GW00360138

VAV Air Conditioning Systems

Keith Shepherd
BEng, MSc

Blackwell
Science

Editorial Offices:
Osney Mead, Oxford OX2 0EL
25 John Street, London WC1N 2BL
23 Ainslie Place, Edinburgh EH3 6AJ
350 Main Street, Malden
MA 02148 5018, USA
54 University Street, Carlton
Victoria 3053, Australia
10, rue Casimir Delavigne
75006 Paris, France

Other Editorial Offices:

Blackwell Wissenschafts-Verlag GmbH
Kurfürstendamm 57
10707 Berlin, Germany

Blackwell Science KK
MG Kodenmacho Building
7–10 Kodenmacho Nihombashi
Chuo-ku, Tokyo 104, Japan

First published 1999

Set in 10/13 pt Times
by DP Photosetting, Aylesbury, Bucks
Printed and bound in the United Kingdom by
Alden Press, Oxford and Northampton.
Bound by MPG Books Ltd, Bodmin, Cornwall

DISTRIBUTORS

Marston Book Services Ltd
PO Box 269
Abingdon
Oxon OX14 4YN
(*Orders:* Tel: 01235 465500
Fax: 01235 465555)

USA
Blackwell Science, Inc.
Commerce Place
350 Main Street
Malden, MA 02148 5018
(*Orders:* Tel: 800 759 6102
781 388 8250
Fax: 781 388 8255)

Canada
Login Brothers Book Company
324 Saulteaux Crescent
Winnipeg, Manitoba R3J 3T2
(*Orders:* Tel: 204 837 2987
Fax: 204 837 3116)

Australia
Blackwell Science Pty Ltd
54 University Street
Carlton, Victoria 3053
(*Orders:* Tel: 03 9347 0300
Fax: 03 9347 5001)

A catalogue record for this title is available from the British Library

ISBN 0-632-04276-1

Library of Congress
Cataloging-in-Publication Data
Shepherd, Keith.
VAV air conditioning systems/Keith Shepherd.
p. cm.
Includes bibliographical references and index.
ISBN 0-632-04276-1 (hardcover)
1. Variable air volume systems (Air conditioning) I. Title.
TH7687.95.S53 1998
697.9′3–dc21 98-36740
CIP

For further information on
Blackwell Science, visit our website:
www.blackwell-science.com

To Olya

Contents

Introduction

This book provides an in-depth review of the design, performance and control of VAV systems. Among heating, ventilating and air conditioning (HVAC) engineers the acronym VAV universally means variable air volume, although in the strictest sense it is variable air volume *flow rate* that is both implied and understood. It would typically be assumed implicit by these same engineers that such a VAV system were a VAV *air-conditioning* system, since the term 'demand controlled ventilation' is more usually applied to systems in which the volume flow rate of air is designed to be variable purely according to the need or demand for ventilation. The VAV concept covers a wide variety of system and equipment types. While this book takes as its central theme conventional single-duct throttling VAV systems, all principal variations on the VAV theme are introduced and discussed, including, in particular, fan-assisted terminal units (or FATs) and dual-duct VAV.

VAV systems originated over 30 years ago in the USA. Since that time they have been extensively developed and widely applied in multizone air conditioning applications throughout the world. Their development continues to the present day, and the concept remains one of the principal tools available to the HVAC designer considering commercial applications of medium to large scale.

It is neither the intention of this book to champion the cause of VAV systems among HVAC engineers nor to build a case against their use. It will always be fundamental that the best system, if all options are equally well designed and engineered, will be the one that is most suitable for the particular application. Rather the present work aims to re-examine the characteristics, design and engineering of VAV systems. It offers no magical insights. Instead it seeks to re-evaluate fundamental characteristics, and to include within a single book at least some of the wealth of information that exists on these systems within an extensive research literature that, while openly available, may not be readily accessible to busy practising HVAC engineers.

Chapter 1 introduces the VAV approach to air conditioning in buildings, and traces its origins and development. Chapter 2 considers system layout and components, while Chapter 3 looks at the fundamentals of VAV system design and operation. Chapter 4 deals with the principal aspects of VAV design, including heating, humidity control and room air distribution, and Chapter 5 considers ventilation, indoor air quality and building pressurisation. Chapter 6 looks at the loads on VAV systems and their impact on VAV design and operation. The equipment at the heart of VAV operation – the terminal units themselves – are considered in detail in Chapter 7. The supply fan and air distribution system are the subjects of Chapter 8, which considers fan duty modulation and control in detail. System return fans, or the alternative options to these, are similarly treated in Chapter 9. The work concludes with comparisons of capital and running costs, energy use and environmental impact between VAV systems and the various alternatives available to the system designer. An extensive bibliography is provided as an aid to engineers and researchers who may need to investigate particular aspects of VAV systems in more detail than a work of this nature can provide.

Keith Shepherd

Acknowledgements

My thanks to friends and former colleagues at the School of Engineering Systems and Design, South Bank University, and the Department of Building Engineering, UMIST, for their help and assistance during both good and difficult times. In particular, I wish to acknowledge the professional and personal contributions of Dr Tassos Karayiannis, Martin Ratcliffe and Tony Day at South Bank. These are the people who really make education and research work. Credit for the idea for this book belongs to Professor Kenneth Letherman of UMIST, although he carefully omitted to mention what the magnitude of the task would be! Very special thanks and appreciation are given to my wife, Olga Konstantinovna, for absorbing the stresses and strains of this during the first year of our marriage, in new and unfamiliar surroundings and for invaluable help in the preparation of the illustrations. Special thanks also to Mike Archer for opening a door back into the real world of Building Services Engineering. Finally, a general thanks to all who have freely given of their advice and experience in the preparation of this book. All errors and omissions are, of course, mine alone.

Chapter 1
Air Conditioning and VAV Systems

1.1 The thermal environment within buildings

In the UK, the vagaries of the climate mean that it is normal for the thermal environment to be regulated within all but the most rudimentary of buildings. In industrial buildings the justification for this often stems from the needs of equipment or processes. In domestic, residential and commercial applications the prime justification is typically seen as the health and comfort of the occupants. Historically in the UK, as elsewhere in Western Europe and North America, the social expectation of a progressive increase in living standards has lead to the wider use of air conditioning in the built environment being principally justified on the grounds of achieving an improved level of occupant comfort. In commercial buildings this is typically linked to a predicted (or hoped-for) increase in occupant productivity, although consideration of the way in which the sensation of thermal comfort interacts with and may influence the psychology of the work place is beyond the scope of this book.

Within buildings the agents of thermal regulation may operate passively or actively. In the face of climatic effects and internal heat gains, much can be done to maintain an acceptable internal environment by careful attention to the design of the building envelope. Where this passive regulation is insufficient to meet the requirements of thermal comfort, its effects must be supplemented actively by the use of purpose-designed mechanical and/or electrical engineering systems. Collectively grouped together under the heading of *heating, ventilating and air conditioning* (HVAC) systems, their operation consumes energy that is typically derived from the combustion of fossil fuels. Natural gas or fuel oil is respectively piped or delivered to the site to fire heating or absorption cooling plants. The motors that drive fans, pumps and refrigeration compressors, the actuators that position valves and dampers, and the control systems are all powered by electricity. Whether generated on-site or taken from the public supply, production of this electricity typically relies on generators driven by gas or diesel engines, gas turbines, or turbines fed with steam raised in boilers fired with coal, oil, natural gas or domestic/industrial waste.

All combustion releases pollutants into the atmosphere, notably carbon dioxide (CO_2), and widespread concern has been voiced that significant and irreversible climatic and environmental changes may occur if the scale of this pollution continues unchecked globally. In the United Kingdom it is generally considered that as much as 50% of the annual consumption of primary energy nationally may be accounted for by building services use, split approximately equally between domestic/residential and commercial/industrial consumers. A major portion of this is attributable to HVAC systems, which also represent significant elements in both the capital and operating costs of buildings. While relatively low energy prices have continued to prevail, in many quarters of the construction industry and the market it serves concern for energy use may have remained the subject of more talk than action. However, concern over the environmental implications of fossil fuel combustion is causing a serious reassessment of the accepted ways of achieving thermal regulation of buildings for occupant comfort, including reducing reliance on active systems, particularly air conditioning.

1.2 Air conditioning in buildings

One of the first numerically important applications for air conditioning arose with the proliferation of purpose-built 'movie theatres' in the United States (and to a lesser extent elsewhere) during the early decades of this century[1]. These provided an impetus for the development of equipment and design techniques, and introduced both public and business to the uses and economics of air conditioning.

In its strictest sense air conditioning involves the control of air temperature, humidity, purity and movement. In some applications where specialist building use or industrial processes are involved there may also be a requirement to maintain control over air pressure. As far as comfort is concerned, the relative insensitivity of human beings to changes in humidity within a space means that close control of the latter is typically not necessary. Whilst a system that cannot provide full year-round control of space humidity strictly represents *partial* air conditioning, such a distinction is rarely drawn in practice. A distinction that may usefully be drawn is that between so-called *comfort conditioning* and industrial air conditioning. The former is intended to provide a comfortable environment for human beings, whereas the latter will typically have as its aim the provision of an environment in which the maximum performance (output or quality) of a scientific or industrial process can be achieved. In either case the techniques employed by the HVAC designer do not differ markedly, the principal difference being the tolerance within which the specified environmental design criteria are to be held. Tolerances on either side of the nominal design value specified of ± 1 or 2°C in dry-bulb temperature and $\pm 10\%$ in relative humidity are typically adequate for comfort conditioning. The yield or quality of an industrial process may be adversely affected if dry-bulb temperature or relative humidity deviate from the specified design values by more than ± 0.5°C and $\pm 2.5\%$, respectively. Comfort conditioning covers most of the commercial (office-type) applications to which VAV systems have routinely been applied.

The functions of heating, ventilating and air conditioning are frequently closely related. Where an air conditioning system is employed, it almost invariably also satisfies ventilation needs, and may also provide space heating. Technically and operationally air conditioning is distinguished by the ability to provide cooling and dehumidification of the air supply to the spaces served. Since this requires the use of precision-engineered refrigeration plant, the provision of space cooling is expensive both in terms of capital and operating costs. Based upon practice within the HVAC industry, information published by the Building Services Information and Research Association[2] suggests that capital costs for mechanical services in typical fully air conditioned office space may be expected to be between two and three times those for heated and naturally ventilated offices. Based upon data gathered on approximately 400 office buildings, figures given by the Energy Efficiency Office[3] suggest that the annual energy costs for HVAC systems in the air conditioned offices, expressed in £/m^2 of treated area at 1990–91 values, was similarly between two and three times that in the heated and naturally ventilated offices. This ratio was broadly independent of whether buildings were considered which were typical of the national stock of office space generally, or those whose design and management could be considered to represent, for the time, good current practice with respect to energy efficiency. The comparison included, where applicable, energy used for space and hot water heating, refrigeration, and the operation of fans, pumps and controls. Energy used for lighting, office equipment, mainframe computers (and their dedicated HVAC services) and catering was excluded.

1.3 The use of air conditioning in the UK

Historically, climatic and economic considerations, combined with traditional construction methods, have meant that in the UK the needs of the thermal environment in domestic and residential buildings have been met almost exclusively by natural ventilation and space heating alone. In commercial buildings, however, the situation may be very different. City locations frequently necessitate the adoption of a sealed building envelope to minimize the ingress of dirt and noise from the external environment. Over the last 30 years, design trends and construction economics have generally favoured the use of extensively glazed and thermally lightweight building envelopes. In commercial buildings the years from 1980 have seen an increasingly sophisticated use of the internal space, driven by the influence which advances in information technology (IT) have exerted on the way in which businesses operate. These factors have been accompanied by increasing expectations of comfort on the part of building occupants. The overall result has been that in many commercial applications space cooling has become the dominant consideration in the regulation of the thermal environment all year round.

Results from a Department of the Environment research project suggest that approximately 10% of the floor area of existing commercial and industrial buildings is air conditioned, including 24% of office floorspace and 14% of retail floorspace[4]. On this basis, total floorspaces for these building categories are approximately 90 M m^2 and 107 M m^2 respectively. Correlations were shown to exist between air conditioning and the size and age of buildings. Only 10% of floorspace is air conditioned in offices of less than 1000 m^2, rising to 50% for offices over 1000 m^2 and to 70% of offices over 10 000 m^2. As far as building age is concerned, only approximately 5% of office floorspace pre-1900 is fully air conditioned, while 43% of that post-1990 is. Latitude was shown to have little to do with the incidence of air conditioning, although the greater number of large buildings in south-east England is certainly reflected in its greater proportion of air conditioned floorspace. At approximately 33%, this is almost double the national average of 18%.

1.4 Air conditioning systems

A wide variety of air conditioning system types and sub-types are available to the HVAC designer. Hybrids combining elements of more than one type of system further increase the range of potential design solutions for a given application. Each system has characteristics of a technical, cost or operational nature that may render it more or less suitable for a particular application. Comparing the performance of alternative system types is a complex business for all but the simplest applications. There will rarely be a uniquely ‘right’ solution in HVAC design.

System selection is at the very heart of the design task, involving:

(1) appreciation of the possible design solutions;
(2) knowledge of their technical, economic and operational characteristics;
(3) weighting of their relative characteristics according to the requirements of the application, resulting in identification of the most suitable design for that particular application.

It may well be correct to suggest that a system that is well designed, installed, commissioned, maintained and operated will always prove superior, regardless of type. It is equally true, however, that there is ultimately no technical 'fix' for poor initial system selection.

As might be expected, there is a natural progression of increasing equipment size and system complexity as the number of conditioned zones served by a single system increases. This progression starts at the most basic level with the self-contained individual room air conditioner, whose mechanical characteristics have earned it the epithet of the 'window rattler', and develops through traditional split systems serving several zones to culminate in large central plant systems conditioning upwards of a hundred individual zones (Fig. 1.1).

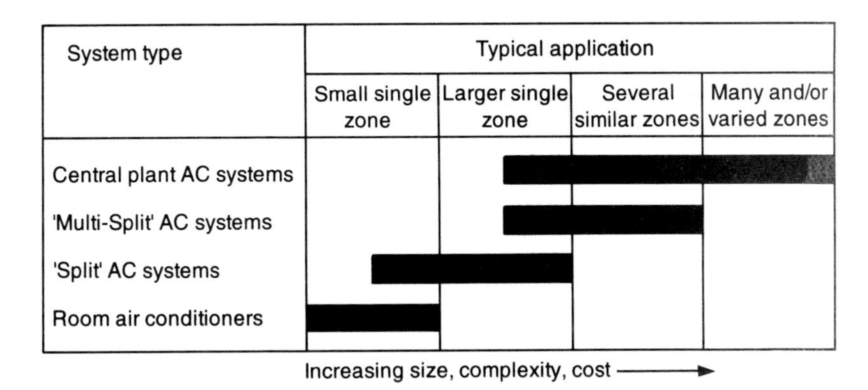

Fig. 1.1 Air conditioning systems: the progression in size and complexity.

1.5 Central plant systems and multi-zone air conditioning

VAV systems invariably come under this category, being used to condition groups of spaces requiring individual environmental control. The classic multi-zone air conditioning application of the post-war period, and one that will be referred to throughout subsequent chapters, is the modern office block. It is estimated that in 1990 the UK stock of office space in the private sector alone amounted to approximately 72.5 M m^2, and included some 275 000 individual buildings[5], and this is supported by subsequent research[4]. With up to 24% of this floor space air conditioned, this represents a very significant use in buildings, and makes it easy to understand why this class of building is of particular interest to HVAC designers and building owners, developers, contractors, tenants and occupiers alike.

A large office building may be air conditioned by one or more central plant systems. The functions of the central plant are typically (but not exclusively) divided between an air handling unit (AHU) and a refrigeration plant. The size and access requirements of the components of a central plant system generally mean that they are located in dedicated plant rooms remote from the conditioned areas – commonly at roof or basement level in multi-storey buildings, with sheltered roof areas being used to site air-cooled refrigeration plants. Whilst central plant systems may employ *DX* cooling (the cooling coil in the air handling unit functioning as the evaporator in the refrigerant circuit), large refrigeration plants typically produce chilled water which is then piped to cooling coils in the AHUs served. In very large refrigeration plants the condensers may be water-cooled types, the waste heat from space cooling being ultimately rejected to atmosphere through a cooling tower.

Central plant systems are commonly classified according to the method by which their cooling effect ('coolth') is distributed to the individual spaces that are to be conditioned. In all-air systems a supply of cooled and dehumidified air is ducted to all zones served. The volume flow rates of air required for the cooling of each zone are typically several times those required for ventilation. VAV is the principal type of all-air system for new multi-zone applications (Fig. 1.2). In air/water systems all zones receive from the central plant a supply of both cooled and dehumidified air and chilled water. The primary air from the central plant is ducted to each zone for ventilation and humidity control. Whilst it also does some sensible cooling, the bulk of this is done by circulating the chilled water through fan coils (Fig. 1.3) or chilled ceiling panels in the zone. If the ventilation requirement for each zone is low enough, outside air may be introduced locally to each fan coil. The central AHU can then be dispensed with, but the fan coils must then perform any dehumidification required for control of space humidity. Chilled ceilings and their development, chilled beams, are an increasingly popular design option (Fig. 1.4).

Although superceded for new-build applications, in the once-popular induction system the primary air is supplied at relatively high pressure to terminal units in each zone. Discharging through special nozzles, the primary air induces room air to flow over cooling coils in each of these induction units (Fig. 1.5). The chilled water supplied to the zones circulates through these coils to provide most of the sensible cooling required. In the case of fan coils the primary air is typically 100% outside air, with no recirculation. For induction systems it is also ideally the case that the primary air supply is limited to the outside air required for ventilation. However, the sensible cooling capacity of induction units is also a function of the primary air volume flow rate. If the required system volume flow rate for the primary air supply exceeds the total ventilation requirement, for economy of energy use and running costs the balance of the primary supply will normally be provided by recirculation. Fan coil systems are the principal type of air/water system for new multi-zone applications, whether a new building is involved or refurbishment of an existing one.

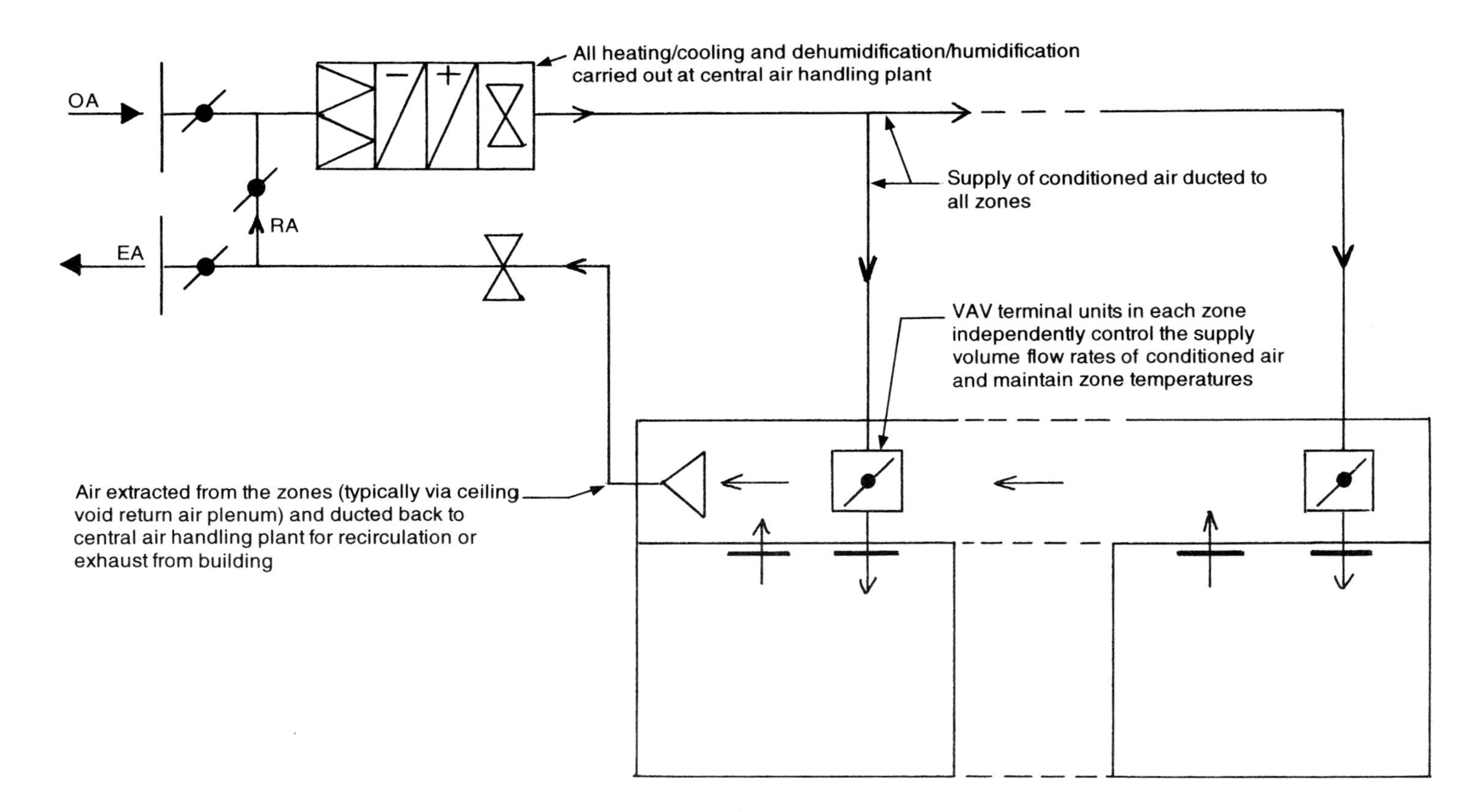

Fig. 1.2 Multizone air conditioning with variable air volume.

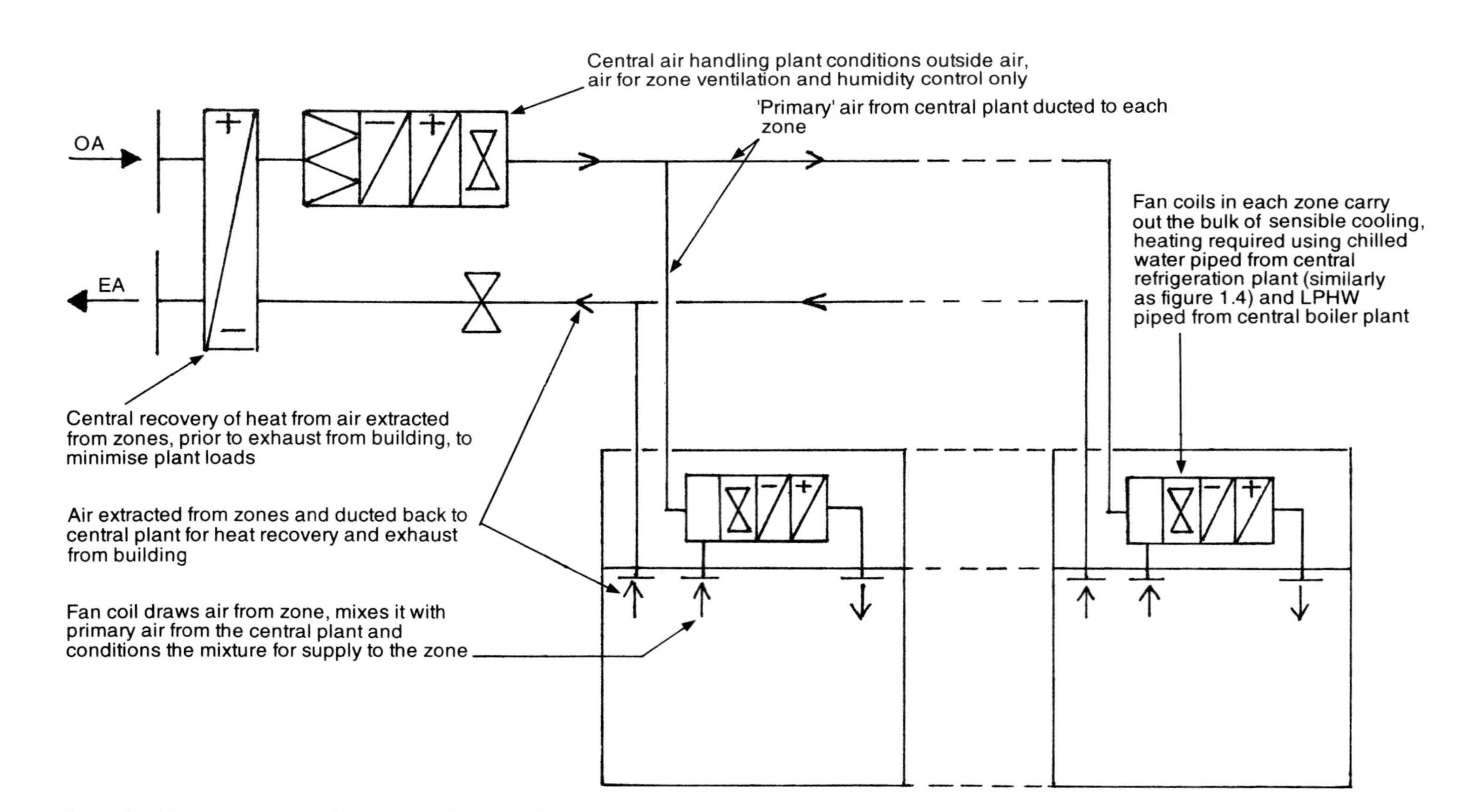

Fig. 1.3 Multizone air conditioning with fan coils.

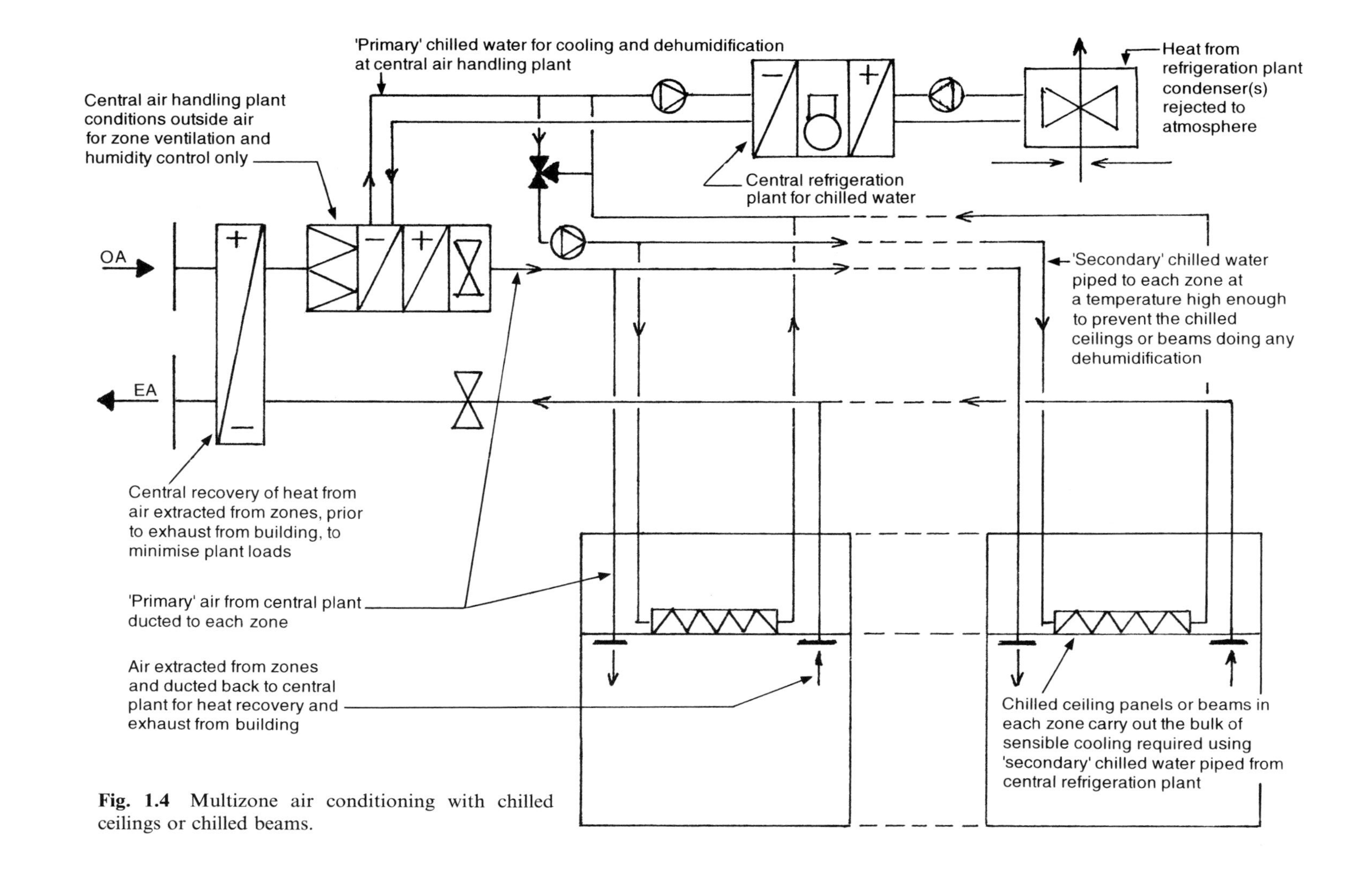

Fig. 1.4 Multizone air conditioning with chilled ceilings or chilled beams.

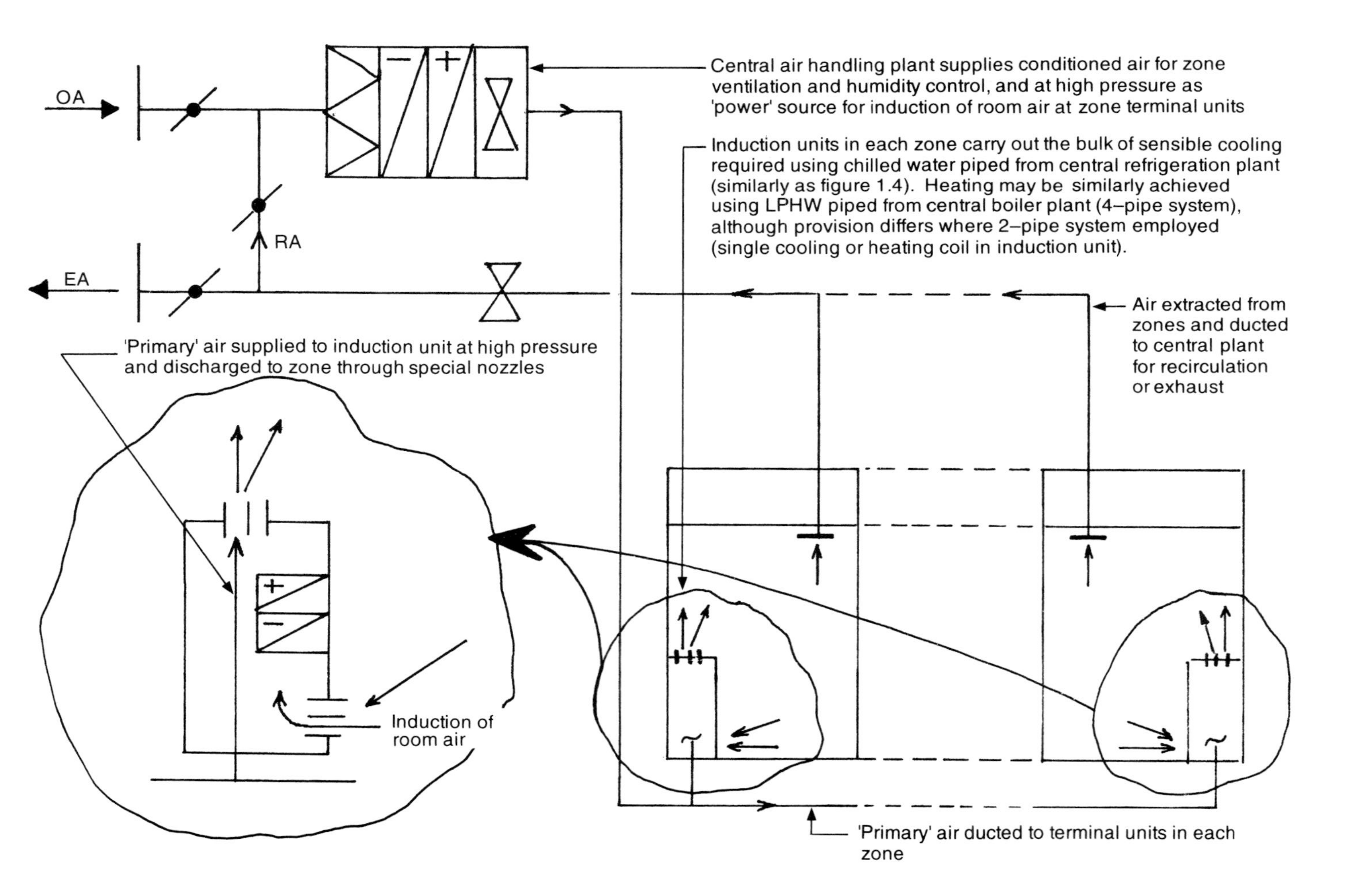

Fig. 1.5 Multizone air conditioning with induction units.

In a variation of the basic air/water system, terminal heat recovery units in each zone use a common water circuit as either a heat sink during cooling operation or a heat source during heating operation (Fig. 1.6). These reversible heat pumps are effectively water-cooled room air conditioners with automatic or manual changeover to heat pump operation. Internal valving allows the roles of both evaporator and condenser to be reversed according to whether cooling or heating of the space is required. Zone units perform both cooling and dehumidification. Conditioned outside air for ventilation may be ducted from a central plant. Alternately, if zone ventilation requirements are low enough, outside air may be introduced directly to each unit.

Developments in refrigeration systems technology resulted in the appearance, in the late 1980s, of *variable-refrigerant-flow* (VRF) systems. In VRF the 'split' concept has developed into a central plant system for multi-zone applications. DX fan coils in each zone are linked by refrigerant pipework to a common central condensing unit with variable-speed compressor (Fig. 1.7). Zone units perform both cooling and dehumidification. Again outside air for ventilation may be ducted from a central plant or introduced locally, according to the level of the requirement.

1.6 All-air systems and the VAV approach

All-air sytems have been, and remain, widely used in commercial air conditioning applications, and within this class of system the VAV approach has come to predominate. By the early 1990s an industry assessment showed that these systems represented approximately 51% by value of the UK market for terminal units (uniquely required for multi-zone applications), then valued at approximately £44 million for basic terminal units themselves, rising to approximately £59 million with factory-fitted valves, controls and diffusers[6]. Virtually all the remainder was accounted for by fan coils. Over the corresponding period in the USA it was suggested[7] that VAV systems might account for about 22% by floor area of all commercial buildings, and about 41% of those both heated and cooled. Since then, however, VAV's share of the market for terminal units has been drastically cut by the economic climate which has prevailed within the construction industry in the UK, and the overall effect of this will be returned to in the following section.

In common with other HVAC systems, the design of an air conditioning system must start with an assessment of the likely cooling load with which the system may reasonably have to cope. This load will arise through some combination of the heat gains originating both outside and within the spaces served by the system, the former typically of climatic origin (e.g. direct solar gain) and the latter typically originating from people, lighting, equipment

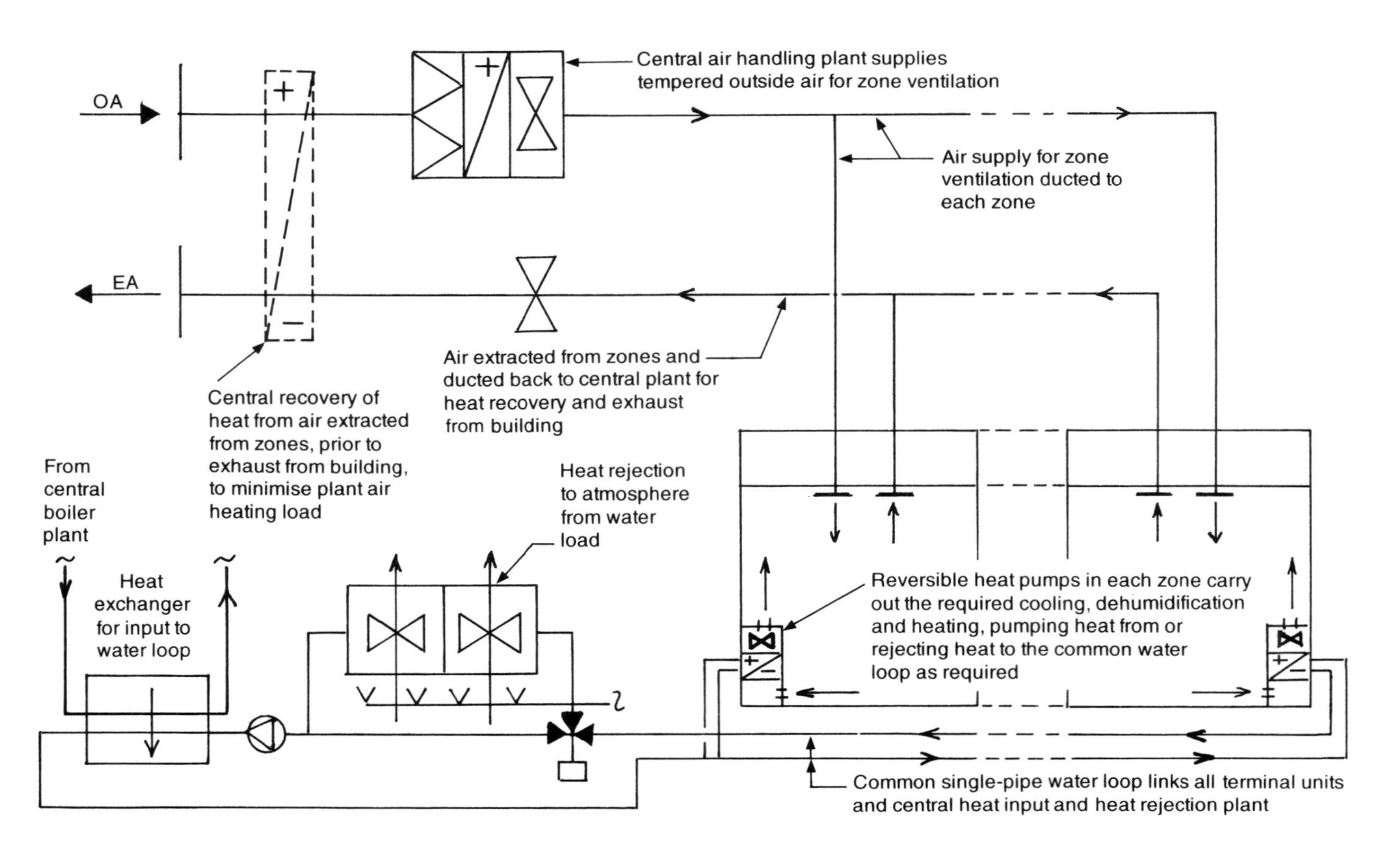

Fig. 1.6 Multizone air conditioning with terminal heat recovery.

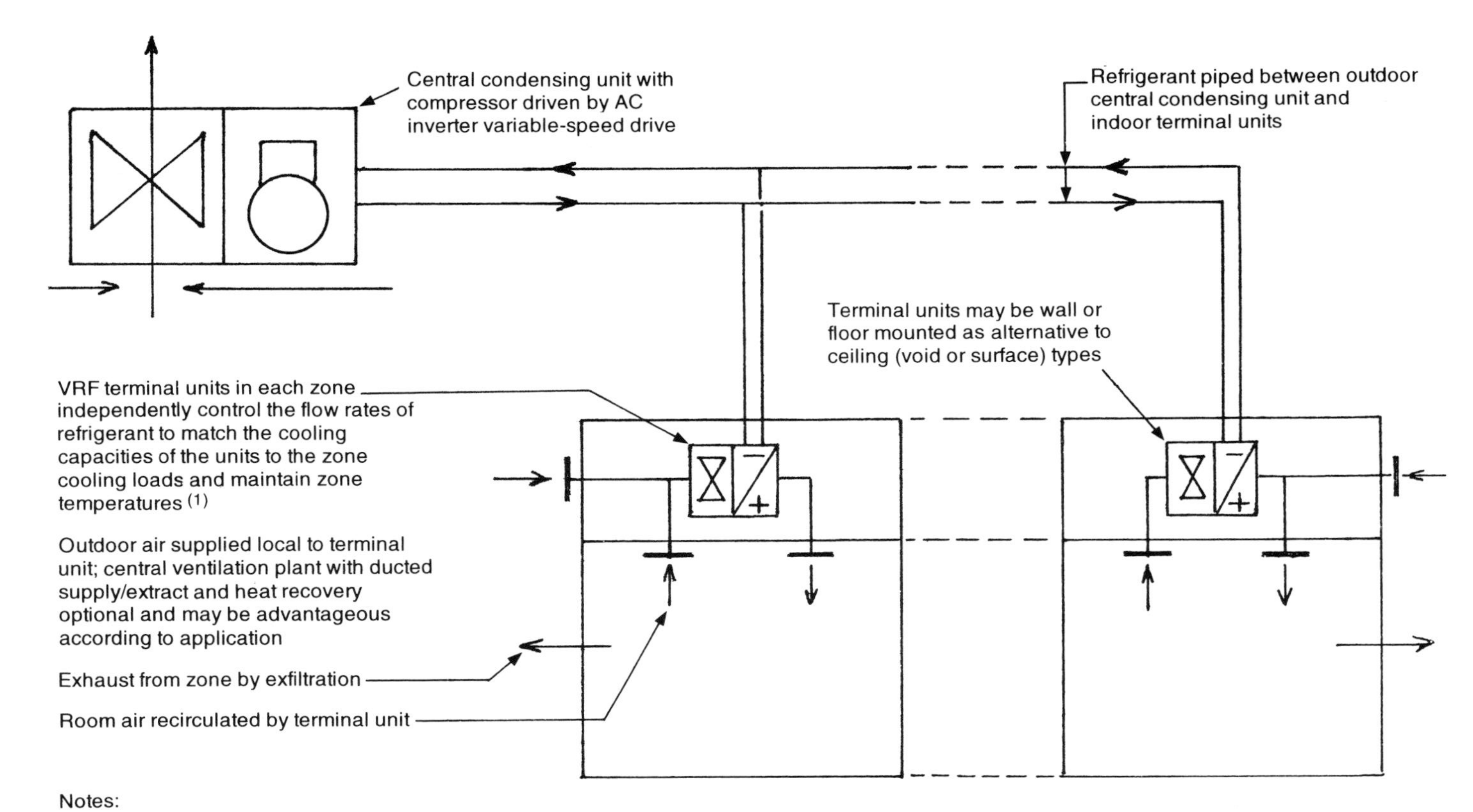

Fig. 1.7 Multizone air conditioning with variable refrigerant flow.

and processes within the spaces. Cooling loads and cooling load calculations are adequately discussed in almost all air conditioning textbooks, but perhaps the best known and most thorough treatment of the subject in the UK is that of Jones[8]. While an in-depth appreciation of cooling load calculations is certainly not a prerequisite for understanding the remaining chapters of the present work, any reader worried at discovering a mental block at this point is referred to either their own or the office copy of the aforementioned text.

The maximum cooling load that a system may reasonably be expected to cope with is its design cooling load, and this is important in sizing the system components. However, the most significant characteristic of cooling loads is their complex variability – according to weather conditions and time of day for the building as a whole and its individual facades, and from zone to zone according to fluctuating occupancy and the frequently variable nature of the processes being carried on within the space (Fig. 1.8). This variability of cooling load gives rise to a need to control the cooling output of an air conditioning system, both at the central plant and at each zone, to match the prevailing load requirements. This leads to two fundamentally different design approaches for all-air systems serving the typical multi-zone applications in which this book is primarily interested.

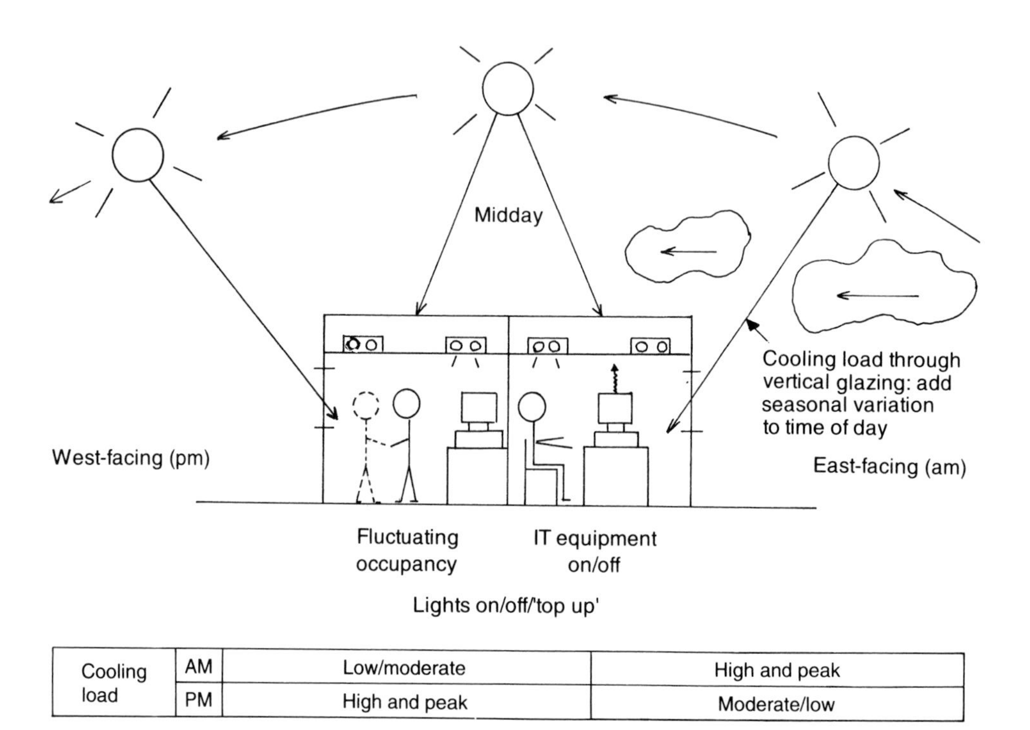

Cooling load		West-facing	East-facing
	AM	Low/moderate	High and peak
	PM	High and peak	Moderate/low

Fig. 1.8 Asynchronous and variable nature of cooling loads.

In the first, the volume flow rates of conditioned air are maintained constant to each zone, the supply air temperature to each individual zone being continuously varied to match the prevailing zone cooling load. This is the constant-volume/variable-temperature or CVVT approach – variously referred to simply as constant volume (CV) or constant air volume (CAV). In the alternative approach conditioned air is supplied to each zone at an identical, constant temperature, the volume flow rate to each individual zone being continuosly varied to match the prevailing zone cooling load. This is the variable-volume/constant-temperature or VVCT approach – commonly referred to simply as variable air volume or VAV. In deference to the purists it is, of course, volume flow rate that ought, strictly-speaking, to be signified in both cases. However, it is universally accepted that this is *implied.* A hybrid approach is, in fact, possible, resulting in the variable-volume/variable-temperature or VVVT approach – usually simplified to VVT.

Applied to multiple zones, the constant-volume approach can lead to the wasteful reheating of cooled air for zone temperature control. The VAV approach typically avoids any need for this, and offers in addition potential savings in both fan energy consumption and capital cost. It is these three considerations which have been largely responsible for the popularity and wide application of VAV systems in recent years. Inevitably, however, since there is no perfect air conditioning system, these inherent attractions of the VAV system are accompanied by some equally inherent problem areas for the HVAC designer. In particular, the provision of space heating, achieving satisfactory room air distribution performance, and ensuring that adequate ventilation is maintained under all normal operating conditions require careful consideration if the potential benefits offered by the VAV approach are not to be diminished or devalued by poor operating performance.

By comparison, CVVT techniques can be employed in air/water systems without the risk of wasteful reheating because in these systems the variable suppy air temperature is achieved by the use of zone cooling coils, rather than by central plant cooling and zone reheating.

1.7 The origins and development of VAV systems

In a very real sense the opening window was the original variable-air-volume device[9]. However, the earliest references to VAV systems that would be recognised as such by engineers accustomed to modern HVAC systems appeared in the mid-1960s in professional and trade journals serving the building services industry in the USA.

A case in point is the study of a 17-storey office block in the Canada Square development in Toronto[10]. Completed in the summer of 1963, the periphery of the building was conditioned using a low-pressure VAV system incorporating terminal reheat. The terminal units were mounted at low level between the structural floor framing and discharged directly into the spaces served. Double-inlet, mixed-flow fans with flat performance curves supplied the system from equipment rooms on each floor, the excess pressure developed at reduced fan duty being absorbed by in-duct throttling devices which maintained constant static pressure in the duct branches. The interior zones on each floor were served at constant volume flow rate using the self same equipment that served the zones on the periphery with variable volume flow rates. Relief air dampers were used, rather than return air fans.

The completion date of the 'Canada Square' project implies that the preliminary design work may have been in hand prior to 1960. Allowing for the timescale required to design and develop the terminal units, the further implication is that VAV systems such as this (and others which may have gone unreported) were actively under consideration by engineers for application to large commercial air conditioning projects in the second half of the 1950s. Indeed crude VAV installations with limited flow modulation may have been unsuccessfully attempted as far back as the 1940s, using dampers modulating under the control of local space temperature sensors as pressure-dependent flow regulating devices[11].

While the concept of applying variable-volume techniques to the air conditioning of multiple zones had been recognised for many years, it was more than anything else the characteristics of conventional air terminal devices that held up its successful application. These were such that dumping and inadequate air motion prevailed under the reduced volume flow rates associated with partial-load conditions. With the evolution, in the early 1960s, of a suitable ceiling diffuser capable of maintaining satisfactory room air distribution at low air volume flow rates, the way was clear for the rapid growth, acceptance and variety of VAV systems. As an illustration of this rapid growth, the records of one prominent US manufacturer show an increase in sales of VAV terminal units from *c*.500 in 1964 to *c*.40 000 in 1968.[12]

Over the next ten years a number of further applications of VAV systems were described (including Collins 1966, Brehm 1968, Heck 1970, Sehgal 1972, Vivien 1972, Watt 1972 and Levine 1974, amongst others). For the interested reader full details of these, and other published references to early applications of VAV systems, are provided in the Bibliography. Still in the USA, the early 1970s saw the first application of a VAV system in the Manhattan area of New York, where some 3800 system-powered terminal units were used to condition the interior zones of the 15-storey police headquarters. However, perimeter loads were still handled by an induction system. By 1972 the VAV applications of one US consultancy practice had totalled over ten million square feet of office space, covering individual projects ranging from 50-storey buildings right down to relatively small-scale applications[13]. However, it was not until September 1972 that the information intended for inclusion in the then-forthcoming 1973 Systems volume of the ASHRAE Handbook and Product Directory could be previewed[14]. The editor's note to this stated: 'For the first time, the 1973 volume will explore the ramifications of the often controversial variable air volume (VAV) system'.

As should be apparent from this, the origins and early development of VAV systems were firmly rooted in the USA. Prior to the mid-1970s there are few references to VAV systems outside North America, and almost without exception these are confined to Australia.

While the early VAV systems and their component equipment might appear crude to engineers familiar with sophisticated, state-of-the-art VAV engineering, it is clearly a misconception to suppose that VAV systems were originated *in answer* to the energy crises which followed in the wake of the Arab Oil Embargo of 1973–74. Enlightened engineers and building clients already appreciated the potential of VAV techniques:

(1) to eliminate the energy waste and operating cost penalty of reheating for control purposes;
(2) to take advantage of load diversity in system sizing, with consequent benefits in reduced capital costs and architectural space requirements;
(3) to reduce the energy use and operating costs associated with ducted air distribution.

The extent to which reheating for zone temperature control was a significant consideration in multi-zone air conditioning applications prior to the availability of variable-volume techniques is perhaps open to some debate. It is, however, a fundamental principle of air conditioning theory that in such applications constant-volume all-air systems must inevitably depend, to a greater or lesser extent, on reheating for control of zone temperatures *whenever zone loads are not at their design values*. It may not always be appreciated by system designers that this provision of reheat at constant-volume air system terminals is a true *year-round* requirement, spanning not only the acknowledged 'off-peak' (spring and autumn) months of the cooling season, but also virtually the whole of the peak summer-time months, since actual (full-load) design conditions may exist for only a relatively few operating hours (if any!) during this period. This is equally true whether terminal reheat, dual-duct or 'multi-zone' systems are considered. It is undoubtedly in whether the extent of reheating is greater or lesser that the leeway for debate exists. This certainly also depended, at least in part, on the particular application, and on how well the system was designed and controlled. Prior to the age of information technology, and in permanently lit interior zones where cooling load varied principally according to changes in occupancy, the annual requirement for reheating might have been modest. In any event, to dwell further on this aspect is unproductive, since for many years now HVAC engineers have worked and studied in a professional and social environment that instinctively, and quite rightly, rejects the use of wasteful reheating on principles of energy efficiency, operating cost and environmental impact. As there can be few HVAC engineers who would wish to see less emphasis on these principles, it can remain a matter for academic debate whether the *bête noire* reputation of reheating may, as is sometimes the case, have achieved a notoriety that overshadows the reality.

The benefits of load diversity are straightforward to confirm from an inspection of even the most basic of cooling load calculations, and need no further comment at this point. Of particular interest in the context of this book, and the commonly-perceived justification for the application of a VAV system, is item (3), reducing the energy use and operating costs associated with ducted air distribution. Indeed it may be fair comment to suppose that the cause of VAV systems would have been better served if this perception had not been so strongly seized upon, both inside and outside the HVAC field.

However, if not *originated* in response to the energy crises of the early/mid-1970s, the effects of the latter certainly stimulated extensive *development* of a VAV concept which was conveniently at hand to meet the new urgency for more energy-efficient air conditioning systems. New equipment was developed, including terminal units, fixed and variable-geometry air terminal devices, methods of fan duty regulation, and controls. Design and control techniques were refined, and system variants proliferated in the latter half of the 1970s and throughout the 1980s.

In the second half of the 1970s, references to VAV systems began to appear in small but significant numbers in professional and trade journals in the UK and continental Europe. In the ten-year period between 1975 and 1985, the application of VAV systems became widespread in the UK. As a measure of the growth in their use, industry statistics suggest that the proportion of the UK market for terminal units accounted for by VAV systems grew to about 50% by 1989[15,6]. This growth in market share was at the expense of constant-volume and induction units, the last-mentioned virtually disappearing from the UK market. It is probably not unreasonable to suggest that the increase in popularity of VAV systems was such that, for many designers, clients and developers alike, VAV became virtually the automatic choice for air conditioning of multi-zone commercial applications of medium to large size. In particular the speculative development market saw the VAV approach as highly flexible in the so-called 'shell and core' type of development where a building shell was developed with a core provision of HVAC and domestic services, the shell being subsequently fitted out to suit the requirements of the eventual tenants.

However, strong dependence on commercial applications, and on 'new-build' projects in particular, has rendered the market for VAV systems and equipment in the UK sensitive to influences in the economy as a whole. In consequence the VAV market was particularly badly hit by the recession in new commercial building projects during the first half of the 1990s, and the continuing stagnation of this market which followed. In the case of fan coils, their popularity in the refurbishment market, where the size of existing services ducts and voids may dictate an air/water design approach, has helped to alleviate some of this market sensitivity, since a decline in new commercial building projects may subsequently result in an increase in refurbishment projects. As a consequence, over the last few years both the number and value of VAV terminal units sold (forecast at approximately 17 000 units and £7.5 million respectively for 1997) have both remained fairly steady at little more than one-third of the corresponding numbers and values of fan coils (forecast at approximtely 48 000 and £19.5 million for 1997), and indeed comparable to the 'mixed bag' of 'other' terminal units which includes constant-volume terminals, induction units and water-source consoles[16]. In other countries, such as the USA, the need for air conditioning in a wider range of building types has lead to consideration of the VAV approach in a range of educational, healthcare and public buildings, as well as in the 'conventional' office block applications.

References

1. Nagengast, B.A. (1993) The 1920's: the first realizationn of public air conditioning. *ASHRAE Journal*, **35**, January, S49-56.
2. Hayward, R.H. (1988) *Rules of Thumb. Examples for the Design of Air Systems*, Building Services Research and Information Association, Bracknell, Technical note 5/88.
3. Building Research Energy Conservation Support Unit (BRECSU). (1993) *Energy Efficiency in Offices. A Technical Guide for Owners and Single Tenants*, BRECSU, Garston. *Energy Efficiency Office Best Practice Programme*, Energy consumption guide 19.
4. Rickaby, P.A., Bruhns, H.R., Mortimer, N.D., Clark, E. and Steadman, J.P. (1995) Mechanical Ventilation and Air Conditioning Systems in UK Buildings: Volumes 1–3, Project Report of the Department of the Environment Scooping Study for HVAC Systems. DoE (1995), Extracts in *Building Services Journal*, **18**, June, 12–13.
5. Samuelsson-Brown, G. and Whittome, S. (1992) *The Office Sector – A Review of Floorspace, Building Type, Market Potential and Construction Activity in Great Britain*, Building Services Research and Information Association, Bracknell, April, Report no. MS 7/91.
6. Building Services Research and Information Association (1990) Product profile – terminal units (2), *BSRIA Statistics Bulletin*, **15**, June, i–iv.

7. Englander, S.L. and Norford, L.K. (1990) VAV system simulation, Part 1: Development and experimental validation of a DDC terminal box model. In *Proceedings of the Third International Conference on System Simulation in Buildings*, held in Liege, 2–5 December 1990. Laboratory of Thermodynamics at the University of Liege and the International Building Performance Simulation Association, Inc. (1991) Liege, 553–580.
8. Jones, W.P. (1994) *Air Conditioning Engineering*. 4th edn, Edward Arnold, London.
9. Daryanani, S. *et al.* (1966) Variable air volume air conditioning. *Air Conditioning, Heating and Ventilating*, **63**, March, 56–78.
10. Shuper, A. (1964) A study in variable volume air conditioning. *Air Conditioning, Heating and Ventilating*, **61**, November, 57–60.
11. Steketee, N.F. (1972) Variable volume air conditioning systems. *Australian Refrigeration, Air Conditioning and Heating*, **26**, January, 39–42.
12. Terry, K.B. (1969) Variable volume air conditioning systems and devices. *Australian Refrigeration, Air Conditioning and Heating*, **23**, October, 34–48.
13. Kettleman, J.E. (1972) Variable air volume systems and hardware: the experience record. *Building Systems Design*, **69**, August 29–32.
14. Rickelton, D. and Becker, H.P. (1972) Variable air volume. *ASHRAE Journal*, **14**, September, 31–55.
15. Building Services Research and Information Association (1982) Product profile no. 14: terminal units, *BSRIA Statistics Bulletin*, **7**, October, i–iv.
16. Building Services Research and Information Association (1997) *BSRIA Statistics Bulletin*, **22**, March, 13.

Chapter 2
System Layout and Components

2.1 The classic VAV system

Quite simply, there is no such thing as the classic VAV system, nor ever has been. VAV is a generic term which covers a wide variety of system configurations and equipment types. As long ago as 1974, the HVAC engineer looking for 'the classic variable-air-volume system' was likened to Diogenes searching, lantern in hand, for 'the honest man'[1]. The variety of VAV systems and equipment may actually have increased since then.

A 1990 report by the Building Services Research and Information Association revealed 22 companies active in the UK market for VAV terminal equipment[2]. Several years later, despite the effects of economic recession on the construction industry, it was still possible to obtain technical literature from 23 manufacturers or suppliers of different VAV terminal units. Since then at least one new design of terminal unit has appeared and achieved commercial application in the UK. While the design and operation of terminal units is covered in some detail in later chapters, for the present it will be sufficient to regard a terminal unit as a complete item of equipment which, when installed in a VAV system, is capable of controlling the volume flow rate of conditioned air supplied to a zone in response to an input from a zone controller. By this definition component manufacturers or suppliers are excluded. The 23 manufacturers of terminal units offered a total of almost 50 different designs, including 14 of the fan-assisted type. Allowing that commercial considerations would undoubtedly have precluded some of these designs from ever achieving a UK application, this still represents a considerable variety of equipment. This variety, however, can be both an advantage and a disadvantage to the HVAC designer. The advantage is that it permits, in theory at least, a VAV system to be tailored to a particular (but fundamentally suitable) application. The disadvantage is that even fairly minor variations in system design may have largely unpredictable and far-reaching effects on system behaviour.

Depending on the size of the application, an individual VAV system may serve a whole building, a section of a building, a single floor, or a group of floors. The size of the air handling plant and the associated air distribution ducts will vary widely according to the approach adopted. The extent to which an application is served by an individual VAV system has significant implications for the latter's design and operation, by virtue of the potential for load diversity that it may offer.

The principal variations and combinations of system types and equipment are shown in Fig. 2.1. With the exception of dual-duct VAV systems, which have never been popular in the UK, all may be more-or-less commonly encountered in typical 'mainstream' commercial applications in the UK. The lack of enthusiasm shown by HVAC designers in the UK towards the dual-duct approach to VAV system design stems from the cost and space requirements of the dual networks of supply air distribution ducts required in this configuration – one each for chilled and heated air. The approach does, however, have its supporters in the USA.

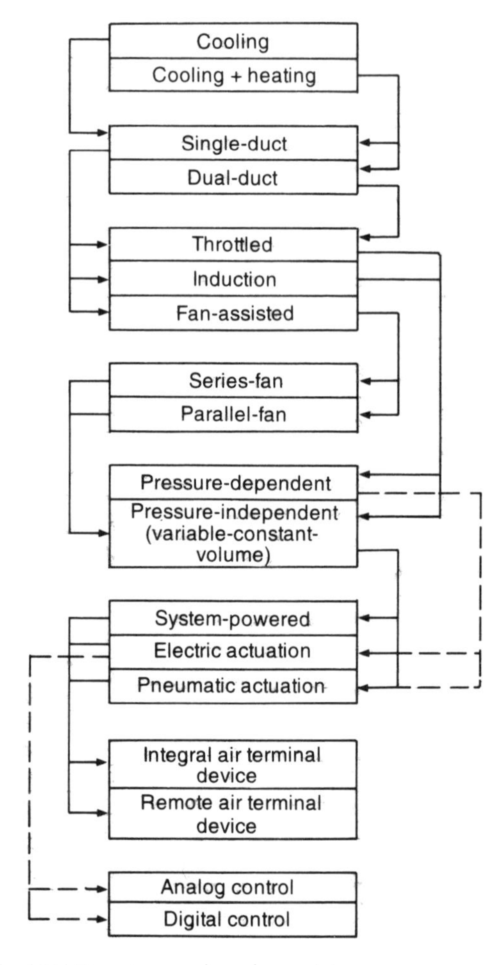

Fig. 2.1 Principal VAV system and equipment types.

From Fig. 2.1 the two system combinations that have historically been most widely applied in the UK would be:

(1) cooling-only, single-duct systems with throttling terminal units giving a pressure-independent supply of conditioned air using a volume control device actuated by system static pressure, and delivering the conditioned air to the space served via an integral supply air diffuser;
(2) cooling-only, single-duct systems with throttling terminal units giving a pressure-independent supply of conditioned air using a volume control device actuated electrically under analogue electrical control, and delivering the conditioned air to the space served via remote supply air diffusers.

Either of these combinations is as close to the 'classic VAV system', in any event historically, as we are likely to get, and a study of their basic configuration and components will form the foundation for both the rest of this chapter and, indeed, of the book as a whole.

2.2 Basic configuration: single-duct VAV systems

Whatever the variety that exists in detail among single-duct VAV systems, all examples of the genre share certain fundamental elements of configuration. They may all be considered to consist of four primary subsystems – psychrometric, supply, zone and exhaust. These are illustrated in Fig. 2.2 for a basic cooling-only VAV system. Each of these primary sub-systems in turn comprises various secondary elements, some of which may rightly be regarded as sub-systems in their own right.

The individual elements of the *psychrometric subsystem* are the plant items which are involved in the air conditioning process itself. In modern practice these are typically packaged together in an air handling unit that filters the supply air stream and cools, dehumidifies and heats it as necessary. Although the AHU invariably includes the supply air fan, and frequently the extract/return air fan, it will be more useful to regard these as elements of, respectively, the supply and exhaust sub-systems. As in any other all-air system, an *outside air economy cycle* is typically employed, with outside air, return air and exhaust air dampers linked to maximise the use of outside air for 'free cooling' – free, that is, in the sense of not requiring the use of mechanical refrigeration. In the UK climate the potential for this is highly significant, as for much of the annual operating period of a VAV system in a typical commercial application the outside air temperature is low enough for the required supply air temperature to be achieved by mixing variable amounts of outside and return air (up to 100% outside air) without the use of

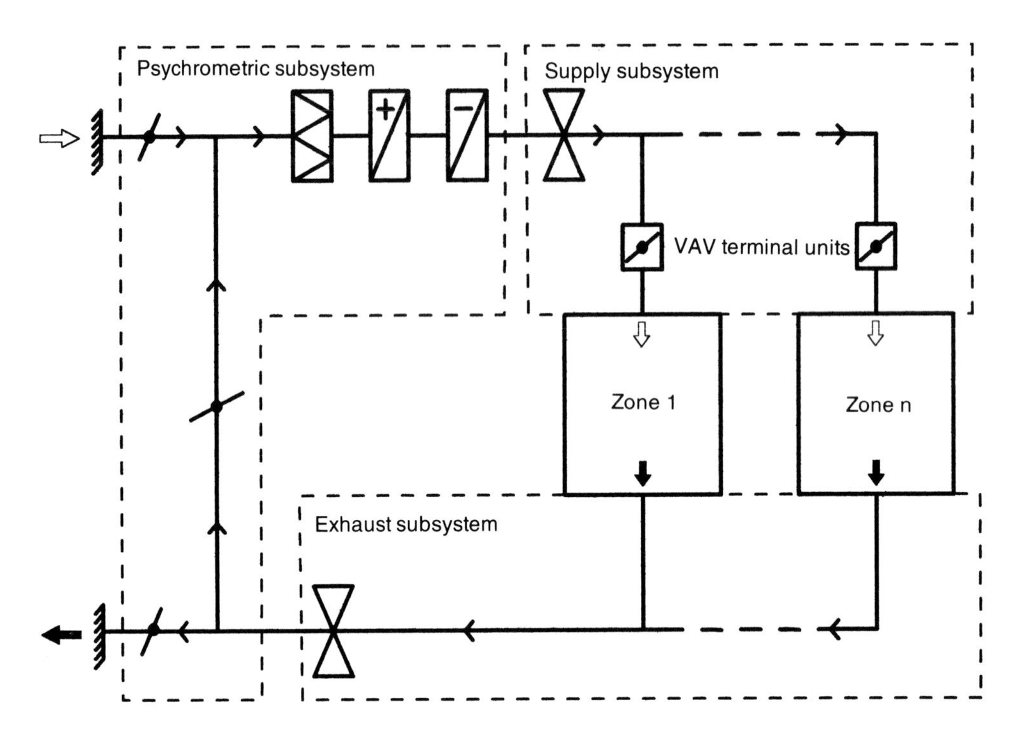

Fig. 2.2 Basic cooling-only single-duct VAV system.

mechanical cooling plant. The ability to do this is a major advantage of all-air systems in temperate climates. The outside, return and exhaust dampers are again typically part of the packaged AHU.

In the UK, even a relatively low level of return air recirculation is typically adequate to prevent freezing of the filter during winter operation. However, an additional heating coil (not shown) may be provided before the filter to protect it in the event of the recirculation air damper failing in the fully closed position. The main heating coil may be close enough to provide this protection.

Within the AHU heating and cooling/dehumidification of the air is carried out by individual heating and cooling coils, which are simply finned-tube heat exchangers. In UK practice the heating coil, or heater battery, is typically supplied with LPHW. The cooling coil may be supplied either with chilled water or liquid refrigerant. In the latter case it functions as the evaporator of the refrigeration plant, and is generally referred to as a *direct-expansion* (DX) coil. Various enhancements to the basic plant may be incorporated to suit the performance and budget of the particular application, including various equipment options for transferring heat either from or to exhaust air prior to its discharge from the building. The most common addition to the basic psychrometric sub-system is a humidifier to maintain relative humidity in the zones served at a minimum level of about 40% during periods of low outside air temperature (and consequent low moisture content). Although the avoidance of very low relative humidity in occupied zones is most definitely a comfort consideration, the increased occurrence of static electric discharges as air becomes drier is of concern to users of IT equipment, especially the now-ubiquitous personal computer. This humidification is typically achieved using steam generated local to the AHU by one or more purpose-made packaged electrode boilers. This is injected directly into the supply air duct(s) downstream of the fan. In some designs of AHU the steam humidifier may be included as a component part.

In general principle the operation, performance and control of the heating and cooling coils and humidifiers in the central plant of a VAV system differs little from that in any central air conditioning plant. This subject is dealt with at length in standard air conditioning textbooks, and the present treatment of these issues will be limited to consideration, in later chapters, of the specific implications of VAV operation. Among the latter the outside air economy cycle will be looked at in some depth, because its correct control and operation is not only a strong influence on the energy consumption of the refrigeration and heating plant, but is crucial to the achievement of adequate ventilation performance with VAV systems, and may have a strong influence on system stability. Beyond this it will generally be sufficient to regard the psychrometric subsystem as a simple block item which delivers a supply of suitably conditioned air (however this is achieved and controlled) to subsequent components of the system.

The elements of the *supply subsystem* are the variable-duty supply fan and a network of air distribution ducts, VAV terminal units, and air terminal devices. The supply fan receives conditioned air from the psychrometric subsystem, and supplies the energy to maintain the flow of this air through the ducted air distribution network to the points where it is discharged into the zones served. Controlling the duty of the supply fan lies at the heart of VAV system engineering, and the subject is dealt with in Chapter 8. As in any air system, noise generated by the supply fan will be propagated in both upstream and downstream directions. Attenuation of the former may or may not be necessary according to the nature of the application and the location both of the building and of the AHU itself, while the latter is invariably necessary. The level of attenuation required may typically be achieved by the provision of acoustic attenuators either within the main duct(s) of the air distribution system or as part of the AHU itself.

The ducted air distribution typically forms a branched-tree configuration (Fig. 2.3), decreasing in size between the fan and the furthest unit as air is progressively fed off through branch ducts to individual zones. The initial duct section downstream of the fan discharge is termed the root section. Various levels of branching of the distribution network may be used to suit

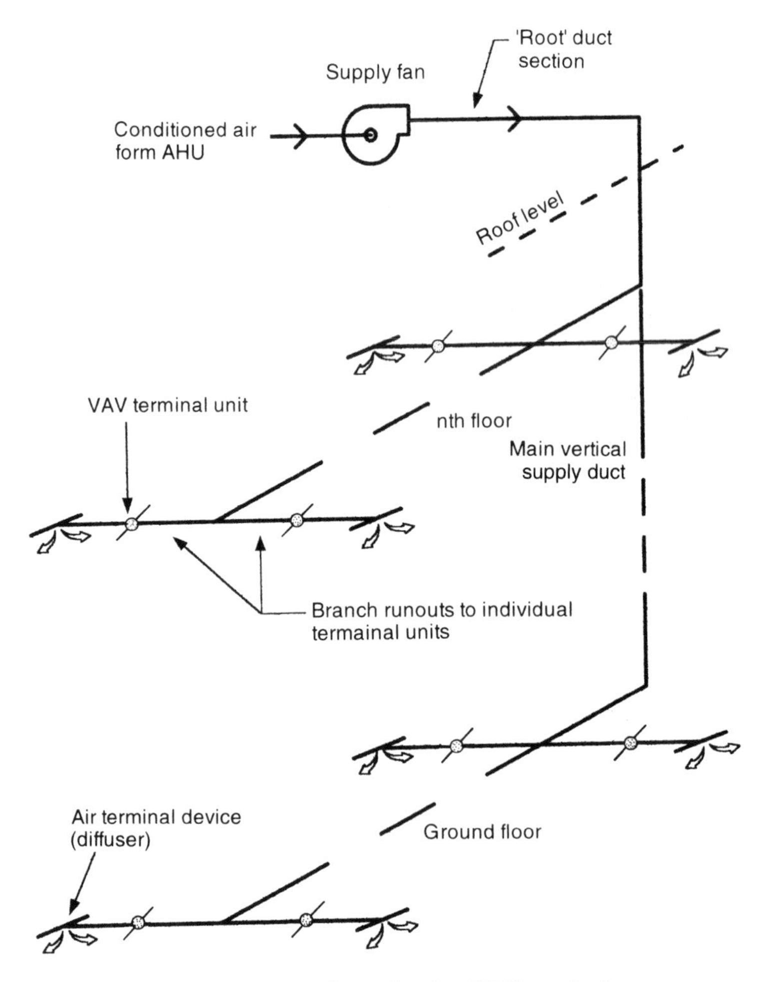

Fig. 2.3 Simple branched tree configuration for VAV supply ducts.

the physical layout and zoning of the application. Vertical branching to serve individual floors of a multi-storey building is a typical example of this. At the lowest level branches serve individual *VAV terminal units*. At the present time, air distribution ducts in commercial applications are still, almost without exception, manufactured from galvanised sheet steel. This material is readily available, economic, and well suited to both manual and machine fabrication techniques. Ducts are most commonly of simple rectangular or circular cross-section, although the more complex flat oval shape has its proponents. Once installed, they are clad externally with thermal insulation. *Ductwork* is the generic term for air distribution ducts and their associated fittings and accessories. Duct fittings include bends and changes of shape or cross-sectional area. Accessories principally include fire, balancing and control dampers. A ducted distribution system, or a part of a system complete in itself, is sometimes described as a range of ductwork. If necessary to achieve design noise levels in the spaces served, acoustic attenuators or silencers may be connected into the ductwork, or suitable lengths of duct may be lined with acoustic insulation.

It is not uncommon for individual zones to be served by single terminal units, although large zones may require several identical units. The terminal units control the volume flow rate of conditioned air that is finally introduced into each zone. In variants of the basic cooling-only system this flow may be heated by fitting terminal units with individual LPHW or electric heating coils. Noise generated by the (throttling) flow control action of the terminal units, and propagated downstream through the discharge ductwork, is typically attenuated by fitting purpose-made acoustic attenuators to the terminal units themselves. Where necessary, this may be combined with lining of the ductwork on the discharge side of the terminal unit with acoustic insulation.

The supply air stream is introduced into each zone via one or more *air terminal devices* (ATDs). The function of the ATD is to distribute the supply of conditioned air within each space in such a way that the environmental design criteria are achieved. In the commercial applications with which the present work is primarily concerned, this means the provision of thermal comfort. This is the role of room air distribution (RAD). In VAV systems the air terminal devices are typically *diffusers*. These may be of rectangular or circular planform, although the characteristics of linear slot diffusers have made them a popular choice with designers of VAV systems. Some designs of terminal unit incorporate an integral diffuser, usually of the linear slot or circular 'swirl' type. Other types are physically separate from the diffusers, which they supply via short lengths (or ranges, if the layout is more complicated) of ductwork downstream of the terminal unit.

Traditionally VAV terminal units were linked with downward air distribution from high level using ceiling-mounted diffusers. The terminal units were themselves thus inevitably mounted in the ceiling void. However, at least one manufacturer did for a time offer a design of terminal unit for perimeter installation, and in recent years a growing number of installations have placed terminal units within raised floor voids using circular swirl diffusers to distribute air from floor level. The physical design and operating characteristics of terminal units themselves are considered in Chapter 7.

Varying the volume flow rate of supply air to each zone has strong implications for the performance of air terminal devices, the successful operation of the room air distribution design, and thus ultimately on the level of thermal comfort achieved in the space served. This is particularly so where air terminal devices are ceiling mounted. The principal factor of interest to the designer of VAV systems is the limitation that a particular type of air terminal device has on the degree to which its supply air volume flow rate may be reduced from its maximum design value without leading to an unacceptable deterioration in room air distribution performance. Just how far the volume flow rate may be safely reduced is defined in terms of a *turndown ratio* or *percentage turndown*. Turndown ratio is the ratio of maximum design volume flow rate divided by a minimum value which some degree of testing has shown can maintain acceptable room air distribution in a typical application. It is thus always greater than unity. Percentage turndown, on the other hand, is the percentage of its maximum design value by which the supply volume flow may be reduced. A turndown ratio of 5:1 and a percentage turndown of 80% are thus equivalent.

The *zone subsystem* comprises the individual control zones. Where there is only one terminal unit per zone, there are obviously as many control zones as there are terminal units. Individual VAV systems may comprise hundreds of terminal units, and large commercial applications with upwards of 2000 terminal units spread between a number of zoned VAV systems are not unknown.

It is in the zones that the cooling load on the VAV system arises. Here the cool supply air is distributed, and mixing with room air takes place to result in a certain level of thermal comfort within the occupied zone. It is the aim of the design and operation of the VAV system that this level should be acceptable, in terms of the design brief, in every zone.

In industrial VAV applications, pressure control within and between zones may be important. However, for typical commercial applications it may be assumed that the VAV system is operating to maintain a dry-bulb temperature set-point within each control zone. The implications of VAV operation on the control of humidity and air movement within individual zones are discussed in the following chapters. Both are significant factors in achieving thermal comfort. The thermal behaviour of individual zones will not be considered in detail.

Air is extracted from the individual zones, by the fourth and final element of Fig. 2.2. This is the *exhaust subsystem*, the individual elements of which are a variable-duty extract/return air fan, a network of extract ducts, and the extract air terminal devices in the zones themselves. The latter may be diffusers or grilles, although air-handling luminaires are also widely used with VAV systems. The use of air-handling luminaires for extract has definite implications for VAV design and operation, and these are considered in the following chapter.

Where, as shown in Fig. 2.2, a proportion of the air extracted from the zones may be recirculated through the air conditioning plant, the components of the exhaust subsystem may equally be termed the *return* air fan, return air ducts and return air diffusers, etc. While UK designers have traditionally favoured the use of an extract/return fan, alternative approaches are possible, and these are considered further in Chapter 9. In particular the use of a relief fan, rather than a return fan, has had its fair share of proponents in the USA.

Previous comments regarding ductwork in the supply subsystem are in general equally applicable to VAV extract ducts. However, it is often the case where zones are open plan that air is extracted generally from the ceiling void via simple stub ducts terminating in the ceiling void itself local to a common vertical riser, rather than via fully ducted connections to individual zones. The unducted return air diffusers or air handling luminaires then simply provide an air transfer path from individual zones into the common return air plenum formed by the ceiling void. Extract ductwork in VAV systems is thus typically less extensive than the supply ductwork, with consequent benefit to the capital cost of the installation.

VAV terminal units for regulating the volume flow rate of air extracted from a zone are commercially available, and are indeed little different from their supply-side counterparts. However, they will typically only be encountered in special applications requiring the accurate matching of air extract to supply at zone level. This is usually associated with a need for close pressurisation control and is more likely to arise in industrial air conditioning applications.

Capital cost considerations alone mean that in UK commercial VAV applications terminal units are invariably restricted to the supply side only. Only the total volume flow rate handled by the extract/return air fan is regulated.

HVAC designers commonly use positive pressurisation of air conditioned areas to minimise infiltration and the associated sensible and latent cooling loads. The extent of this pressurisation is invariably slight, since it must not interfere with door opening or closing. This effectively limits it to a level of less than 50 Pa above adjacent unconditioned areas or relative to 'free air' (outside the building). The pressurisation is achieved by operating the air conditioning system with an excess of supply air volume flow rate over extract, a differential of between 10 and 20% of the supply value being typical. It is also quite common for some air to be drawn from the zones and exhausted directly from the building by local mechanical extract ventilation systems. Toilet extract systems offer the most common example of this in commercial applications. Matching the total extract volume flow rate to the varying supply volume flow rate, whilst making allowance for these additional effects, is a major element in the control of VAV systems – and one that has generated much discussion in the subject literature. Controlling the duty of the extract/return fan is thus again at the heart of VAV system engineering. Without terminal units, the behaviour of the exhaust subsystem is fundamentally different from that of the supply sub-system. Effectively it operates as a single-zone VAV system.

2.3 Increased complexity: dual-duct VAV systems

Dual-duct VAV systems may be categorised according to the number of supply air fans that are employed. In *dual-duct, single-fan* systems a common supply fan serves both hot and cold duct systems. In *dual-duct, dual-fan* systems the hot and cold duct systems are each served by a dedicated supply fan. The former approach is shown schematically in Fig. 2.4, the latter in Fig. 2.5.

2.4 Hybrid and pseudo-VAV systems

Although rarely encountered in modern VAV applications, a design approach is available that provides a variable-volume supply to the spaces served by bypassing or 'dumping' excess conditioned air into the ceiling void return air plenum under part-load conditions. While individual zone control is achieved, the approach generates neither fan energy savings nor load diversity benefits to sizing of the system. The cool ceiling that results may contribute to an overcooling problem at low load. Instead of 'dumping' unwanted conditioned air into the ceiling void, it may be bypassed directly into the return air duct system (Fig. 2.6). While this may avoid the potential overcooling problem associated with creation of a cooled ceiling, it still generates neither fan energy savings nor load diversity benefits to sizing of the system.

A modern approach to bypass VAV systems is shown in Fig. 2.7. Here each zone has its own throttling damper which delivers a variable supply rate of conditioned air according to the requirement of its zone thermostat. A bypass connection between the main supply and extract ducts allows constant-volume system fans (for low capital cost) to continue to operate at constant volume while the cumulative effect of the throttling action by the zone dampers reduces the total rate of supply to the zones. To allow the use of constant-volume system fans, the volume flow rate in the bypass duct is controlled to maintain a constant static pressure just downstream of the bypass duct. A master thermostat controls changeover of the central plant from supplying heated air to chilled air, and vice versa, based on analysis of feedback from the individual zone thermostats. Individual zone control is again provided, and the system certainly depends on VAV techniques for its operation and control. Furthermore it may also allow some load diversity for system sizing. However, its mode of operation does not yield the fan energy savings characteristic of true VAV systems.

Other more exotic hybrids of constant-volume and variable-volume technology have on occasion been proposed, including combining dual-duct VAV with a limited constant-volume function based on a supplementary

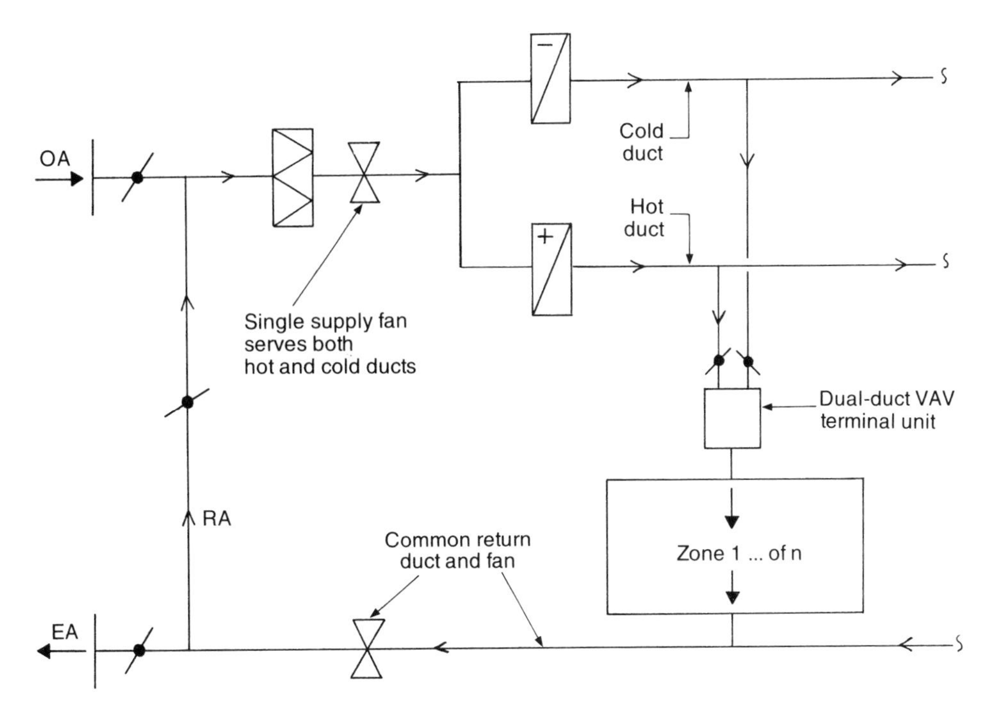

Fig. 2.4 Dual-duct, single-fan VAV system.

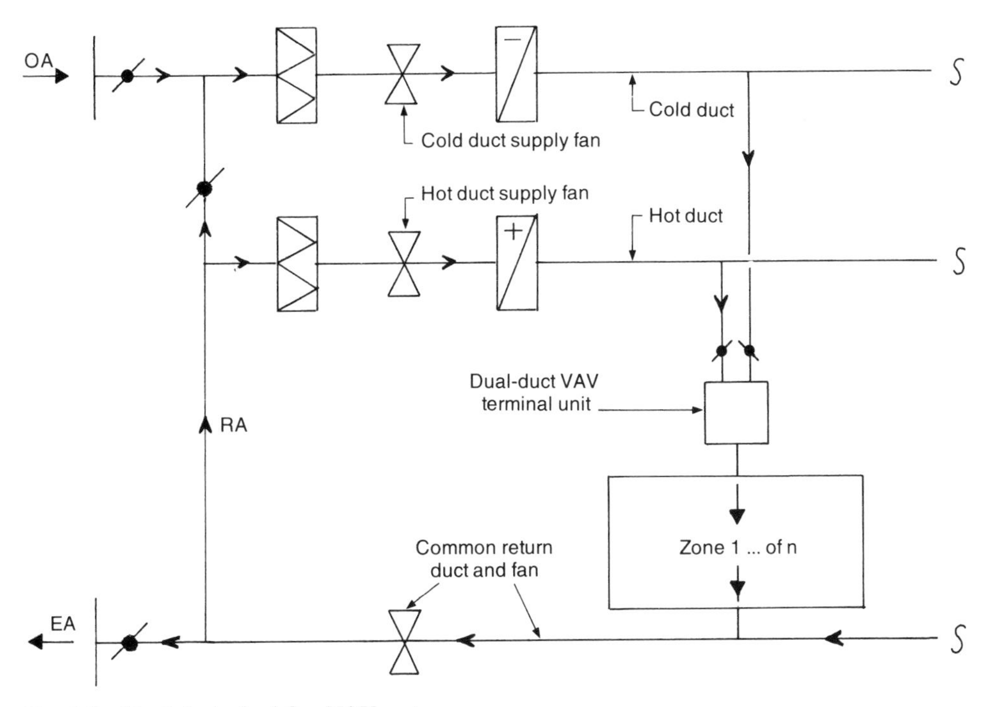

Fig. 2.5 Dual-duct, dual-fan VAV system.

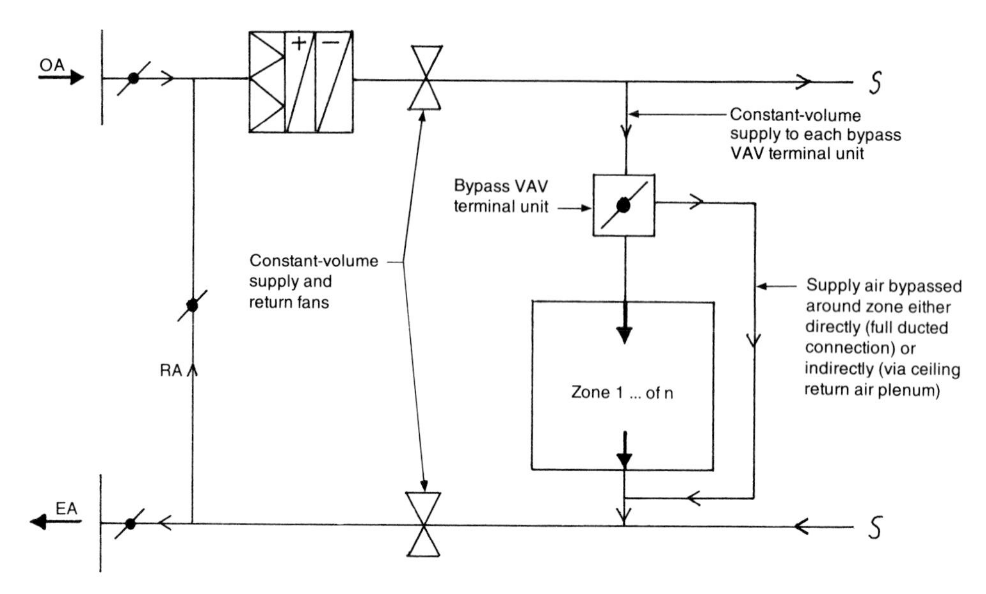

Fig. 2.6 Bypass VAV system with bypass at zone level.

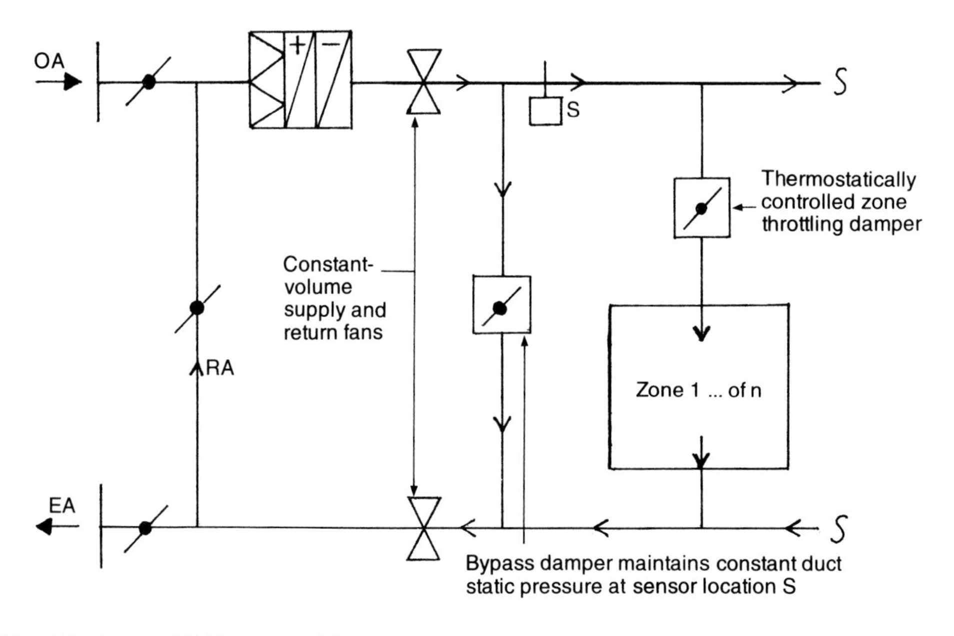

Fig. 2.7 Bypass VAV system with bypass at system level.

cooling coil in the 'hot' duct. However, the application of such hybrids has generally been so limited as to render any discussion of their characteristics superfluous here. While perfectly valid and valuable applications of HVAC technology (including aspects of VAV systems engineering) within themselves, systems supplying large single zones with a variable volume flow rate of conditioned air, or multi-zone systems which are controlled as a single zone, are not regarded as true VAV systems for the purpose of this book.

References

1. Procel, C.J. (1974) Variable air volume systems: loads and psychrometrics. *ASHRAE Transactions*, **80**, Part 1, 473–479.
2. Building Services Research and Information Association (1990) Product profile – terminal units (2), *BSRIA Statistics Bulletin*, **15** June, i–iv.

Chapter 3
VAV System Fundamentals

3.1 The psychrometry of VAV systems

Psychrometry, the physics of air/water vapour mixtures, is fundamental to the design of any air conditioning system. It has been suggested to more than one student of air conditioning, perhaps tongue-in-cheek, that one reason for the popularity of VAV systems with HVAC designers is that they are simple psychrometrically! Since HVAC designers are only human, there may of course be some element of truth in this suggestion. The Psychrometric Chart, that traditional engineers' tool for psychrometric design, deals only with specific properties of the air being conditioned, i.e. enthalpy, moisture content and volume as expressed per kilogramme of dry air. Any process shown on the chart is thus independent of volume flow rate, and the treatment of a VAV system, whether in single-duct or dual-duct configuration, is no different from that of corresponding constant-volume all-air systems.

3.2 The summer cooling cycle for single-duct VAV

Figure 3.1 shows the generalised summer and winter cooling cycles for the basic single-duct, cooling-only system of Fig. 2.2 with steam humidifier added. The components of the psychrometric subsystem are shown inset top left. Since, for convenience and ready comparison, Fig. 3.1 shows both summer and winter cooling cycles, the state points of each are defined using suffixes S and W respectively.

Under summer design conditions, outside air at state O_s is mixed with a proportion of the air extracted from the zones which is recirculated at state R_s', and a mixed-air state M_s results. State R_s is the *mean* of the states of the air in all zones served.

It is characteristic of VAV systems that dry-bulb temperature (t_R) is controlled on a zonal basis, and in typical commercial applications of VAV systems a common zone design value may be assumed. However, even under summer design conditions, variations between zones (t_{RS1}, t_{RS2}, ... t_{RSn}) will result from differing *offsets* from the design value where simple proportional

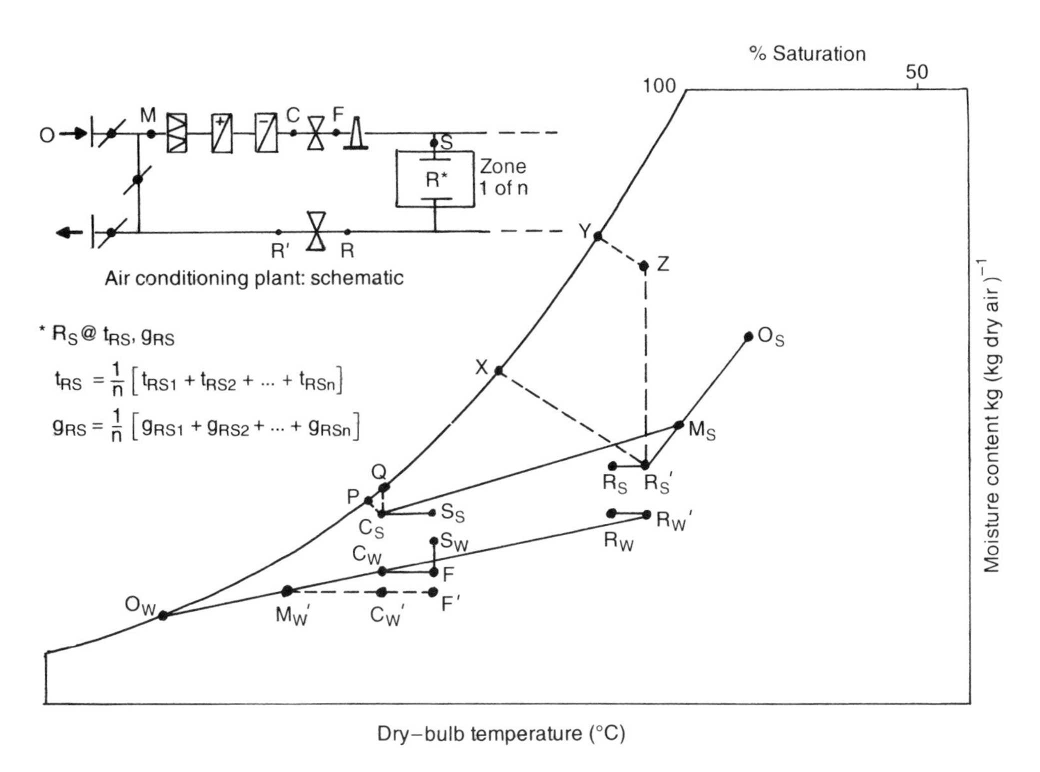

Fig. 3.1 Psychrometric chart for year-round cooling with VAV.

control is employed (since load diversity exists between the zones), and from imperfect mixing between supply and room air in each zone. Proportional control offset can, however, be removed (albeit typically at the expense of some increase in capital and operating costs) by the use of *proportional-plus-integral* (P + I) control of zone temperature. Furthermore, while dry-bulb temperature is typically controlled on a zonal basis, *humidity is not*, leading to variations in zone moisture content ($g_{RS1}, g_{RS2}, \ldots g_{RSn}$). There will thus be a range of zone air states ($R_{S1}, R_{S2}, \ldots R_{Sn}$), each combining the respective zone dry-bulb temperature and moisture content ($t_{RS1}/g_{RS1}, t_{RS2}/g_{RS2}, \ldots t_{RSn}/g_{RSn}$), although under system design conditions some of these stages may well be common. Humidity control in VAV systems is considered in Chapter 4. For the present it is sufficient to note that the adequacy of humidity control afforded by a VAV system should be assessed by estimating the greatest potential deviation of any zone moisture content from the average value. This requires the zone to be identified which experiences both the lowest design value of sensible heat ratio (SHR), i.e. the ratio of sensible to total (sensible plus latent) cooling load, and the lowest proportion of its zone design sensible cooling load under system design conditions.

Between the point of extraction of the air from the zones and mixing with outside air at O_s the mean state of the air changes to R_s'. Since in the return air path there are no plant items capable of performing intentional dehumidification or humidification, the moisture content is assumed to remain unchanged at the mean of the zone values. The difference between states R_s and R_s' is thus one of dry-bulb temperature. It is typically the case in VAV systems serving open-plan office areas that the ceiling void is used as a *return air plenum*. The heat gain to this from recessed light fittings and through the perimeter fabric is removed by the return air drawn from the plenum and returned to the central AHU. It is generally accepted that up to 30% of the total potential sensible heat gain from simple recessed luminaires may be removed via the return-air plenum in this way, and rather more than this (*c*.40%) where the return air is drawn into the ceiling plenum through air-handling luminaires. Air drawn from the ceiling plenum and exhausted from the building carries with it this heat. Air that is recirculated through the central AHU carries it back to the main cooling coil, where it is removed without requiring a cooling air flow to the zones. Under steady-state conditions the heat emission from the lighting is balanced by a sensible gain to the zone and the heat carried away into the return-air plenum. The temperature rise of an extract air flow which removes X% of the total sensible heat gain from lighting is given by:

$$\text{Temperature rise (K)} = (0.01\, X\, \dot{Q}_L) \div (\rho\, c_p\, \dot{V}_E) \tag{3.1}$$

where

$\dot{Q}_L$ = sensible heat gain from lighting (W)
ρ = density of air ($kg\,m^{-3}$),
c_p = specific heat of air at constant pressure ($kJ\,kg^{-1}\,K^{-1}$)
$\dot{V}_E$ = extract volume flow rate (litre s^{-1}).

Substituting nominal values of $1.2\,kg\,m^{-3}$ and $1.02\,kJ\,kg^{-1}\,K^{-1}$ for density and specific heat respectively:

$$\text{Temperature rise (K)} = 0.00817\, X\, \dot{Q}_L \div \dot{V}_E \tag{3.2}$$

Between leaving the ceiling void and reaching the return fan, heat transfer occurs across the walls of the return ductwork as it passes through any space or void where the air temperature is different from that inside the duct. Certainly as far as the summer cooling cycle is concerned, a net rise in the dry-bulb temperature of the return air may be expected. Any increase in the temperature of air recirculated through the central AHU means that more heat must be removed at the main cooling coil. If ventilation is inadequate or restricted by location, the combination of high outside temperature under design conditions and heat gains resulting from plant operation can make HVAC plant rooms particular problem areas in this respect. Notwithstanding specific problem areas, however, the cost effectiveness of applying thermal insulation to return air ducts is likely to be marginal unless the air-to-air temperature difference between the return air duct and the surrounding service void is significant. In any event, calculation of heat transfer to or from ductwork requires the prior definition of duct layout, sizes, and design volume flow rates. Hence during the early stages of design it is customary to use an allowance based on experience and the estimated extent of the duct system. It is of little consequence if different paths through the return ductwork give rise to different amounts of heat transfer to the return air, because they all ultimately converge together in the root section of the network.

FInally, the return air experiences a further increase in dry-bulb temperature due to fan gain. This represents two components:

(1) the power actually transferred to the air stream by the fan (the *air power*);
(2) that portion of the difference between the air power and the power input to the fan motor itself, representing the energy dissipated by inefficiencies in the fan/drive/motor combination, that is actually transferred to the airstream.

Air power is dissipated as heat, through the frictional losses experienced by an airstream as it flows through ductwork and any associated items of plant and equipment. However, for convenience in defining the psychrometric processes of the cooling cycle, the temperature rise associated with this gradual conversion of air power into heat is conventionally lumped together with the temperature rise due to inefficiencies in the fan/drive/motor combination (which does *in reality* occur across the fan), and both are regarded as occurring across the fan. Whatever the mechanism and location of the heat transfer associated with the fan gain, the end result is the same – an increase in the dry-bulb temperature of the airstream which must be taken into consideration when plotting or calculating the psychrometric processes of the cooling cycle.

The fan gain to the return airstream is naturally affected by the efficiency of the fan, and the location of the fan motor (whether it is within the airstream or not). However, a temperature rise of the order of 1 K per kPa of fan total pressure can be shown from a simple heat balance on the airstream, and is a widely accepted basis for general psychrometric design. Fan gain is directly proportional to fan power, and both fan power and the capacity of the airstream to absorb heat are directly proportional to volume flow rate. Temperature rise across the fan is, therefore, *independent* of volume flow rate, and the difference between system supply and extract air volume flow rates is of no consequence in this respect. The absence of terminal units and psychrometric plant items in the return air path, combined with the limited extract ductwork systems made possible by the use of ceiling void return-air plenums, means that the total pressure requirements of VAV return fans are typically modest. Temperature rise of the return airstream across the fan is thus unlikely to exceed 1 K under design conditions, with 0.5 K being a more typical estimate. Frequently, as in Fig. 3.1, a combined allowance is made for temperature rise of the return airstream due to both fan and duct sensible heat gains.

The mixture of outside and recirculated return air at state M_S passes through the main cooling coil in the central AHU. This cools and dehumidifies the mixed airstream to state C_S, which depends upon the performance of the cooling coil. The principal interest of the HVAC designer is to determine the design load which must be satisfied by the cooling coil. This requires only a knowledge of the mixed-air state at which the airstream enters the coil, the state at which it leaves the coil, and the mass flow rate of the airstream. For this reason it is customary to show the psychrometric process occurring as the airstream passes through the cooling coil as a straight line between the entering and leaving states. In reality the heat and mass transfer mechanisms occurring mean that, even for an ideal coil, charting the changing state of the air stream as it passes through the cooling coil is considerably more complex than this simple process line would suggest. Here it is sufficient to note that the performance of the coil is a function both of its physical design (number of rows of tubes, tube spacing, fin design and pitch, etc.) and of the operating parameters of the cooling medium used, commonly chilled water or refrigerant (in a direct expansion or DX coil).

Real heat transfer processes require the existence of a finite temperature difference between cooled and cooling media. Neglecting any considerations of operating economy or efficiency, using conventional refrigerating plant places a lower limit to chilled water flow temperature of *c*.4–5°C to avoid the risk of freezing the evaporator in the refrigeration plant. The use of ice-storage in conjunction with mechanical refrigeration may allow chilled water flow temperatures of only a few degrees to be achieved in practice, with a consequent reduction in both temperature and moisture content of the airstream off-coil. Although the application of DX cooling coils to VAV systems may be considered inherently problematic, evaporating temperature is in any event ultimately limited by the need to avoid freezing of the moisture condensing out of the airstream.

A cooling coil consists of a number of small-bore copper tubes fixed horizontally across the air stream. Tubes are spaced vertically above each other in banks, and each such bank of tubes is termed a row. Chilled water (or refrigerant) flows through the tubes, while the the air stream that is to be cooled flows over them. The heat and mass transfer achieved by the coil is improved by increasing the area of chilled surface available for contact with the air stream. A great increase in area is typically achieved by fixing close-pitched copper or aluminium fins to the outside of the tubes. For a given cross-sectional (or face) area, coil performance may be improved by increasing the depth of the coil, in the direction of air flow, by adding further rows. The effect of this on both manufacturing cost and the pressure drop of the airstream can be appreciated by likening the process to adding another single-row coil in series for each additional row. A lower off-coil state, C_S, thus typically requires a deeper cooling coil, which is more expensive and has a higher air-side pressure drop.

In air-conditioning applications, the temperature differences driving the heat transfer in cooling coils are much lower than in heater batteries, with a *log-mean temperature difference* (LMTD) of the order of 6–8 K (in comparison with 55–60 K for a heater battery). Chilled water flow temperatures in practice may be up to 7°C, returning from the coil at about 13°C. To achieve a maximum LMTD, cooling coils are typically piped in contra-flow, so that the coolest chilled water always meets the coolest air.

In the basic cooling-only VAV system the dry-bulb temperature of the airstream leaving the cooling coil is maintained constant throughout the annual operating period, a technique commonly termed *off-coil control.* However, because the airstream leaving the cooling coil is close to saturation (remembering that for an ideal coil it would be saturated), and at saturation dry-bulb and dew-point temperatures are identical, this has led to the technique also being described as *dew-point control.* The design value of the off-coil temperature is determined at the design cooling load.

As far as VAV operation is concerned, adequate mixing of outside air with recirculated air from the return airstream is essential, since the reduction in total supply airflow under part-load conditions could result in stratification of the two airstreams, leading to control problems. To achieve satisfactory mixing of the outside and recirculated airstreams requires control dampers with good characteristics, and the existence of equal air pressures upstream of the dampers[1]. Furthermore, such diverse factors as wind pressure acting on the outside air intake and exhaust air discharge openings on the building face may alter flow patterns in the fresh and exhaust air duct connections, and affect operation of an outside-air economy cycle.

Any stratification of the mixed airstream has the potential to cause a deterioration in the heat and mass transfer performance of the main cooling and heating coils, and associated problems in their control. The greater log-mean temperature difference between the fluid streams when heating means that the performance of heater batteries is typically more robust in this respect than that of cooling coils. Flow separation in the AHU is thus to be avoided, as is the positioning of the supply fan inlet or discharge in close proximity to the face of the cooling coil. Establishing and maintaining even airflow patterns and ensuring thorough mixing of the airstream before reaching the cooling coil are thus fundamental considerations in the design of the AHU.

The airstream leaving the cooling coil is handled by the supply fan. The fan gain from this results in an increase in dry-bulb temperature that may again be taken to be of the order of 1 K per kPa of supply fan total pressure. Pressure drops across the plant items in the central AHU, the more extensive supply ductwork, and the pressure drop across VAV terminal units results in a higher design total pressure requirement for the supply fan. The consequence is a temperature rise of between two to three times that associated with the return fan. Since the early 1970s, an increasing awareness of the need to reduce the energy consumption of HVAC systems has lead to the conscious avoidance of high-velocity air distribution systems with their associated high fan total pressure requirements. VAV systems, therefore, are typically designed as low- to medium-pressure systems, with highest mean duct velocities of less than $15\,m\,s^{-1}$ and supply fan total pressures up to approximately 1.5 kPa. This implies that a temperature rise of between 1 and 2 K must be allowed for fan gain to the supply airstream under design conditions.

Packaged central-station AHUs may be configured as either *draw-through* or *blow-through* units, according to whether the supply fan either sucks or blows air across the main heating and cooling coils. In the former case the fan gain reheats the supply airstream, in the latter it preheats it. In either case the sensible heat gain must be offset by the the main cooling coil. The blow-through arrangement has the potential to achieve a lower supply air temperature, because the dry-bulb temperature of the supply airstream leaving the AHU is then a function only of the lowest temperature which can be achieved by the cooling coil. Naturally the implications for achieving a low off-coil temperature apply equally here. However, three disadvantages of the blow-through configuration are serious enough to mean that it is rarely seen. First, the uneven flow distribution in the discharge from centrifugal supply fans (the type most commonly used in packaged AHUs) is likely to result both in a deterioration in the performance of the main cooling coil and control problems. Second, a potential *hygiene* problem exists from an increased risk of carry-over of condensate from the cooling coil into the main supply duct. While this may be minimised by the use of suitably designed moisture eliminator plates, albeit at the expense of introducing an additional pressure loss into the AHU, the introduction of the Construction (Design and Management) Regulations 1994 can only add to the significance of this consideration. Third, in the blow-through arrangement the positive pressure differential which exists between the AHU and its surroundings results in a wasteful leakage of conditioned or partially-conditioned air into plant rooms. It also typically results in a need for a deeper trap to be provided on the drain from the cooling coil.

The supply airstream leaving the central AHU is distributed through the network of supply ducts to the conditioned zones. Between leaving the central AHU and arriving at the zones, a further increase in dry-bulb temperature results from heat transfer through the walls of the ducts as they pass through warm plant rooms, service ducts and ceiling or roof voids. This is another effective, but undesired, reheating of the supply airstream that must be offset by the main cooling coil. The extent of the duct gain is reduced as far as economically practical by thermally insulating the supply ductwork. From consideration of a simple heat balance over unit length of duct, the heat transfer to the supply airstream is proportional to the wetted perimeter, while the consequent temperature rise is proportional to duct cross-sectional area. The ratio of wetted perimeter to duct cross-sectional area is inversely proportional to duct size, and the temperature rise of the supply airstream, for a constant air temperature around the duct, is thus disproportionately greater in the smaller ducts closer to the zones. In the early stages of design an estimate must usually be made of the temperature rise due to this duct gain. In Fig. 3.1 the effects of both fan and duct gain have been combined.

Air is supplied to the zones at state S_S, and experiences increases in both dry-bulb temperature and moisture content, to states R_{S1} ... R_{Sn} respectively, in offsetting the cooling loads on each zone. State S_S is a nominal value with respect to dry-bulb temperature. The moisture content of the supply airstream does not change between the central AHU and all zones. However, in the supply network the duct gain to the individual supply airstreams which finally reach the zones may differ. Because of this, the actual supply air temperature may be slightly different to each zone. In a VAV system the individual terminal units will automatically compensate for this, an action that will be indistinguishable from the general volume flow rate control function of the terminal units. The greater the nominal supply air temperature differential (design values of $t_{RS} - t_{SS}$) at which air can be supplied to the zones, the lower can the system volume flow rate be. This benefits capital and operating costs and system space requirements. The permissible supply air temperature differential is ultimately limited by the performance of the room air distribution design, and avoidance of draughts in the occupied space. Traditionally a differential of the order of 8–10 K has been regarded as a practical maximum for conventional mixing air distribution from high level using ceiling-mounted diffusers. The magnitude of both fan and duct gain must, therefore, be seen in terms of representing 10–12.5% of the total cooling capacity of the supply airstream leaving the main cooling coil, per 1 K rise in temperature.

The actual design supply air temperature itself must also represent a practical cooling coil performance. For cooling-only VAV systems using mixing air supply techniques in the zones, it is typically *c*.13–14°C for a design zone air temperature of 22°C. A principal interest in the use of fan-assisted VAV terminal units is their ability to receive air at a lower temperature from the central plant. The terminal unit fan allows mixing of the low-temperature primary supply with air drawn from the ceiling return plenum to maintain a combined volume flow rate (and thus avoid the risk of draughts) with reducing zone cooling load. A reduction in design supply air temperature of 1 K, with a corresponding increase in the supply air temperature differential, represents a potential for reducing system volume flow rate by between 11% (relative to an 8 K supply differential) and 9% (relative to a 10 K differential). A reduction of 3 K in supply air temperature thus offers a potential reduction in system volume flow rate of the order of one-third. The use of such low-temperature supply airstreams is generally termed cold air distribution, and is typically associated with the use of ice storage techniques. In addition to the duct gain to such a supply airstream, its potential to cause condensation must also be considered, as its temperature will typically be below the dew-point temperature of air around the ducts. (At a nominal room design condition of 22°C dry-bulb temperature and 50% saturation, air has a dew-point temperature of approximately 11.5°C.) Even at conventional supply air temperatures, thermal insulation to supply ducts typically incorporates a vapour barrier to prevent the risk of moisture passing through the insulation material to condense on the cold surface of the duct, or interstitially within the insulation itself. At a low supply temperature the consequences of any discontinuities, imperfections, or deficiencies in workmanship during installation of the vapour barrier can only be more serious.

3.3 The winter cooling cycle for single-duct VAV

Figure 3.1 also shows the corresponding generalised winter cooling cycle for the single-duct, cooling-only system of Fig. 2.2 with steam humidifier added. The state points of the winter cycle are indicated by suffix W. For the same zone design dry-bulb temperature the moisture content of state R_W, still the mean of all zone states (R_{W1}, R_{W2}, ... R_{Wn}), and hence of all zone dry-bulb temperatures (t_{RW1}, t_{RW2}, ... t_{RWn}) and moisture contents (g_{RW1}, g_{RW2}, ... g_{RWn}), ... will be lower than for the summer cycle. Extracted air returning to the central AHU from the zones still experiences a rise in dry-bulb temperature resulting from fan and duct gain, to state R_W'. In reality the magnitude of the duct gain may be expected to differ from the summer cycle, depending on the winter temperature of the plant rooms and service voids through which the ducts pass. Since the winter cooling cycle will inevitably represent part-load operation, the heat transfer between the duct and its surroundings may also be expected to change (increase) as volume flow rates and duct velocities reduce.

In VAV systems the temperature rise due to fan gain reduces under part-load operation. This may be confirmed by consideration of the basic heat balance which equates fan gain to the temperature rise of the airstream handled by the fan. The proportionality of the temperature rise to fan total pressure, already noted under design conditions (Section 3.2), remains equally valid under part-load operation. If, as is typically the case, only the overall system extract volume flow rate is controlled, fan total pressure is a simple function of volume flow rate. Where fan efficiency is reduced under part-load, the rate of temperature rise per unit of fan total pressure will be correspondingly increased. The efficiency of variable-speed fan drives may also be expected to drop off with decreasing fan duty, with implications where heat dissipated by the drive passes into the airstream. However, having noted the theoretical effects, it must be said that there is little (if any) justification for the designer of a VAV system to consider them further, because the temperature rise due to fan gain for the return fan is typically low even under design conditions. Furthermore, system processes will inherently compensate for minor changes.

A proportion of the return airstream is recirculated, mixing with outside air at state O_W to achieve the year-round design temperature off the main cooling coil (state C_W), and is then supplied to the zones. The dry-bulb temperature of the mixed airstream is again increased from state C_W to state F by fan and duct gain on the supply side of the system. Again as far as VAV operation is concerned, supply fan power reduces under part-load conditions, although the supply fan total pressure is not in this case a simple function of the system volume flow rate. Certainly traditional control strategies for VAV supply fans were evolved to maintain duct static pressure nominally constant at some specified location along the main supply duct. Even so, supply fan total pressure will tend to decrease under part-load operation, and with it the temperature rise of the supply airstream due to fan gain. Again also, both fan total efficiency and the efficicency of a variable-speed drive may also be expected to decrease under part-load conditions. For practical purposes there is again little justification for the designer of a VAV system to regard the temperature rise of the supply airstream due to fan gain as other than constant. Duct gain may similarly be expected to show some variation with volume flow rate. At the zones the VAV terminal units will in any case attempt to adjust the supply volume flow rates to compensate for any variations in supply air temperature. Minor effects of changes in fan and duct gain under part-load conditions will thus be indistinguishable from normal operation of the terminal units.

During the winter cycle the moisture content of the supply airstream is increased by direct steam injection to keep the mean value of all zone relative humidities above a minimum of approximately 40%. For practical purposes the increase in dry-bulb temperature of the airstream during steam humidification may be ignored, and the process treated as one of moisture addition at constant dry-bulb temperature. Fan and duct gain are again shown as a combined allowance, prior to humidification. As in the summer cycle, the extent of duct gain along each path through the network may differ, and with it the precise dry-bulb temperature of the supply air to each zone. Air is supplied to the zones at a nominal state S_W, and experiences increases in both dry-bulb temperature and moisture content, to states R_{W1}, R_{W2}, ... R_{Wn} respectively, in offsetting the reduced cooling loads on each zone. As shown, state R_W is a mean value. If state C_W cannot be achieved by mixing of outside and return air alone, due perhaps to the proportion of outside air required to assure adequate ventilation of each zone, the mixed airstream at the new state (shown as M_W') is pre-heated by the heater battery in the central AHU to state C_W'. This differs from state C_W only in moisture content. The effect of fan and duct gain on the supply side of the system then result in state F′, and an increased rate of steam injection must be provided to achieve the design supply air condition, S_W.

UK winter design outside air dry-bulb temperatures are of the order of −1 to −3°C for the southern part of the country. It requires only simple arithmetic to show that over this range of values of O_W an off-coil temperature of up to 12°C should be possible using up to 60% recirculation. The daily mean outside air dry-bulb temperature, averaged over the period October to April inclusive, is quoted[2] as being 6°C or higher for most of England and Wales (7°C for the London area).

3.4 The outside air economy cycle and 'free cooling'

In an all-air system the supply airstream typically performs both ventilation of the zones served and the transfer to them of the cooling and dehumidifying effect produced by the central plant. The supply air volume flow rate required for comfort conditioning is typically several times greater than the volume flow rate of outside air required for ventilation. The actual ratio depends on the characteristics of the particular application. For commercial applications under design conditions it will typically be between three and six times the required ventilation rate. In a VAV system, as the supply volume flow rate to a zone reduces under part-load condiditons, this ratio decreases to a minimum value that is typically between one and two. As long as the volume flow rate of outside air required for ventilation is maintained, the balance of the supply airstream may be either outside air, recirculated return air, or a mixture of both, according to which will require the least amount of mechanical cooling to bring it to the supply air state. This is the purpose of the outside air (OA) economy cycle, and is achieved by synchronised modulation of the outside air, recirculation and exhaust air dampers.

The OA economy cycle originated for use with constant-volume systems. Its application to VAV systems presents additional problems of design and control if adequate ventilation of all the zones served is to be ensured. However, the cycle has the potential to reduce the use of mechanical refrigeration considerably, with consequent energy savings, during much of the annual operating period for air conditioning systems in temperate climates. For this reason it is widely used with VAV systems in the UK. However, in other climatic conditions its use may require careful consideration of load and weather profiles, energy utility rates and capital and investment costs. In particular hot and humid climates are unlikely to benefit from an OA economy cycle.

The cycle has the following three distinct operating regimes:

(1) During the summer cooling cycle of Fig. 3.1, when the specific enthalpy of the outside air is greater than that of air at state R_S', the minimum proportion of outside air which will satisfy the ventilation requirements of the zones is used.
(2) When the specific enthalpy of the outside air lies between this return air state and the design specific enthalpy for the supply leaving the cooling coil at state C_S, 100% outside air is used.
(3) Once the outside air temperature reduces below that required for the airstream leaving the main cooling coil, the off-coil dry-bulb temperature is maintained by progressively mixing an increasing proportion of recirculated return air with a decreasing proportion of outside air. This process is continued to avoid use of the main heater battery in the central AHU down to as low a value of outside air temperature as possible. However, it can do so only as long as sufficient outside air continues to be drawn into the central AHU to provide adequate ventilation of all zones. Depending on the ventilation requirements and the system load profile, it may be necessary to override the process to achieve this, increasing the proportion of outside air and heating the mixture to maintain the off-coil temperature.

Practical considerations of sensor cost, performance and reliability have traditionally meant that the control of an OA economy cycle is typically based on a comparison between the dry-bulb temperatures of the outside and return airstreams, rather than between their specific enthalpies. There are, however, two areas of the psychrometric chart where a given outside air state will have a different effect on the operation of the OA economy cycle, depending on whether it is dry-bulb temperature or specific enthalpy that are compared.

The first, and largest, area is that bounded by points $R_S'XYZ$ in Fig. 3.1. In this area the dry-bulb temperature of outside air is less than that of the return airstream, but its specific enthalpy is actually higher. The line YZ is drawn along the line of specific enthalpy at the summer design outside air condition. In this region it would be advantageous to maximise the recirculation of return air, but this cannot be achieved with control based on sensing dry-bulb temperature only. However, while the result is an increase in the annual energy consumption for a typical system, realistic outside air states likely to be encountered in this region of the chart during actual system operation may be expected to represent rather less than 5% of the potential annual operating hours for air conditioning systems in UK commercial applications.

The second, and much smaller, area is lower on the chart, bounded by points C_SPQ. In this area the dry-bulb temperature of outside air is less than the off-coil temperature, while the specific enthalpy is greater than at the off-coil condition. In this region it is thus advantageous to operate under a regime of dry-bulb comparison, since this will not call for any output from the cooling coil. However, the frequency of occurrence of outside air states in this region of the psychrometric chart is so low that it has little practical significance.

Overall, while wet-bulb temperature sensors have been used to give an approximation to specific enthalpy of the outside airstream, analysis suggests that for the UK climate dry-bulb sensors may typically be expected to yield a coincident effect for 95% of the time.

The term 'free cooling' is frequently associated with the operation of an OA economy cycle. This is based on the absence of any need for mechanical refrigeration when the outside air temperature is equal to or less than the design off-coil temperature. It is thus 'free' only with regard to mechanical refrigeration, as fan energy must still be bought and paid for.

3.5 The winter heating cycle for single-duct VAV

Where a VAV system performs space heating as well as cooling, this is typically carried out by the VAV terminal units in each zone. Each terminal unit must, therefore, be given a heating capability in the form of an air heater battery. This results in complete flexibility to cool some zones while heating others. In psychrometric terms, the only modification to the winter cooling cycle of Fig. 3.1 for a zone requiring space heating is that the dry-bulb temperature of the supply airstream received, at state S_W, must be increased at the terminal unit (at constant moisture content) to a value sufficiently higher than that of the design room condition, which for the present purpose we may take as R_W, to offset the fabric and infiltration heat losses from the zone.

It should perhaps be stressed at this point that the provision of air heater batteries will not necessarily be restricted to terminal units serving zones which are subject to fabric and infiltration heat losses (generally perimeter and 'top floor' zones), as there may be a real risk of overcooling at minimum turndown in internal zones.

3.6 Year-round cooling and winter heating with dual-duct VAV

The psychrometric differences between air conditioning systems are typically reflected in differences in the sequence in which the processes are linked together, rather than by differences in the processes themselves. Cost and space requirements mean that dual-duct VAV systems are unlikely to be encountered in the mainstream of new-build and refurbishment applications in the UK. Hence it is not proposed to consider the psychrometric aspects of this major variation on the VAV theme, save to note that:

(1) The hot duct supply to each zone can be modulated to full shutoff.
(2) Mixing of the hot duct and cold duct supplies to any zone is only permitted once the cold duct supply has been modulated down to a fixed minimum setting, which is based either on ventilation requirements or room air distribution performance.
(3) The use of dual fans allows maximum economy of fan energy consumption, and also permits maximum recovery of the heat transferred from light fittings to a ceiling return air plenum.

3.7 Supply volume flow rates

Design supply air volume flow rates for each zone served by a VAV system are determined conventionally from the sensible heat balance equation:

$$\dot{Q}_{SE} = \dot{m}_S \, c_P \, (t_S - t_R) \tag{3.3}$$

where
$\dot{Q}_{SE}$ = sensible heat gain (kW)
$\dot{m}_S$ = supply air mass flow rate ($kg\,s^{-1}$)
c_P = specific heat of air at constant pressure ($kJ\,kg^{-1}\,K^{-1}$)
t_S, t_R = supply, room air temperatures (°C).

Substituting for mass flow rate and transposing yields the supply air volume flow rate:

$$\dot{V}_S = \dot{Q}_{SE} \div [\rho \, c_P \, (t_S - t_R)] \tag{3.4}$$

where
$\dot{V}_S$ = supply air volume flow rate ($m^3\,s^{-1}$)
ρ = density of air ($kg\,m^{-3}$)

and all other terms are as defined for equation 3.3 above.

This illustrates the first justification for the use of VAV systems. Under part-load operation, and with steady values of t_S and t_R, volume flow rate is proportional to sensible cooling load.

By comparison, in an all-air system with a constant volume flow rate the design zone air temperature can only be maintained under part-load conditions by reheating the supply air to reduce the temperature differential ($t_S - t_R$). Reheating, as opposed to simply heating, implies the heating back up of air which has previously been cooled. It is totally wasteful of energy, and is thus to be avoided. In constant-volume systems scheduling of the supply air temperature from the central plant can be employed to reduce the energy required for this reheating. For example, in a constant-volume system operating with terminal reheat, the air temperature off the main cooling coil would ideally be continuously adjusted to supply air just cool enough to satisfy, without reheating, whichever zone was experiencing the maximum proportion of its design cooling load at that instant of time. In this way wasteful reheating by the terminal reheater batteries serving each zone would be minimised. In practice, without a feedback from each zone to enable identification of the critical zone, the only practical way to approach this condition is to relate both plant load and operation to a common variable. Fortunately outside air temperature is generally suitable for this, and the off-coil temperature may be scheduled (or compensated) according to its value. Such a technique can at best only approach the minimum use of reheat, and defining the control schedule itself can be quite involved for all but the most straightforward systems.

Even a perfect scheduling technique can only minimise the reheat requirement. It cannot avoid it altogether. The different times of the day at which the various facades of a building receive peak solar gain implies that, to be really effective, scheduling must be combined with zoning of constant-volume systems according to the orientation of the zones served.

For design purposes it is typically adequate to take the density of air as $1.2\,\mathrm{kg\,m^{-3}}$ at 20°C, 43% saturation and 101.325 kPa,[3] and the specific heat of humid air as $1.02\,\mathrm{KJ\,kg^{-1}K^{-1}}$. A Charles's Law correction can then applied for the actual supply air temperature:

$$\dot{V}_S = [\dot{Q}_{SE}/(t_S - t_R)]\,[(273 + t_S)/359] \tag{3.5}$$

where

$\dot{V}_S$ = supply air volume flow rate ($\mathrm{m^3\,s^{-1}}$)
$\dot{Q}_{SE}$ = sensible heat gain (kW)
t_S, t_R = supply, room air temperatures (°C).

The *simultaneous* cooling load on the VAV system is the sum of the zone cooling loads at a specified instant in time. The highest such load is the simultaneous peak value, and is taken as the design sensible cooling load for the system. Plant and equipment are selected and sized on the basis of this peak simultaneous sensible cooling load.

Where the zones served by the VAV system have various orientations (for example, some with eastward-facing glazing and others with westward-facing), it is evident that the design sensible cooling loads for the different orientations cannot be experienced simultaneously. This gives rise to the second principal justification for using VAV systems – design load diversity. The peak simultaneous sensible cooling load for a VAV system may typically be expected to be between 70 and 80% of the sum of the connected zone peak design loads. This is reflected in smaller air-handling and refrigeration plants than would be required by an all-air constant-volume system serving the same zone loads. Air distribution ducts are smaller, capital cost is reduced, and less space is required for plant and ductwork. The lower space requirement may have some knock-on effect in reducing the cost of the building structure, although the magnitude of this is difficult to quantify.

3.8 Variable supply temperature options

In its basic cooling-only form a single-duct VAV system supplies air to the conditioned spaces at a nominally constant dry-bulb temperature throughout the year. In theory VAV operation removes the need for scheduling of the supply air temperature from the central AHU. However, in the control strategies adopted for some systems this may in fact be varied, usually according to system volume flow rate. The justifications for this may be the provision of adequate ventilation, the maintenance of room air distribution performance, or the balancing of fan and refrigeration plant operation for minimum overall energy consumption. The latter aspect is typically more appropriate to hot and humid climates, where the potential for 'free cooling' with an outside-air economiser cycle is more limited. In the UK's temperate climate, on the other hand, this potential is extensive, and fan operation tends to be the dominant factor in the annual energy consumption of all-air systems.

In any event a VAV control strategy employing variable supply temperature must give full consideration to its implications for internal zones, because here the cooling loads are independent of outside conditions, and may thus peak at any time of the year. Furthermore, a change in the supply air temperature requires a change in the temperature of the airstream coming off the main cooling coil. This may be expected to have an effect on the dehumidification performed by the main cooling coil, on the moisture content of the air supplied to the zones, and thus on humidity control within the zones.

3.9 Latent cooling loads and humidity control

A latent heat balance may be taken for any zone:

$$\dot{Q}_L = \dot{m}_S \, h_{fg} \, (g_R - g_S) \tag{3.6}$$

where

$\dot{Q}_L$ = latent heat gain (kW),
$\dot{m}_S$ = supply air mas flow rate (kg s^{-1})
h_{fg} = latent heat of evaporation of water (kJ kg^{-1})
g_R, g_S = room, supply air moisture contents (kg $kg^{-1}_{DRY\ AIR}$).

This may again be expressed in terms of the supply volume flow rate to the zone, V_S $m^3\ s^{-1}$:

$$\dot{Q}_L = 352 \, \dot{V}_S \, h_{fg} \, (g_R - g_S) \div (273 + t_S) \tag{3.7}$$

This expression also assumes a density for air of 1.2 kg m^{-3} at 20°C, 43% saturation and 101.325 kPa, and includes a Charles's law correction to adjust the volume flow rate for the actual supply air temperature. It is typically adequate for design purposes to take a value of 2454 kJ kg^{-1} for h_{fg}, being the latent heat of evaporation of water at 20°C and 101.325 kPa.[4] An expression for the moisture content in the zone is obtained by rearranging the balance equation:

$$g_R = g_S + [1.159 \times 10^{-6} \times \dot{Q}_L \, (273 + t_S) \div \dot{V}_S] \tag{3.8}$$

From this it is evident that, for a constant latent cooling load and supply air moisture content, g_S, the zone moisture content increases with reducing supply air volume flow rate.

All dehumidification is carried out by the cooling coil in the central air handling unit. In the basic VAV system this is controlled in sequence with the outside and return air mixing dampers to provide air at a nominally constant supply air condition to the supply subsystem. The value of g_S is based on satisfying the design latent cooling load on the zone with the lowest Sensible Heat Ratio – at that zone's design supply volume flow rate and the nominal design relative humidity.

3.10 Chilled water vs. DX cooling with VAV

Economy in capital cost typically makes DX cooling an attractive option as the size of an air conditioning system decreases. However, the use of DX cooling raises a fundamental problem when combined with VAV design. At the heart of this problem is the matching of cooling capacity to system load. If the two do not match, then fluctuations will occur in the temperature of the air stream leaving the cooling coil, which will translate into fluctuations in the supply air temperature to the zones. The range and period of these fluctuations are a function of the *turndown* and *resolution* available from the cooling plant. Turndown is the ratio of maximum capacity to minimum capacity without cycling on/off, while resolution is the smallest achievable capacity change, expressed as a percentage of full capacity.

If the supply air temperature changes continuously, zone air temperatures may never be stable. A zone temperature sensor will not respond instantaneously to a change in supply temperature. It will typically be some minutes (perhaps up to 20) before the VAV terminal unit damper is repositioned to increase or decrease the supply volume flow rate to the zone in a corrective manner. If control of the temperature off the cooling coil is poor, during this time the supply temperature will probably have changed again, and another cycle of zone and terminal unit response will be initiated. Such cyclic changes in supply volume flow rate will be accompanied by a cyclic variation in noise from the supply diffuser.

A DX refrigeration plant typically relies primarily on cycling of the refrigeration compressor(s) to control the off-coil temperature. Provided that the compressors are fitted with cylinder unloaders, a unit with two refrigeration circuits, each with its own two-cylinder compressor, thus has four stages of capacity control (100%, 75%, 50% and 25%), plus zero. This gives a turndown of 4:1, and a resolution of 25%. Unless the cooling load exactly matches one of these capacity stages, cycling of the compressors and the off-coil temperature will occur. If the load is less than the staged cooling capacity, then the off-coil temperature will drop until the highest stage cycles off (or unloads in the case of a cylinder of the compressor). However, the load is then greater than the capacity of the remaining stages, and the off-coil temperature rises until the next stage of capcity cycles back on. Following cycling off of the final stage of control, there is typically a delay before restart of the machine is possible. While this delay is designed to protect the machine as a whole against damage from excessive on/off cycling, the off-coil temperature will rise during the delay period. In any event, excessive cycling of a compressor can lead to a dramatic reduction in its life.

One option for increasing the load-matching capability of a DX machine is to incorporate hot gas bypass from the first-stage compressor to the associated evaporator at low load, although conventional hot gas bypass still leaves capacity control reliant on compressor cycling/cylinder unloading above 50% of full capacity. Modern implementation of the technique using electronic control may offer improved flexibility of control across the capacity range.

In contrast, a chilled water cooling coil with a three-port modulating control valve can achieve a turndown in the range 20:1–50:1, combined with a resolution of the order of 0.1%, resulting in minimal cycling of the supply air temperature.

In addition to fundamental considerations of their load-matching ability, the use of DX cooling coils can pose a further potential control problem when associated with VAV systems, because the type is sensitive to uneven thermal loading, i.e. the rate of heat transfer per unit area of the coil face[5]. This may arise not only from the existence of a temperature-stratified air-stream, but also from a disturbed flow profile at the coil face. With the reasonably high face velocity that would be expected under system design conditions, a DX coil may be robust to moderately uneven thermal loading. As the face velocity reduces during part-load operation, however, the effects of any uneven loading are magnified, and unstable operation may result. If this tendency towards unstable thermal operation occurs in a critical part of the operating range of the thermostatic expansion valve (TEV), which controls the flow of refrigerant to the coil, the valve may become prone to hunting. Thus if it is intended to use DX cooling with a VAV system, great care is required in designing the AHU to ensure even airflow patterns and thorough mixing of the airstream before reaching the DX coil. Direct liquid overfeed refrigeration has been suggested as offering a potential for improved efficiency and low costs when associated with VAV systems[6].

References

1. Arnold, D. (1986) VAV and the economy cycle. *Building Services CIBSE Journal*, **8**, May, 68–69.
2. Chartered Institution of Building Services Engineers (CIBSE) (1982) Weather and solar data. *CIBSE Guide Section A2*, 7, CIBSE, London.
3. Chartered Institution of Building Services Engineers (CIBSE) (1977) Flow of fluids in pipes and ducts. *CIBSE Guide Section C4*, 58, CIBSE, London.
4. Rogers, G.F.C. and Mayhew, Y.R. (1981) *Thermodynamic and Transport Properties of Fluids*, 3rd edn (SI units), Basil Blackwell, Oxford.
5. Coad, W.J. (1984) DX problems with VAV. *Heating, Piping and Air Conditioning*, **56**, January, 134–139.
6. Scofield, C.M. and Fields, G. (1989) Joining VAV and direct refrigeration. *Heating, Piping and Air Conditioning*, **61**, September 137–140, 147–152.

Chapter 4

Heating, Humidity Control and Room Air Distribution

4.1 Space heating: design options for VAV systems

Selection of a suitable approach to space heating is perhaps the most fundamental decision that an HVAC designer must take when proposing to employ a VAV system for a particular application. The options conventionally available to the system designer are identified in Fig. 4.1. Their principle characteristics are compared in Table 4.1, and each is discussed in subsequent sections of this chapter. For HVAC engineers in the UK the options most likely to be encountered in current new-build or refurbishment applications are VAV with wet perimeter heating and VAV with fan-assisted terminals.

4.2 Space heating with single-duct VAV systems

Two distinct approaches are possible to the provision of space heating where single-duct VAV air conditioning systems are employed. In the first the VAV system itself satisfies the heating requirements of the spaces served. In the second, space heating is provided by a separate system. With specific and very limited exceptions, VAV systems are not generally suitable for changeover techniques of the kind that may be employed with air/water systems (perimeter induction and fan coils) in their two-pipe configurations. The changeover principle implies a change from a constant cooling supply air temperature to a constant heating supply air temperature, and vice versa, according to whether the building concerned requires cooling or heating. Ignoring the practicalities of achieving reversal of the control action at the terminal units, this might conceivably be adequate for a building with single solar exposure, not subject to moving shadows, and with substantially constant internal sensible heat gains. This is hardly the typical large multi-zone commercial air conditioning application. In any event, being best suited to climates with well-defined cooling and heating seasons (hot summers and

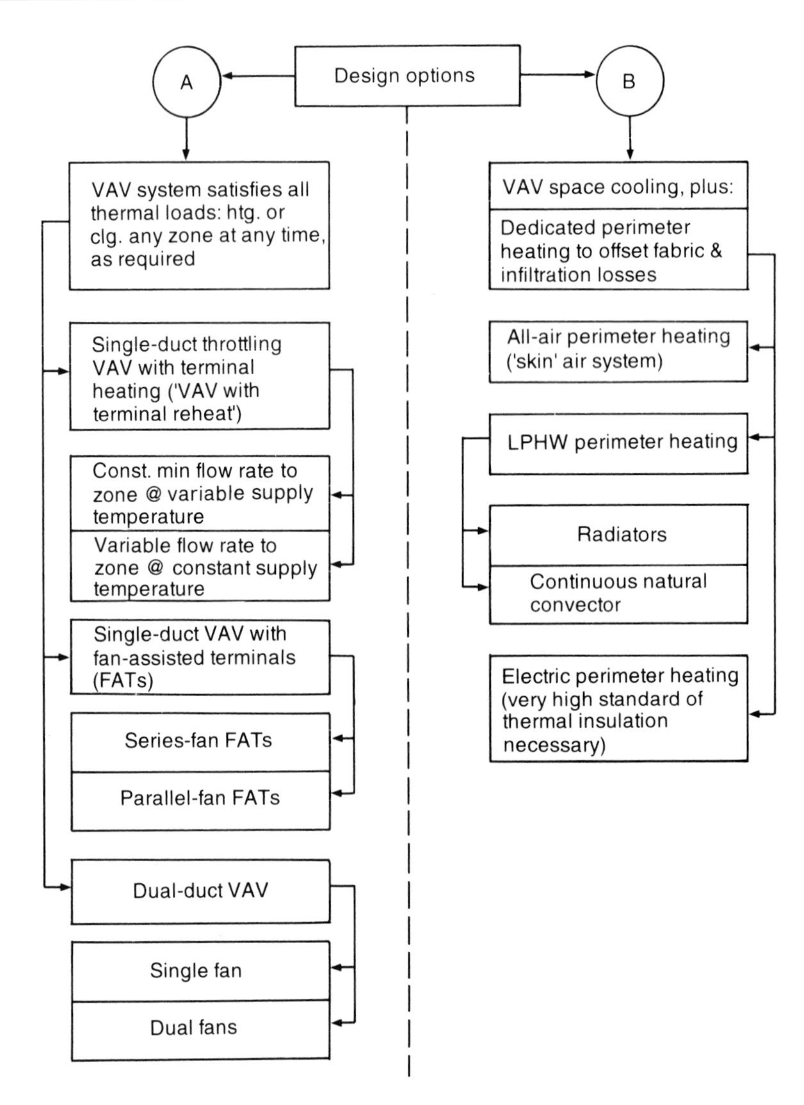

Fig. 4.1 Design options for space heating with VAV systems.

cold winters) and brief spring/autumn interludes – such as are indeed found in parts of the USA – the changeover principle is out of place in the UK. Here the temperate climate needs a fully flexible system that can cope with changes from heating to cooling and back again for any zone during the same day, and between zones at the same time. This is commonly, and perhaps somewhat confusingly, described as a requirement for *simultaneous* cooling and heating. The 'simultaneous' refers to the system, which can at any time be heating some zones and cooling others. Individual zones can only ever be either heated or cooled.

Single-duct VAV systems can provide simultaneous heating and cooling by the addition of heater batteries to the terminal units serving perimeter zones and those on the top floor of a building. While it is feasible for these to employ electric resistance heating elements, in the UK this is effectively prohibited both by the high cost of peak-rate electricity and by the very high electrical Maximum Demand Tariffs which apply during the peak winter heating months. Heating at the terminal units thus typically utilises conventional air heater batteries supplied with LPHW from a central boiler plant. VAV terminal units serving internal zones, or away from the perimeter in large open-plan areas, provide cooling only.

Assuming a zone served by a single VAV terminal unit, the maximum potential design load for the terminal unit heater battery is the sum, for the zone, of the transmission heat loss through the building fabric, any infiltration heat loss, and a part of the ventilation heat requirement. In the basic VAV system the supply airstream leaves the central AHU at a nominally constant dry-bulb temperature and moisture content throughout the year. During the heating season if this cannot be achieved by mixing of outside and recirculated return air or, in the absence of recirculation, by the use of heat recovered from air exhausted from the building, the necessary heating is provided by the main heater battery in the AHU. The temperature of the airstream received by the VAV terminal unit serving a zone is the constant supply condition of the air leaving the central AHU, plus any temperature rise that occurs along intervening ductwork. It is that portion of the ventilation heat requirement between the temperature at inlet to the terminal unit and the zone air temperature which constitutes a heating load on the zone.

The heating provided at the VAV terminal units is typically carried out at the zone's minimum supply volume flow rate. The heating capacity of the supply air is matched to the zone load by varying the discharge air temperature from the terminal unit at the constant minimum volume flow rate. This is achieved conventionally by modulating the volume flow rate of LPHW through the terminal unit heater battery, at constant flow temperature, using the conventional combination of a three-port mixing control valve in a diverting circuit. Lower water temperatures than the conventional 82/71°C flow and return may be practical where a source of low-grade recovered heat is available – as, for example, when using a double-bundle water-cooled condenser in the refrigeration plant.

Incorporating a 'dead band' (or 'zero energy band') between reaching the minimum zone supply air volume flow rate and increasing the supply air temperature ensures that heating can take place only at the minimum zone supply air volume flow rate. The design value of discharge air temperature from the VAV terminal unit is ultimately limited by the performance of the room air distribution design. This will typically be optimised for cooling rather than heating. At a given volume flow rate the risk of stratification of

Table 4.1 Comparison of options for space heating with VAV systems.

	Single-duct VAV with terminal heating		Single-duct VAV with fan-assisted terminals		Dual-duct VAV		VAV cooling with perimeter heating by:		
Characteristic	Constant min flow to zone	Variable flow rate to zone	Series FATs	Parallel FATs	Single fan	Dual fans	All-air ('skin' air)	LPHW	Electric
References to Fig. 4.1	A1.1	A1.2	A2.1	A2.2	A3.1	A3.2	B1	B2	B3
Simultaneous cooling or heating of adjacent zones?	Yes	Yes	Yes	Yes	Yes	Yes	No	Yes if local emitter control	Yes if local emitter control
Heat transfer medium (to zone)	Air	Air	Air	Air	Air	Air	Air	LPHW	–
Heat output provided by:	LPHW/ electric heater battery	LPHW/ electric heater battery	LPHW/ electric heater battery	LPHW/ electric heater battery	Hot duct supply	Hot duct supply	Central LPHW heater battery	Radiators, continuous nat. convector	Panel heaters, continuous nat. convector
Use of VAV system fans for 'unoccupied' heating?	Yes, full recirculation possible	Yes, full recirculation possible	Terminal fans only; full recirculation	Terminal fans only; full recirculation	Yes, full recirculation possible	Yes, full recirculation possible	No. CAV 'skin' fans on full recirculation	No	No
Zone flow rate during heating operation	Constant at minimum level	Variable; but < cooling design flow	Constant at cooling design flow	Constant at $\leq 67\%$ cooling design flow	Constant or variable, at $\leq$ 50% design	Constant or variable, at $\leq$ 50% design	'Skin' constant; VAV (ideal) at minimum in heated zone	VAV (ideal) at minimum in heated zone	VAV (ideal) at minimum in heated zone
Recovery of lighting gain to ceiling void	At central plant	At central plant	At zone	At zone	At central plant	Full recovery if full recirc at central plant	Full recovery if full recirc at 'skin' plant	At central plant	At central plant
Risk of stratification?	Yes, unless VG diffusers	Yes, but less than A1.1	Low if supply Δt moderate	Low if supply Δt moderate	As for single-duct systems	As for single-duct systems	Low if supply Δt moderate	No	No
Risk of downdrafts from glazing, etc.?	Yes	Yes	Yes	Yes	Yes	Yes	Yes	Minimised if emitters correctly positioned	Minimised if emitters correctly positioned

Acoustic implications during heating	None	Changing flow rates	None	Cycling of terminal fan	None	None	None	None	None
Local control of heat output in VAV zones?	Yes	Yes	Yes	Yes	Yes	Yes	No	Yes, if TRVs fitted to emitters	Yes, if local emitter control
Risk of simultaneous heating/cooling in same zone?	Minimised by adequate 'dead band'	Minimised by adequate 'dead band'	Minimised by adequate 'dead band'	Minimised by adequate 'dead band'	Minimised by adequate 'dead band'	Minimised by adequate 'dead band'	Yes	Yes, high if no local emitter control	Yes, high if no local emitter control
Additional space requirements	LPHW circuit + minimal at terminals	LPHW circuit + minimal at terminals	LPHW circuit + larger terminals	LPHW circuit + larger terminals	High, due 'hot' and 'cold' ductwork	High, 'hot' and 'cold' ductwork + dual system fans	Highest: additional CAV ductwork	LPHW circuit + emitters on outside walls in zones	Lowest: emitters on outside walls in zones
Pipework in ceiling voids?	Yes	Yes	Yes	Yes	No	No	Yes	Yes	No
Capital cost implications	LPHW circuit, TU heating coils and control valves	As A1.1 plus zone controls	As A1.1, plus higher cost of FATs	As A1.1, plus higher cost of FATs (but less than 'series' type)	Dual ducts, higher cost of TUs, system and zone controls	As A3.2, plus dual fans	Highest; zoned CAV system	Zoned LPHW VT circuits, heat emitters, local controls	Lowest: emitters and wiring, local controls
Running cost, energy use implications	LPHW pumps, FTP increased by heater battery ΔP	As A1.1 plus increased flow to zone during heating (fan power)	LPHW pumps, terminal unit fans (but reduced system FTP)	LPHW pumps, terminal unit fans (but run less than 'series' at lower ΔP	Complex, mixing of hot and cold duct supplies at terminals	Complex, dual system supply fans, mixing of hot and cold duct supplies at terminals	LPHW pumps, 'skin' CAV fans	Zoned LPHW VT circuit pumps	Highest unit energy cost and environmental impact; needs very high standard of insulation
Maintenance implications	LPHW circuit zone control valves	As A1.1	As A1.1, plus terminal unit fans	As A2.1	Increased complexity of terminal units, system/ zone control	As A3.1, plus dual supply fans and controls	Greatest, CAV system plant and controls	Zoned LPHW VT circuit pumps and controls	Least, emitters and controls

the supply air stream occurring under winter heating conditions becomes greater as the supply air temperature differential increases.

Conversely, for a given temperature difference the risk of stratification occurring increases as the supply volume flow rate reduces. In a sense this gives the VAV terminal unit with heater battery the worst of both worlds, because the discharge velocity from the air terminal devices (ATDs) is at a minimum when operating at the minimum zone supply volume flow rate – unless the discharge area of the ATD is decreased for heating.

An alternative strategy to that of variable-temperature air supply at constant (minimum) volume flow rate is possible using variable-volume air supply at a constant temperature. This arrangement requires two local control loops. In the first the terminal unit three-port control valve modulates the volume flow rate of LPHW through the heater battery to maintain a constant discharge air temperature. In the second the terminal unit controller varies the supply volume flow rate to the zone to match the heating requirement. Volume flow rate at the zone design heating load need not, of course, be as high as at the design cooling load. However, system volume flow rate and fan energy consumption are increased under just those conditions contributing to the large proportion of VAV system annual operating hours typically spent at high system turndown. This mode of heating operation also requires a reversal of the control response of the VAV terminal unit to a change in zone air temperature. This must change from closing the terminal unit flow control damper with a decrease in zone air temperature during cooling to opening it with a decrease in temperature during heating. A 'dead band' is again desirable between cooling and heating operation.

Heating of a building during unoccupied hours may be accomplished using full recirculation. Using variable-temperature supply at the minimum zone supply volume flow rate, it may even be possible to switch off the system return fan. Except in those relatively rare cases where thermal inertia is deliberately built in, modern commercial buildings typically employ lightweight construction techniques, and for economy of operation space heating is intermittent. A reduced (or 'setback') set-point temperature is adopted overnight, and may also be used at weekends if the building is routinely unoccupied. This night temperature is based on the need to afford protection against damage by low temperatures to the building fabric and fittings, as well as the now ever-present IT equipment which is such an important feature of modern offices. The building is brought back up to its working temperature just in time for occupancy by a morning warmup (preheat) period. Individual zone control is typically not required under night setback conditions, and both night heating and preheating may be carried out using the heater battery in the central AHU. However, the system supply fan must always run when heating is required, and its overall efficiency is at a minimum in this region of its operating envelope. Without zone control, overheating of individual zones may occur, and sufficient time must be allowed to correct this under normal cooling operation prior to morning occupancy. Such 're-cooling' is inherently wasteful of energy.

A significant advantage often put forward for the use of all-air HVAC systems in commercial buildings is that no pipework is required in the ceiling voids. This avoids the potential for leakage, and reduces access requirements for maintenance. Unless the heating load can be met solely by the heater battery in the central AHU, this advantage for a VAV system *vis-à-vis* fan coil units is lost if heating is provided at the terminal units. This assumes that the use of electric heating has been dismissed on grounds of energy cost and environmental impact. VAV systems with terminal unit heater batteries are often described as VAV with terminal reheat. However, unless the object of heating is purely to prevent overcooling at very low cooling loads below the capacity of terminal unit minimum supply volume flow rates, no reheating as such is involved. By increasing the pressure drop across a fully-open terminal unit, the use of terminal unit heater batteries increases the fan total pressure required under all operating conditions – both during heating and cooling. This increases the energy consumption of the VAV supply fan in comparison to that in an otherwise identical cooling-only system. The increase in pressure drop across the terminal unit may be low during heating operation at minimum zone supply air volume flow rate. However, for one popular range of VAV terminal units the manufacturer's technical performance data suggests that the addition of a single-row LPHW heating coil can increase the pressure loss across a fully-open terminal unit by between 35 and 45 Pa (between 70 and 90%) at 80% of the maximum volume flow rating of the unit. Use of a four-row heating coil could increase the fully-open pressure loss at this volume flow rate by between 140 and 180 Pa (between 233 and 360%). Using the same source, corresponding figures for a volume flow rate of 20% of the maximum flow rating would be between 3 and 4 Pa for a single-row coil, and between 12 and 16 Pa for a four-row coil. If a VAV terminal unit were selected so that the zone design supply air volume flow rate was approximately 80% of the unit's maximum volume flow rating, 20% of this rating would represent approximately 25% of the zone's design volume flow rate. The minimum supply air volume flow rate to a zone might typically be expected to be specified at approximately one-third of its design value. A coil of several rows might be necessary when a source of low-grade recovered heat is to be used, and this has definite implications for the operating economics of any such proposal.

4.3 Space heating with fan-assisted terminals

VAV systems employing *fan-assisted terminals* (FATs) are also able to provide full heating and cooling flexibility. With this type of terminal unit, room air distribution is improved during heating operation. There are two types of FAT, the series type and the parallel type.

In the series type the terminal unit fan is in series with the damper controlling the volume flow rate of the primary supply airstream. All air flow to the zone is handled by this fan, which runs continuously. The terminal unit effectively acts as a local constant-volume AHU for the zone. With reducing sensible cooling load, return air from the zone or ceiling void is mixed in progressively increasing proportion with the reducing quantity of primary supply air from the central AHU. The total supply air volume flow rate to the zone is maintained approximately constant. The VAV terminal unit has an LPHW heater battery fitted, and operation of this is sequenced with the cooling provided by the mixture of primary supply and recirculated air. This primary supply to the series FATs is not the all-outside-air primary supply used with fan coils. Use of the term is simply a convenient way to distinguish the supply as that from the central AHU. This will typically use an outside air economy cycle, and is generally as described in Chapter 3.

When the primary supply volume flow rate has reached its minimum permitted level, a demand for heating from the zone is met by increasing the discharge temperature from the terminal unit at the same constant volume flow rate. This benefits the room air distribution in two ways. First, the constant volume flow rate through the supply outlets tends to maintain a constant and steady pattern of air distribution. Second, the high volume flow rate to the zone decreases the required supply air temperature for any heating load. This reduces the risk of stratification occurring when using supply outlets at high-level (typically ceiling-mounted diffusers), and allows higher zone heating loads to be served.

A 'dead band' is desirable between cooling and heating operation. This avoids the risk of simultaneous heating and cooling of a zone through deficiencies in equipment or controller accuracy – or their deterioration through component wear. It also helps to reduce component wear from repeated changeover between cooling and heating under late autumn and early spring conditions. Energy consumption is reduced by allowing zone temperature to 'float' over a small range when the primary supply air volume flow rate is at its minimum permitted value and the output of the terminal unit heater battery is still zero. Operation in the 'dead band' uses the sensible heat from lighting that is picked up by the return air recirculated through the terminal unit. It is important that a 'dead-band' control strategy should take into account the practical accuracy of sensors and controllers, and of their commissioning on site. Performance may be affected by sensor or controller drift, valve wear, and even tampering with control settings.

The terminal unit fan must operate whenever heating is required outside normal occupancy periods. However, this is carried out using full recirculation, and intermittent cycling of the fan may be employed. The system supply and return fans are not required to run. As with any noise source located in close proximity to an occupied space, the acoustic performance of the fan must be carefully considered. Manufacturers of VAV terminal units typically produce secondary acoustic attenuators for use with their equipment. Indeed this is equally true whether terminal units are of fan-assisted type or not. While these secondary attenuators may be easily connected to the terminal units, there is an inherent penalty associated with their use – in terms of equipment and installation cost and fan energy consumption.

The nominally constant zone supply volume flow rate with series-fan terminal units is beneficial to room air distribution under both part-load cooling and heating operation. However, there are significant implications for fan energy consumption. The energy consumption of the main VAV supply fan is reduced through a reduction in fan total pressure. The pressure drop across the index VAV terminal unit, and the losses along the supply ductwork downstream of this, are made good by the terminal unit fan, rather than by the system supply fan. However, there is no load diversity for the terminal unit fans during either heating or cooling operation. The total supply air volume flow rate handled by all of these taken together is the sum of all the individual zone design values. This is significantly greater than that handled by the main VAV supply fan, and remains constant throughout the annual operating period. Furthermore the overall efficiency of the small terminal unit fans may typically be expected to be poorer than that of the main system supply fan for much of the annual operating period.

The use of FATs may not, however, be decided purely on heating considerations. During cooling operation the energy consumption of the main VAV supply and extract fans may be decreased by taking advantage of the mixing capabilities of fan-assisted terminal units to reduce the temperature, and hence the volume flow rate, of the primary supply air. This provides a reduction in the capital cost of the main VAV fans and ductwork to offset against the additional capital cost of fan-assisted terminal units. Additional maintenance requirements of the terminal unit fans must also be considered, and the use of FATs may also be expected to incur significant additional electrical Maximum Demand charges, where this type of tariff applies.

In some designs of fan-assisted terminal the fan operates in parallel with the damper controlling the flow of the primary supply airstream. Only part of the total airstream supplied to the zone is handled by the fan, which runs intermittently. Operation of the fan is sequenced to commence as the primary supply volume flow rate from the central AHU approaches its permitted minimum value for the zone. Heating operation is generally similar to that described for the series-fan type. Total supply volume flow rate to the zone varies during cooling operation, and remains approximately constant during 'dead-band' and heating operation – but at a level roughly mid-way between the minimum and design values (a maximum value of perhaps two-thirds) of the primary supply volume flow rate from the central AHU.

Since in parallel-fan terminal units the fans themselves operate for less time than in series-fan types, and handle a lower volume flow rate, their energy consumption will be correspondingly less, although the system supply fan must now make good the pressure losses across the index terminal unit and along the supply ductwork on its downstream side. With the parallel-fan type of FAT the acoustic effects of intermittent fan operation must be considered. The fluctuating noise level resulting from switching the terminal unit fan on and off will typically be more noticeable to occupants than the constant fan noise level from a series-fan type would be – even if this was actually rather higher.

4.4 VAV with all-air perimeter heating

Two approaches are again possible where a cooling-only VAV system is combined with an independent means of space heating. The first is to use a constant-volume, variable-temperature (CVVT) all-air system to offset the fabric and infiltration heat losses. This is sometimes also referred to as a 'skin' air system. The alternative approach is to serve the perimeter of the building with an LPHW heating system. In both cases the perimeter heating is fully independent of, but should be sequenced with, the cooling-only VAV system.

Where all-air perimeter heating is employed, the system may also offset a proportion of the summer-time cooling load for these areas. In this case the VAV terminal units serving the perimeter zones would be correspondingly down-sized. Following conventional constant-volume system design practice, the building perimeter would be zoned according to the orientation of the facades, with each zone served by its own perimeter system. These can operate on full recirculation continuously, with the VAV system providing all ventilation air. Morning warm-up of the building served can be carried out by the perimeter system(s) without running the main VAV system.

The supply air temperature of the perimeter constant-volume system is scheduled according to outside air temperature (decreasing with increasing outside air temperature), without individual zone control. Full recirculation results in full heat recovery of that portion of the lighting gain which is picked up by the extract air that is recirculated through the perimeter system AHU. Depending upon the design and control of the lighting installation, this may be adequate for the perimeter heating requirements for a substantial part of the heating season. Potential also exists for the use of the perimeter return air as an internal source for a heat pump, although the capital cost of such a scheme is high, so that its economics are unlikely to be sufficiently attractive in speculative developments.

On the negative side, it is difficult to counteract potential downdraughts from cold glazed surfaces, as the need to 'wash' perimeter glazing with warm air supplied at high level by ceiling-mounted diffusers is opposed by the natural buoyancy of the heated air. System fans must still be run to provide heating during unoccupied hours. Internal zones on the top floor of a building, which experience transmission heat losses through the roof, also derive no heating benefit from the perimeter system. Such areas may require space heating even during occupancy. During unoccupied hours they may have the greatest need for protection against frost or condensation. To give this protection with economy in fan operation it may be necessary to heat them independently of the perimeter system. This adds the complexity of yet another separate heating system. Perhaps more fundamentally, however, the coordination and space requirements of the different ductwork and air distribution layouts and air handling plants would seem to be a potential minefield for the HVAC designer – and a major reason why this approach has not been favoured in the UK.

4.5 VAV with wet perimeter heating

The alternative approach to VAV with perimeter heating is to combine a cooling-only VAV system with an independent LPHW heating system serving the perimeter of the building. In the UK this has proved popular with HVAC designers to the extent that VAV with perimeter heating is typically taken to imply this approach. Heat emitters typically take the form of a system of either finned-tube natural convectors or radiators. LPHW is supplied to these from a central boiler plant, and its flow temperature may simply be scheduled according to outside air temperature without individual zone control. Finned-tube natural convectors are particularly suitable for this. These comprise continuous lengths of small-diameter (typically 22-mm OD) copper pipe fitted with intermittent sections of finned heating element. The whole is installed within a continuous casing terminating at the cill line of the glazing. A wide variety of casing styles is available, including (if cost allows) examples which may be integrated into the architectural layout of the floors. Without individual zone control of the perimeter heating, it may be difficult to avoid some conflict between its operation and that of the VAV system. However, this may be minimised if the perimeter heating is zoned according to the orientations of the building facades, in which case LPHW flow temperature may also be compensated for solar radiation (although some HVAC engineers remain sceptical of the efficacy and reliability of other than basic OAT compensation). The treatment of night and weekend heating setback and morning preheating prior to occupancy follow conventional space heating practice, and VAV system fans may be shut down outside normal occupancy.

Conventional location of the heat emitters at low level around the perimeter of the building has the advantage of counteracting potential downdraughts from cold glazed surfaces, which is difficult to achieve with warm air heating. With this approach there is no heating of the supply air at the VAV terminal units. In the absence of any direct solar gain, which cannot be guaranteed under design heating conditions, the portion of the ventilation heat load required to raise the temperature of the supply air stream to the zone space heating set-point can only be offset either by the perimeter heating system emitter or the internal sensible heat gains. Whether any heating credit can be taken for internal gains depends upon the pattern of building operation, and the design and mode of control of the lighting installation.

4.6 Space heating with dual-duct VAV systems

Dual-duct VAV systems inherently provide full heating and cooling flexibility. As already noted, they may be of single-fan or dual-fan type. In both cases all heating takes place at the central AHU, so that no LPHW pipework is required to the VAV terminal units in the ceiling voids. Again full recirculation may be employed for heating during unoccupied hours, although fan energy must still be consumed for this. However, dual-duct, dual-fan systems allow recovery of the heat transferred from lighting to ceiling void return plenums, while for maximum economy in fan energy consumption both fans may be operated on the VAV principle. In the UK the cost and space penalties of dual duct systems have militated against the use of this form of VAV system.

4.7 Room air distribution

4.7.1 Fundamental design considerations

In air conditioned buildings the method adopted for room air distribution is typically governed by the requirements for supplying chilled air for cooling. In commercial applications this has traditionally involved air supply via ceiling-mounted diffusers. It is desired that the chilled supply air should be rapidly mixed with air within the space. Ideally, the temperature of the mixed airstream should have reached the design room temperature by the time that it enters the occupied zone. For office-type applications, where occupants are seated for much of the time, this is nominally defined as an envelope extending from floor level to a height of 1.2 m, and to within 0.15 m of any other room surface.

In addition, by the time it penetrates the occupied zone, air velocities in the mixed airstream should be low enough not to give rise to any sensation of draught by the occupants. Draught conditions have been quoted as equivalent to any sense of cooling felt on a localised portion of the body[1]. This may be caused either by the degree of air movement or by the temperature of the moving air. It is now well recognised that the relationship of human comfort to air movement is not a simple one, with susceptibility to a sensation of draught depending on the part of the body exposed to the air movement. It is also affected by the type of clothing worn. Draught is more readily associated with air movement around the lower legs by women, whilst the back of the neck is generally accepted as being particularly sensitive for both men and women. It also appears that people are more tolerant when the direction of air movement is variable, whereas air movement from a constant direction is more likely to be interpreted as a draught.

However, if too much air movement results in complaints of draughts, the perceived absence of any sensation of air movement at all is likely result in complaints of 'stuffines'. Some air movement is required to maintain a sensation of 'freshness' within a space. Indeed in hot weather conditions a degree of air movement that might be considered draughty in winter may be welcomed by occupants, although in both cases the temperature of the air remains the same. This is of practical importance to the design of room air distribution with VAV systems. Air velocities of less than 0.1 m s^{-1} are typically considered indistinguishable from true still air conditions. Based on studies carried out at the (then) UK Heating and Ventilating Research Association, it was concluded that for a room dry-bulb temperature in the region of 22°C the range of 'comfortable' mean room air velocities will typically lie within the range 0.08–0.24 m s^{-1}.[2] In an early study of air distribution design in open-plan offices the (then) Electricity Council carried out field tests on a number of deep-plan office buildings served by a variety of air-conditioning systems, including in at least one case a VAV system[3]. The study concluded that to maintain room air movement in the range 0.1–0.15 m s^{-1}, 'as required for the comfort of office workers', required constant-volume air supply, but that in the event that the lower limit of 0.1 m s^{-1} – 'which is a subject of much dispute' – were relaxed, a VAV system could achieve adequate comfort control.

4.7.2 Air discharge patterns and the Coanda effect

All the types of air terminal devices that are used to supply air in comfort conditioning installations produce a discharge pattern in the form of a *jet* (or jets). The jet may be broad or narrow, and may radiate in all directions or issue on one or more sides of a rectangle. Around its periphery the jet entrains room air and turbulent mixing takes place. The initial momentum of the supply air as it is discharged from the air terminal device is progressively transferred to the much greater mass of room air. While the total momentum is conserved, the velocity of the initial jet of supply air is ultimately reduced to that of the general level of room air movement. Similarly the turbulent mixing process results in the transfer of heat from the room air to the cooler supply air, which consequently increases in temperature, ultimately reaching the general room temperature. This is the mechanism by which the sensible cooling load on the space is absorbed by the supply air, and it is the designer's intention that the mixing process should be effectively complete by the time the (expanded, slowed, warmed) supply air jet reaches the occupied zone.

At any point along the length of the supply air jet the maximum air velocity ideally occurs on the centreline of the jet. The mean air velocity over its whole cross-section may typically be only 20–30% of this. As its velocity decreases, the supply jet becomes more sensitive to local effects, such as natural convection currents, which can affect its behaviour significantly.

Manufacturers' performance data for ceiling diffusers is typically presented in terms of *throw* and *terminal velocity*. Throw (or radius of diffusion) is the horizontal travel of a supply air jet from the diffuser to the point at which the maximum velocity in the jet has decayed to a nominated terminal velocity. Conversely, terminal velocity is the maximum velocity in a supply air jet at a nominated throw. Both the type of air terminal device and the particular requirements of the application influence what constitutes an 'acceptable' terminal velocity. Manufacturers' data often give values of throw to a nominated terminal velocity of $0.5\,\mathrm{m\,s^{-1}}$, and values in the range 0.25–$0.5\,\mathrm{m\,s^{-1}}$ are typically applicable under design cooling loads in applications where occupants are principally sendentary.

Except in high spaces with relatively large vertical distances between supply outlets and the occupied zone, it is not conventional practice to discharge the supply air jets either directly downwards, or even obliquely towards the occupied zone. Supply air diffusers typically turn the airstream received from the supply duct through 90° and discharge it horizontally across the ceiling (Fig. 4.2). The natural tendency of a fluid in motion to cling to an adjacent solid surface is called the *Coanda effect* after Henri-Marie Coanda, the Romanian engineer who discovered the phenomenon while working with the Bristol Aircraft Company in the early 1900s. In Fig. 4.2 the airstream leaving the diffuser is initially thinned into a wide, narrow jet, and its velocity is increased substantially. The Coanda effect is apparent as a negative pressure or suction which pulls each layer of air in the jet towards the ceiling. Turbulent mixing occurs along the lower periphery of the supply jet, away from the ceiling. Although mixing occurs only along this single surface, the low ratio of mass flow rate to surface area of the jet means that it rapidly entrains and mixes with the room air as it moves across the ceiling. As the jet becomes thicker, its velocity falls, and with it the strength of the suction that pulls it towards the ceiling. The chilled supply air is naturally denser than the room air, and experiences a negative

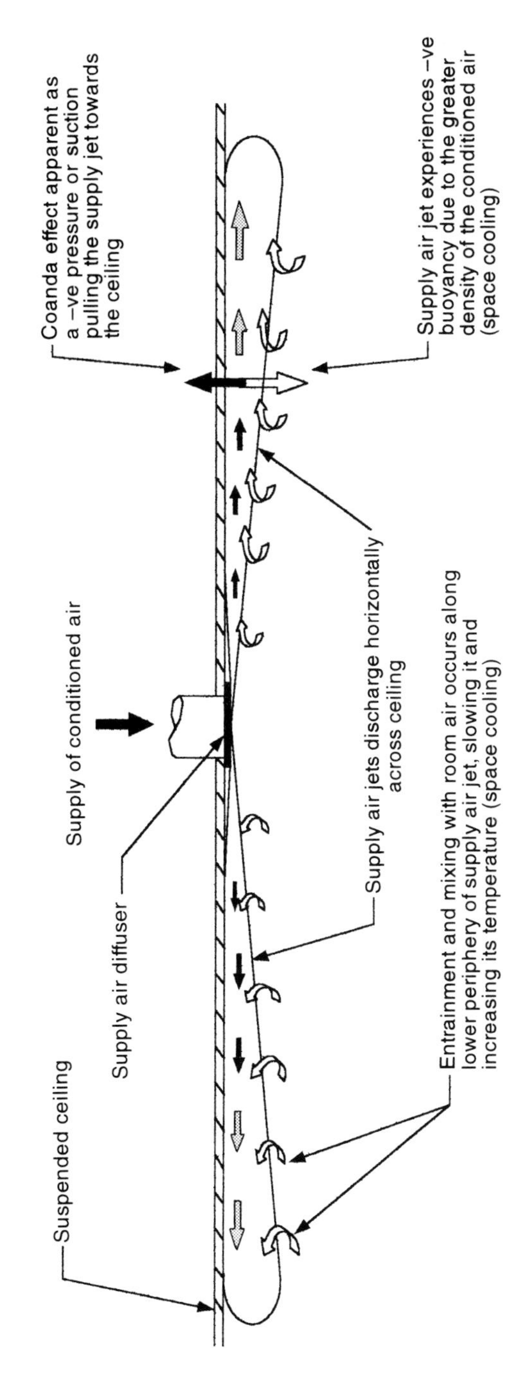

Fig. 4.2 Room air distribution: the Coanda effect for a supply air jet.

buoyancy force tending to cause the supply airstream to fall away from the ceiling towards the occupied zone. This negative buoyancy force is a function of the difference in the absolute temperatures of the supply airstream and the room air. It is greatest at the point of discharge of the supply air jet from the diffuser (when the Coanda effect is at its strongest), and decreases to zero when full mixing of the supply airstream to room air temperature has been achieved. If the negative buoyancy outweighs the strength of the Coanda effect, the supply air jet will 'dump' or fall away from the ceiling. Entering the occupied zone prematurely at too low a temperature and with too high a velocity, it will cause discomfort (a sensation of draught) for any occupant unfortunate enough to be in its path. The design aim is thus to maintain the horizontal mixing motion at ceiling level so that the airstream that finally enters the occupied zone does so with the required temperature and velocity.

Good room air distribution using ceiling-mounted diffusers typically relies on establishing a strong Coanda effect on the supply air jet(s). The effect is only established for supply air jets which are directed across (and close to) the surface of the ceiling, or towards it. A supply air jet that shows no tendency to 'break away' from a surface that it is moving across is commonly referred to as being 'attached'.

In the basic cooling-only VAV system the supply air temperature to the zone is maintained nominally constant, and with it the magnitude of the negative buoyancy force on the supply air jet(s) from a ceiling diffuser. If the area of discharge from the diffuser remains constant, the velocity of the supply airstream reduces in direct proportion to a reduction in supply air volume flow rate to the zone. The strength of the suction force tending to keep the airstream attached to the ceiling is generally regarded as being proportional to the square of the discharge velocity. As the volume flow rate supplied by the ceiling diffuser reduces, a point will be reached when dumping will occur. Unfortunately, once the supply airstream has broken away from the ceiling, its volume flow rate must be increased significantly above the level at which dumping occurred to re-establish attachment of the flow. This may mean a doubling of the supply air volume flow rate, equivalent to an increase of 30–40% of the design value. Research into the design and evaluation of room air distribution suggests that dumping may occur at supply velocities below $1.5\,m\,s^{-1}$.[4] The occurrence of dumping is clearly incompatible with occupant comfort.

Under heating operation, with supply air temperature above room air temperature, the buoyancy force experienced by a supply air jet acts upwards, and is additive to the Coanda effect. This encourages a warm supply air jet to stay close to the ceiling, resulting in a potential for stratification at high level. This may lead to high temperature gradients in the space, uneven temperature distribution and generally poor heating.

4.7.3 'Dumping' and the options for reducing the risk

The risk of dumping can be reduced in a number of ways. The first is by paying careful attention to aerodynamic design. This is the approach adopted in VAV terminal units of the type that incorporate a linear slot diffuser. In such units the design of the flow control damper is such that the velocity of the supply airstream issuing from the diffuser is maintained with decreasing supply air volume flow rate. The geometry of the supply diffuser itself remains unchanged, but its action is still to turn the thinner jet of air and direct it along the ceiling surface in such a way that an initial attachment is made.

Alternatively, the geometry of the supply air diffuser itself may be varied. One design approach (Fig. 4.3) achieves this by dividing the flow passage through the diffuser into two sections. With reducing volume flow rate, a simple weighted flap damper, balanced against the dynamic pressure of the air flow, closes off one flow path to maintain the velocity through the other (and hence over a reduced length of the supply air jet). Supply diffusers of this type are claimed to be able to cope with a reduction in supply air volume flow rate to 25% of the design value. Naturally this performance is obtained at an economic premium.

The variable-geometry approach may similarly be applied to maintain the velocity of a warm supply air jet to avoid the risk of stratification at high level

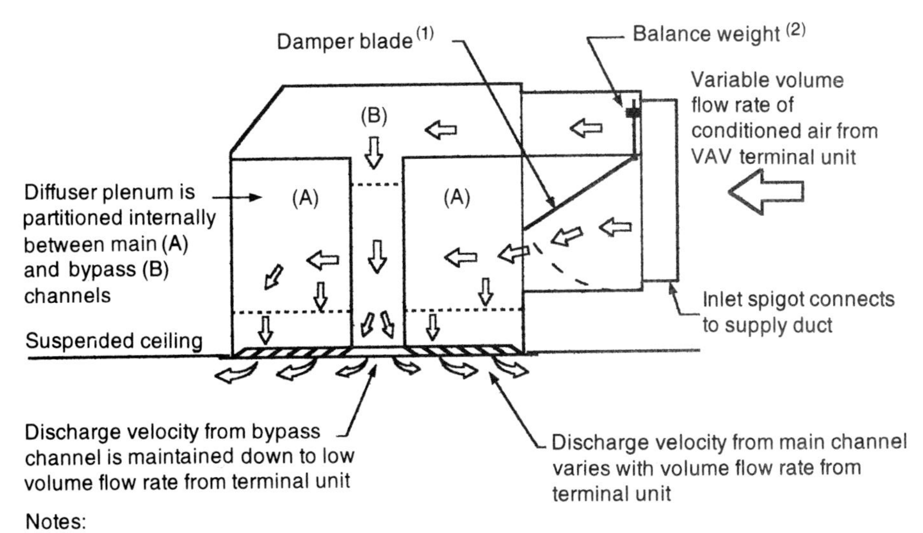

Fig. 4.3 Variable-geometry supply diffusers for VAV operation.

under heating operation. Under these circumstances variable-geometry techniques try, wherever possible, to direct a heated supply air jet towards the area experiencing the heat loss, namely the external wall and glazing (Fig. 4.4). As far as industry wisdom on VAV room air distribution with fixed-geometry supply diffusers is concerned, opinion favours the use of relatively small supply air volume flow rates per diffuser. One respected industry opinion quantifies this in terms of a preferred limitation of approximately $240\,l\,s^{-1}$ (500 cfm) or less per diffuser, whilst acknowledging that acceptable performance has been achieved by designs using three to four times this value[5]. The encouragement to use many small fixed diffusers is based on the premise that in this way no single supply airstream will be large enough to cause an objectionable downdraught. Furthermore the variation in the absolute value of throw with variation in supply volume flow rate will be much less where short-throw, high-entrainment diffusers (such as round and perforated-plate types) are used. If dumping is progressively more likely to occur as outlet velocities drop below *c.* $1.5\,m\,s^{-1}$, then limiting the outlet velocity of circular or perforated-plate supply diffusers to $2.5\,m\,s^{-1}$ under design conditions effectively restricts the permissible reduction in supply volume flow rate to a maximum of 40–50% of the design value. However, performance always carries a price tag, and the use of a greater number of diffusers each handling a relatively small volume flow rate of supply air will typically increase the capital cost of an installation.

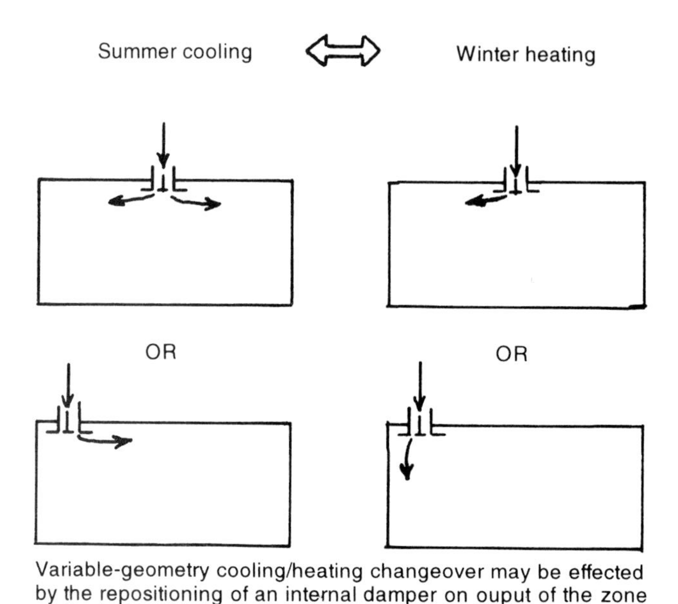

Fig. 4.4 Variable-geometry VAV supply diffuser with heating/cooling changeover.

A further acknowledged technique for delaying the risk of dumping with fixed-geometry diffusers involves *undersizing* the diffusers. These are selected on the basis of 'optimised' volume flow rates which range from 60 to 80% of the required zone design value. This is the region of the operating profile in which experience shows the system will run for a large part of the year, and that during which most of the potential hot weather may be expected to occur. In this way the discharge velocity of the supply air jet(s) can be maintained down to lower supply volume flow rates. The jet velocity will be correspondingly increased under conditions of design sensible cooling load on the zone. However, this typically represents a small potential operating period in which comparatively rare combinations of high outside air temperature and maximum solar gain may result in maximum design cooling loads being experienced. Under such conditions the increased level of room air movement is likely to be welcomed by most occupants.

If the room air distribution scheme is suitably designed, and the air terminal devices properly selected, most types of fixed-geometry ceiling diffuser (Fig. 4.5) can be used with VAV systems. However, linear diffusers have proved to be particularly suitable, and are widely used with VAV systems. In VAV terminal units which incorporate a supply diffuser, this is also typically of the linear type.

Linear diffusers (or linear slot diffusers) supply air in one or two directions (one- or two-way blow). Whilst vertical or oblique discharge patterns are sometimes possible where adjustable discharge vanes are fitted, linear diffusers are typically used to produce a high-aspect-ratio, two-dimensional discharge pattern across a ceiling (Fig. 4.6). Some types divide the air flow into a number of distinct supply air jets spaced along the length of the linear air distribution plenum. The supply air jets from linear diffusers show a strong Coanda effect, and the type is thus well suited to VAV applications. Indeed linear supply diffusers selected for 'optimised' volume flow rates are claimed to deliver acceptable room air distribution performance down to *c*.30% of the design supply volume flow rate to a zone. Certain other advantages accrue from the use of linear air distribution, including a wider coverage of a space from a given number of supply air diffusers. They also offer flexibility in open-plan office layouts. Individual diffusers may be used conventionally, or combined in continuous strips along the length of a ceiling, and from the architectural point of view present a neat (and typically unobtrusive) appearance. Sections not required for active air distribution may simply be blanked off in the ceiling void, or special 'dummy' sections substituted to maintain appearances. Sections not used for active air supply may also be used to provide a return air path to a ceiling extract plenum. Alterntively the necessary return air path may be achieved from an independent continuous length of return air diffuser, which may be coordinated with its supply counterpart. In the event of changes in the use of the space, or

Multi-cone circular ceiling diffusers

Diffuser neck

False ceiling

Diffuser cones

Diffuser cones may be adjusted for horizontal or vertical discharge pattern

Rectangular ceiling diffuser

Diffuser neck (round or square duct connections are possible)

False ceiling

Diffuser core

Core pattern determines air distribution - 1, 2, 3 or 4-way 'blow' (shown) is possible

Perforated-face ceiling diffuser

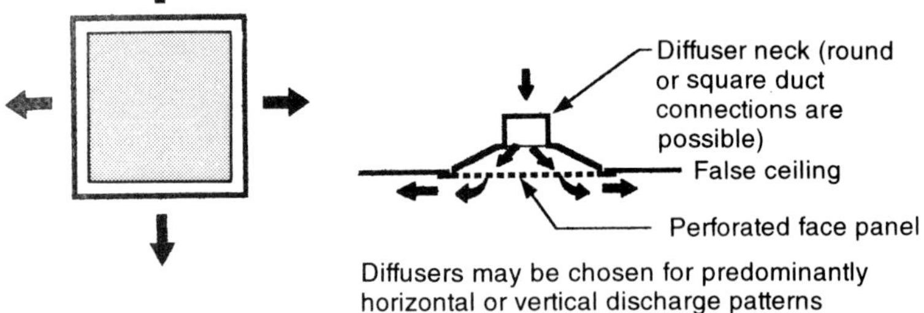

Diffusers may be chosen for predominantly horizontal or vertical discharge patterns

Swirl diffusers

A large number of supply air jets spread out in a combined radial and tangential pattern, resulting in high induction of room air. Angle of the swirl-inducing vanes is adjustable in some designs, and versions are available for use with low-level air distribution via a floor void (ducted or plenum floor)

Fig. 4.5 Standard supply diffuser types suitable for VAV operation.

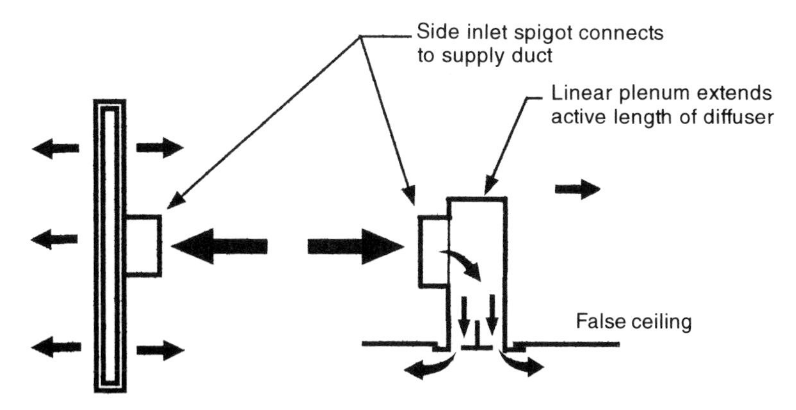

Diffusers may be single- or multi-slot, with fixed vanes or adjustable deflectors. Discharge patterns may be horizontal (to one or both sides), vertical or oblique.

Fig. 4.6 Linear slot supply diffuser.

the introduction of partitioning, there is some flexibility to change the pattern of active supply sections, or to introduce additional active sections without disturbing the overall appearance of the ceiling. This feature is particularly attractive in speculative developments, where the air distribution design may to a limited extent be adapted to the final requirements of the eventual tenant. However, for a given supply volume flow rate, linear slot diffusers are probably the most expensive option relative to other types of fixed-geometry air terminal devices – typically costing about twice as much as rectangular or circular diffusers. The noise level and pressure loss characteristics of linear diffusers also tend to be slightly higher.

Where an adjacent flat ceiling surface is not available, as in the case of sculptured or coffered ceilings, special consideration is typically required where VAV air supply is intended. The presence of downstand beams or other obstructions to the supply air jet presents problems of a fundamentally similar nature to those posed for any air distribution design.

Fan-assisted terminals of the series-fan type supply a constant volume of air to the diffuser(s) under all operating conditions. Room air distribution considerations are thus no different from those with a constant-volume system. With fan-assisted terminals of the parallel-fan type the volume flow rate supplied to the diffuser decreases with reducing sensible cooling load until fan operation is initiated. Thereafter the mixed volume flow rate is increased to a significantly higher level, subsequently reducing again for any further reduction in sensible cooling load, and remaining nominally constant at up to a maximum of perhaps two-thirds of the zone design value throughout heating operation. This should present no particular problems for room air distribution design with conventional fixed-geometry diffusers of linear or other type. Coanda effect horizontal air discharge patterns across the ceiling would again be the rule with both types of fan-assisted terminal.

4.7.4 The risk of low air movement

Thus far, the concern has been with the level of room air movement under design conditions. At the other end of the system operating profile the level of room air movement will decrease as the supply air volume flow rate reduces, and it will not typically be possible to maintain it above a mean value of *c*.0.1 m s^{-1} at the minimum supply volume flow rate. Air movement below *c*.0.05 m s^{-1} is uncomfortable to some people, and it is necessary to consider the risk of stagnant conditions arising in the occupied zone. However, it is questionable whether room air velocities would ever in practice reach such a low level during periods of normal occupancy, although the particular case of a cloudy day in a southwest 'corner' office served by a system which has been oversized (perhaps as a result of the tendency towards overestimation of casual gains from office IT equipment) might conceivably test this statement. In any event good control of space temperature will always allow some leeway in circumstances where other thermal comfort factors must be allowed to 'float'. With regard to room air distribution, as for any other aspect of thermal comfort under working conditions, the psychology of the workplace should never be overlooked.

4.7.5 Acoustics: variable flow rate means variable noise level

Acoustically, as the supply volume flow rate from a diffuser varies, so the noise level generated will vary. For a fixed-geometry diffuser undersized for between 60 and 80% of the design supply volume flow rate, the noise level will be greater under conditions of design sensible cooling load. As with increased air movement, however, occupants are more likely to accept some increase in the 'normal' level of noise from the supply diffuser during the typically short and infrequent periods when maximum cooling is required. Of perhaps greater concern for the HVAC designer is the sensitivity of humans to changes in noise level. People quite quickly become accustomed to constant levels of background noise, and noise that may at first be considered intrusive soon passes unnoticed. However, *changes* between noise levels that are not in themselves intrusive may be perceived as such.

4.8 Humidity control with VAV systems

In VAV systems supply air volume flow rate is typically a function of sensible cooling load *only*, since zone air temperature is the controlled variable. If both sensible and latent cooling loads do not peak simultaneously, space humidity will be less than its nominal design value at the zone design supply volume flow rate, and greater at the design latent cooling load (with zone supply volume flow rate less than its design value).

In commercial applications the latent cooling load is typically low, and derives from two principal sources. The first is the latent heat gain from the occupants. The second is associated with the infiltration of outside air into the building. As noted earlier, HVAC designers typically intend that an air conditioned building should be held at a slight positive pressure, relative to ambient, in order to minimise infiltration. How successfully this can be achieved with a VAV system depends not only on the standard of airtightness of the building envelope but also on the magnitude and duration of static pressure fluctuations within the zones caused by the VAV system itself. The effects of VAV operation on building pressurisation may be complex. The latent heat gain from any infiltration naturally varies with the moisture content of the outside air, and is thus seasonal in nature.

In typical commercial applications, variations in zone latent cooling load are not matched to the variations in sensible cooling load. Indeed, in such applications the latent cooling load may, to all intents and purposes, remain approximately constant for much of the time. We have seen that the moisture content of air in a zone is not only a function of the latent cooling load but also of the supply air volume flow rate. Since this can vary from zone to zone at any given time, and with time for any given zone, VAV systems are typically unable to provide close control of space humidity, which is expected to rise during part-load operation. However, because the latent heat gains in offices are typically low, and humans in general are relatively insensitive to changes in relative humidity (RH) within a broad band between approximately 40% and 70%, this does not usually cause problems.

Any deviation from the assumption of low latent gains needs to be carefully considered. It should also be remembered that in the 'typical commercial application' the occupants are assumed to be sedentary – performing light office work, or its equivalent. Any increase in activity level beyond this will result in an increase in latent heat emission. As an example, at a constant dry-bulb temperature of 22°C an increase in activity from a level corresponding to light office work to that equivalent to light bench work may increase the latent heat emission from an adult male of 2 m^2 body area by a factor of 2.4.[6] Under part-load conditions the supply air volume flow rate to the zone will be increased by the VAV terminal unit(s) to match the associated increase in the sensible heat emission. However, as the sensible heat gain from occupants represents only a part of the total sensible cooling load on the zone, the increase in supply air volume flow rate is unlikely to be of a similar magnitude to the increase in latent heat gain, and the room humidity will rise. Whilst it is not unreasonable to expect occupants to accept the effects of infrequent deviations from generally sedentary activities, the chance of any frequent or more sustained variations of activity level should be considered at the design stage. Furthermore, for a nominally constant total heat emission from the body, the proportion represented by the latent heat component increases with increasing dry-bulb temperature – from *c*.36% at 22°C to 50% at 26°C.[6] In general, changes in the proportion of sensible and latent heat emission from occupants may be expected to have a greater effect on the latent component of the cooling load for a zone than on the sensible component.

At the design stage it is in any event essential to check the extent of the rise in space humidity under part-load operation. In the absence of other information, the absolute worst-case condition may be identified as a zone experiencing design latent cooling load at its minimum supply volume flow rate. If this is 0.3 times the design value, the increase in moisture content of the supply airstream will be 1/0.3 (or 3.33) times that at the zone design volume flow rate. So as not to overstress the significance of an increase of this magnitude, it should be borne in mind that the ratio of sensible to total heat gains under design conditions is unlikely to be much below 0.9 for most zones in typical commercial applications under UK climatic conditions. A zone having high occupancy, such as a meeting room, is likely to be an exception. However, the particular requirements of such a space, including its increased (and intermittent) ventilation needs, may in any event make it more suited to treatment by a dedicated (single-zone) air conditioning system. The significance of the increase in moisture pick-up by the supply air will be greater if, for any reason, the nominal zone design relative humidity is much above 50%, which is again rare in UK commercial applications.

If VAV terminal units also provide space heating to the zones, it may be possible (in theory at least) to employ a zone high-humidity override. The supply air volume flow rate would be increased to limit the rise in space humidity, its excess sensible cooling capacity being cancelled by reheating. Energy consumption would be increased both by the reheat energy and the increased energy consumption of both supply and extract fans at the higher system volume flow rate. Capital cost of the control system would also be significantly higher. No such application appears to have been reported in the subject literature to date.

Where a DX cooling coil is employed in the central AHU, it may be sufficient simply to override evaporating temperature downwards by a small amount on a high-limit humidity signal, reducing the moisture content of the supply air by just enough to maintain space humidity levels. There will inevitably be an associated slight reduction in the dry-bulb temperature of the air leaving the AHU, further reducing (by however little) the supply air flows to the zones. This approach will naturally only be possible where latent loads are typically low. Control strategies are sometimes employed which involve resetting the supply air temperature upward from its design value under decreasing system sensible cooling load. Any such variation results in an increase both in the supply air moisture content and the supply air volume flow rate to all zones.

During winter operation the low moisture content of the outside air typically gives rise to a requirement for humidification to maintain a minimum relative humidity of about 40%. The moisture picked up by as little as $5\,l\,s^{-1}$ of outside air between a condition of -2°C (saturated) and 21°C (40% RH) is sufficient to absorb the typical latent heat emission from one sedentary occupant, while ventilation rates of $8\,l\,s^{-1}$ of outside air per person and above are recommended for open-plan offices[6]. Humidification is now typically based on the direct injection into the supply airstream of steam generated by local packaged electrode boilers. While these avoid the potential hygiene risks which are now generally felt to be associated with humidification based on water sprays or atomisers, the technique is a heavy consumer of electricity. This is significant not only because of electricity's position as the premium energy source, but also because each kW of humidification adds correspondingly to the building's electrical maximum demand. The humidification load is greatest over just those winter months when maximum demand tariffs in the UK typically impose the greatest cost penalties for increasing maximum demand. Hence in commercial applications generally the provision of humidification should be limited to the minimum level required, and indeed avoided altogether if the particular application permits.

References

1. Straub, H.E. (1962) What you should know about room air distribution. *Heating, Piping and Air Conditioning*, **34**, January, 209–220.
2. Holmes, M.J. (1974) *Designing Variable Volume Systems for Room Air Movement*, Building Research and Information Association, Bracknell, Application guide 1/74.
3. Willock, D.A. (1974) *Air Distribution Design in Open Plan Offices*, Electricity Council, Environmental Engineering Section, London.
4. Whittle, G.E. (1986) *Room Air Distribution: Design and Evaluation*, Building Services Research & Information Association, Bracknell, Technical note 4/86.
5. Straub, H.E. (1970) Room air distribution with a variable volume system. *ASHRAE Journal*, **12**, April, 52–58.
6. Chartered Institution of Building Services Engineers (CIBSE) (1986) *Environmental Criteria for Design*, CIBSE Guide, Section A1, CIBSE, London.

Chapter 5

Ventilation, Indoor Air Quality and Building Pressurisation

5.1 Ventilation

Outside air is the medium used for ventilation, and is conditioned at the system's central air handling plant. In the UK, VAV systems typically incorporate an outside-air economy ('free-cooling') cycle. This allows maximum advantage to be taken of the large proportion of the annual operating period during which the required supply air temperature can be achieved simply by mixing outside air and recirculated return air without the need for refrigeration. This may represent a total of between one-third and one-half of the potential system operating hours for a typical commercial application. For a further large part of the potential system operating period, when the outside air temperature is between the return air temperature and the required temperature off the cooling coil in the central AHU, the least amount of mechanical cooling (and hence the greatest economy of operation) is achieved by operating with 100% outside air. This may represent up to a further one-third of the potential system operating time annually.

At other times, when less energy is required to cool or heat recirculated return air, economy of operation requires that the system should use the minimum volume flow rate of outside air necessary for ventilation. For VAV systems, as for any other, this minimum volume flow rate of outside air that is to be drawn into the system is derived by reference to conventional statutory requirements, industry standards or recommendations for ventilation. Principal among the latter are, of course, the recommendations published by CIBSE in the UK and ASHRAE in the USA. It remains the HVAC designer's responsibility to interpret these to arrive at a design allowance of outside air, typically expressed either as a per capita allowance for the design occupancy or per unit floor area of the occupied space, that is appropriate for the particular application. This should take into account particular operating characteristics of the system.

In commercial applications the objectives of ventilation are fourfold:

(1) to provide an adequate supply of oxygen for respiration;
(2) to dilute the level of carbon dioxide produced by respiration to an acceptable value;
(3) to dilute the natural body odours produced by the occupants of the ventilated space;
(4) to dilute artificial contaminants generated within the ventilated space, either by processes carried out within the space, by smoking where this remains permitted, or by outgassing of *volatile organic compounds* (VOCs) from the materials and finishes used in its construction and furnishings.

Although objective number one is undeniably the most fundamental of all, it is invariably also the simplest to fulfil, typically requiring as little as $0.2 \, l \, s^{-1}$ of outside air per capita at typical occupant densities found in office-type working environments. Indeed, where work is sedentary, both oxygen supply and carbon dioxide dilution may be adequately delivered by an allowance of approximately $1 \, l \, s^{-1}$ per capita, although this will be greatly increased where work is of a more physically demanding nature. Achieving an acceptable dilution of natural body odours increases the total ventilation requirement in sedentary environments to a bare minimum level between 5 and $8 \, l \, s^{-1}$ per capita. As levels of smoking have decreased, VOCs have come to dominate modern thinking on artificial contaminants within the indoor environment, and concerns for a healthy environment within buildings, although low boiling point polar compounds ($< 50°C$), carbon monoxide, ozone and nitrogen oxides may also be present. VOCs represent a class of organic compounds typically used as solvents in the manufacture of synthetic materials and products which may be found in the fabric, finishes and furnishings of modern buildings. There are as yet no clear guidelines on limiting levels of long-term exposure, and little is yet known about the level of ventilation required to control their levels effectively within a space. Following years of minimising ventilation rates out of primary concern for energy use, the current trend is towards increasing allowances beyond minimum levels as an important contribution to reducing the occurrence of poor indoor air quality within modern buildings.

5.2 The concept of indoor air quality

The concept of *indoor air quality* (IAQ) extends beyond traditional ventilation considerations. Factors influencing IAQ are complex, and a detailed discussion is beyond the scope of the present work. For this the interested reader is directed to the substantial and growing body of literature on the subject. Suffice it is here to note that the factors include the temperature and relative humidity of the space (especially if RH is consistently low), the amount and quality of the air introduced for ventilation, the efficiency of its distribution, the presence and level of sources of micro-allergens (dust, mould, etc.), the chemical nature and volatility of the materials used in the fabric, finishes and furnishings of a building, and possibly even factors relating to the occupants themselves. At a simple level the latter can include such physical factors as trends in the nature and extent of cosmetic use. At a more complex level it can include undiagnosed illnesses, individual or group psychological factors or the psychiatric problems of particular individuals. Individual or group psychological factors can be triggered by the presence of non-optimal features of the working environment, such as poor lighting, intrusive noise or unsatisfactory workplace ergonomics. Through the effects of air temperature and relative humidity (particularly low RH), IAQ shares strong links with conventional considerations of thermal comfort.

Attempts have been made in recent years to quantify levels of IAQ (using the 'olf' and 'decipol'), although whether this approach will be generally adopted, and can indeed be effectively used by practising HVAC engineers, remains to be seen. As far as VAV systems are concerned, ventilation is the only factor over which such a system can exert any direct influence, although the quality of air filtration naturally has an impact on the level of airborne dust, fibres, and other particulate matter present in the space. Unless activated carbon adsorption-type filters are employed, however, filtration cannot have an impact on gaseous contaminants, including VOCs. As far as the performance of carbon filters is concerned, there are a myriad of potential VOCs which can outgass from carpets, furnishings, cleaning agents, etc. However, evidence suggests that the effect of the typical 'cocktail' of VOCs which is to be found in buildings may be tested on a single compound whose boiling point is roughly the same as the concentration-weighted average boiling point of the VOC mixture. Toluene appears to be such a compound, having a boiling point (110°C) that is close to that of both the mixture of substances in the bioeffluent given off by occupants (106°C) and the concentration-averaged boiling point of 35 selected VOCs (107°C). While activated carbon is very effective on ozone, it has only limited or no effect on nitrogen oxides, carbon monoxide or low boiling point polar compounds[1].

It is the aim of HVAC design to provide a healthy working environment to all occupants of the spaces treated, and one that will allow all persons to feel enough of a sense of personal physical and mental well-being that they can be as productive as they wish to be. IAQ problems have arisen, albeit unintentionally, through the actions of HVAC designers, building operators, architects and builders. While it can be seen that the concept of IAQ extends beyond considerations purely of ventilation, studies carried out in the USA by the National Institute of Occupational Safety and Health[2], and in Canada by Health and Welfare Canada[3] showed that inadequate ventilation was identified as a major contributory factor in cases of 'sick building syndrome' and poor IAQ in general. Within the HVAC industry itself the significant role of ventilation in achieving acceptable levels of IAQ has been clearly underlined in recent years both by the introduction in the USA of ASHRAE Standard 62-1989 ('Ventilation for acceptable indoor air quality') and, perhaps most forcefully, by the potential for future litigation over 'sick buildings'.

Although not inherently incapable of providing adequate ventilation, the complexity of VAV systems does make them more susceptible than their CAV counterparts to contributing to IAQ problems if a strategy for achieving adequate ventilation under all reasonable operating conditions is not fully thrashed out at the design stage and successfully implemented during commissioning. Reductions in the extent of smoking formerly permitted in many buildings may have served to unmask the effects of other contaminants, to which occupants had been effectively desensitised by the effects of smoking. Nor should the effects of changes in construction methods and standards of air-tightness in buildings be overlooked. As a generalisation, commercial buildings were formerly more 'leaky'. It is now probably impossible to confirm or reject the suggestion that high infiltration rates resulting from poor air-tightness may, in many instances, have made up for otherwise inadequate ventilation via VAV systems. Differences in standards of building air-tightness internationally may also reflect in the extent and frequency of occurrence of IAQ problems in the commercial building stock.

The results of failing to achieve acceptable levels of IAQ have become readily apparent, first to the users of many modern buildings, subsequently to their owners or managers and, ultimately, to the architects, HVAC engineers and major clients involved in new building projects. While IAQ itself may have fairly recent roots as a formally recognized concept, the ventilation performance achieved by real VAV systems has certainly been a matter of concern to HVAC engineers since the early 1980s.[4,5] Ultimately IAQ problems can only be solved by those who, inadvertently or not, created them. As in any other well-publicised problem area, IAQ also acts as a magnet for some who see it merely as a suitable vehicle for achieving more personal objectives.

5.3 Environmental effects

It has been customary to refer to the supply of *fresh air* for ventilation, when of course the only thing that can be assured is that it is air from *outside* the building. In the centres of towns or cities this may be far from the idea of 'freshness' typically associated with country, sea or mountain air. Location of outside air intakes away from direct sources of pollution, such as vehicle exhaust fumes or air exhausted from toilets or kitchens, is of course an essential requirement not specific to VAV systems. However, since recommended ventilation rates must inevitably be based on notional average levels of pollution, their adequacy may be adversely affected by particularly poor outside air quality (OAQ). It would seem, after all, logical to regard outside air as having a *capacity to provide ventilation* which bears an inverse relationship to its level of contamination or pollution. The true ventilation capacity of outside air, as far as IAQ is concerned, may be a complex matter. Nor is OAQ a constant factor, being subject to fluctuations and trends over the life of a building.

Outside air that is drawn into an air conditioning system is filtered, but it must be remembered that the purpose of this filtration is to remove dust, dirt and other particulate matter. This of course represents only one aspect of air pollution. Conventional filtration can do nothing to combat gaseous pollutants, and hence in some situations it may not be an oversimplification to suggest 'poor air quality outside, poor air quality inside'. Outside air that infiltrates directly into air conditioned spaces remains totally unfiltered. While such infiltration may be minimised by maintaining a slight positive pressurisation of the spaces relative to outside air, achieving this under all operating conditions is more difficult with a VAV system unless the extract flow rate from each zone is individually controlled to track the zone supply flow rate. If only the total volume flow rate of the VAV extract system is controlled, as is typically the case in UK practice, fluctuations in building pressurisation will undoubtedly occur.

As far as the performance of conventional filter media in bag or panel arrangements is concerned, VAV operation results in a reduction in face velocity at the filter(s) under part-load operation. As far as is known, no published data exists to suggest that this has a detrimental effect on filter efficiency. On a practical note, however, a fixed differential pressure set-point does not give a true indication of filter status for a 'filter blocked' alarm. A differential pressure sensor is required, rather than a differential pressure switch. A more sophisticated approach to determining filter status could reset the differential pressure set-point value according to system volume flow rate.

5.4 Contamination within the system itself

Some of the ventilation capacity of the outside air may be lost by contamination within the central air handling plant itself and the ducted air distribution systems connecting it to the spaces served. As air can be recirculated many times, return air ducts are as serious a source of potential contamination as are the supply ducts. It is important to minimise any such contamination by maintaining acceptable standards of duct hygiene through periodic cleaning. The introduction of new standards and techniques for the cleaning of ductwork have done much to draw attention to this formerly neglected aspect of system maintenance. Visible staining of ceiling tiles around supply diffusers does not promote confidence in either the quality of the air being delivered to the space or the cleanliness of the ducts through which it flows.

5.5 Influences within the space itself

Assuming that the outside air that reaches the zone supply outlets retains adequate capacity to ventilate the spaces served, just how much of this capacity is usefully employed will depend on both the design of the room air distribution scheme and the installed performance of the air terminal devices selected. Some of the supply airstream may simply short-circuit back to the return air system without entering the occupied zone. A corresponding proportion of the outside air thus provides no useful ventilation. It goes without saying that any such short-circuiting is at least to be minimised and, preferably, avoided altogether. It should be stressed that this is not a problem *specific* to VAV systems, but these may be less forgiving to any unpredicted loss of ventilation effectiveness.

Assuming that short-circuiting between supply and extract points is avoided, mixing air distribution is generally quoted as having a typical ventilation effectiveness, η_V, of approximately 0.65, defined as:

$$\eta_V = (C_{EA} - C_{SA}) \div (C_{ZA} - C_{SA}) \quad (5.1)$$

where C is the concentration of contaminant, in parts per million or other consistent units, and subscripts EA, SA and ZA refer to extract air, supply air and zone (room) air respectively. A combination of ventilation and filtration are the conventionally accepted solutions to the problem of removing indoor air contaminants from a space. Achieving satisfactory room air movement is essential to the success of both aspects. Ventilation can only be effective if the outside air reaches all parts of the occupied zone. Filtration of recirculated air can only be effective if airborne contaminants are extracted from the space and transported to the central plant filters. The room air movement that is achieved is dependent on many factors, including room size, shape, layout and type of furnishings, type and level of activity, temperature gradients, location of the supply and extract points (principally the former), and the characteristics of the supply diffusers. Turbulence within the space may play a dominant role in the removal of both particulate and gaseous contaminants.

From the perspective of contaminant removal, the pattern of room air movement may follow that shown in Fig. 5.1.[6] This suggests the possible interplay between jets, currents, subcurrents and turbulent eddies within the space. Larger particles (1 μm and above) held in suspension in the air depend mainly on room air movement for transport to the extract point(s) and removal from the space. The level of room air movement in turn depends on the outlet velocity of the supply diffusers and the physical characteristics of the supply air jets. Particles of submicron size and gases transfer by diffusion, even in the absence of air motion. However, these submicron particles and gas molecules may attach to larger particles, making them also dependent on air movement for transfer to the exhaust points – from eddy to eddy, to subcurrent, and via subcurrents and currents. Any particles that settle out and are deposited on surfaces within the space are not returned and filtered out at the central plant. When disturbed, they may be re-entrained into the room air movement, or remain a residual source of contaminants that can be breathed in by the occupants of the space. Proper air movement is thus critical to the effective removal of particulate and gaseous contaminants from the air within buildings. This will inevitably also benefit thermal comfort by minimising temperature differences and gradients.

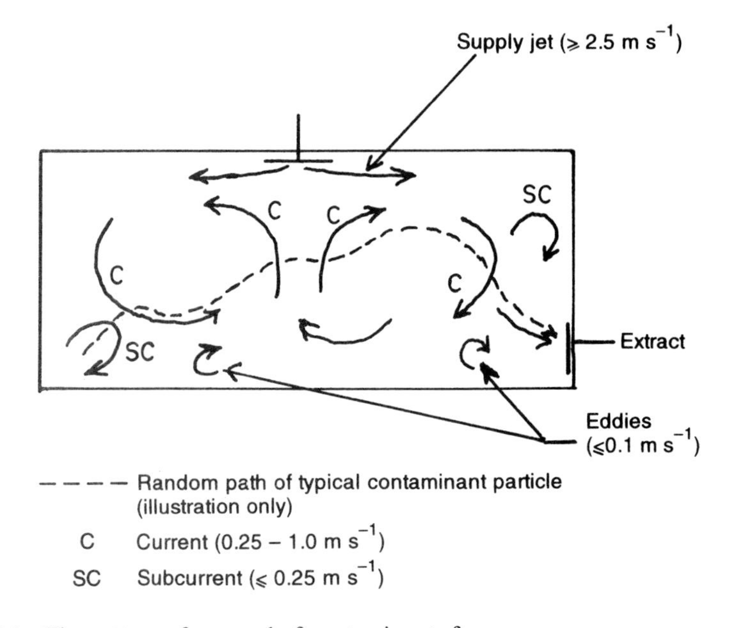

Fig. 5.1 The pattern of removal of contaminants from a space.

5.6 The effect of VAV operation

Under reducing sensible cooling load the VAV terminal units reduce the volume flow rate of air supplied to the zones. As this is a mixture of outside (ventilation) air and recirculated return air, zone ventilation rate is also reduced unless the proportion of outside air (the outside air fraction) is increased. The outside air fraction of any zone cannot be increased independently of that for the whole system.

For a given application and design occupancy, the ventilation requirement of a single zone, or of the whole building, is the same whether provided by a constant- or variable-volume system. However, because of design diversity in the sensible cooling load, the system design supply air volume flow rate is less with the VAV system – typically 70–80% of that of the constant-volume system. For the VAV system the same building ventilation requirement thus represents a higher outside air fraction at the system design supply air volume flow rate – between 1.25 and 1.5 times that of the equivalent constant-volume all-air design.

In the office-type applications that are most commonly served by VAV systems, zone ventilation requirements are typically regarded as being independent of variations in sensible cooling load. To ensure adequate ventilation when the system is operating with 100% outside air, it is only necessary to ensure that the minimum supply volume flow rate to any zone does not fall below its ventilation requirement. In office-type applications this is unlikely to exceed the minimum volume flow rate that can be accommodated while retaining an acceptable performance at either the VAV terminal units or the supply diffusers themselves. It is not normally possible to identify instantaneous values of the ventilation requirements of individual zones, and these are typically assumed to remain constant throughout the occupied period at their design values.

In terms of the effect of VAV operation on ventilation effectiveness, under cooling operation a VAV system differs from other air systems only in so far as 'dumping' may occur. Ignoring the obvious comfort considerations, this is unlikely to be compatible with achieving maximum ventilation effectiveness. Under heating operation, at minimum zone supply volume flow rates, the risk of stratification may be greater with a VAV system. However, the discussion in Chapter 4 shows that neither of these aspects needs be an inherent feature of a VAV installation, as both are amenable to design treatment and the selection of appropriate equipment.

Adequate ventilation must be provided under all operating conditions. Strictly speaking, however, the problem is not one of ensuring adequate ventilation, since this can always be achieved by the simple expedient of operating throughout the year with 100% outside air, but rather of ensuring adequate ventilation economically – in terms of capital cost and energy consumption.

5.7 Year-round VAV operation with 100% outside air

When operating year-round with 100% outside air, adequate ventilation of all zones can be assured if minimum volume flow rate settings are maintained at the terminal units. These should be equal to the required zone ventilation rates. With appropriate design and selection of equipment it is feasible to achieve this in practice. However, the ventilation advantages of year-round operation with 100% outside air do not come without penalties to both system capital and running costs. Consider two VAV systems serving an identical commercial application (A and B in Fig. 5.2). System A operates throughout the year with 100% outside air, while system B uses an outside air economy cycle controlled according to outside/return air dry-bulb temperatures. In zone 2 both systems operate with 100% outside air. For system B in the UK's temperate climate this operating regime may account for up to perhaps one-third of the potential annual operating hours. It is at the extremes of Fig. 5.2, in zones 1 and 3, that the effects of a change to full outside air are felt. Since these regions contain both summer and winter outside air design conditions, there are implications for the design and sizing of the heating, air handling and refrigeration plant. In the case of zone 1 the effects may be somewhat disproportionate, considering that in the UK it typically accounts for a minimal proportion of potential annual operating hours. The more extreme the summer and winter outside air design conditions, the greater will be the capital cost penalty associated with full outside air operation under those conditions. Similarly, the more time spent operating on full outside air either in zone 1 or near the low end of zone 3, the greater will be the associated penalty in annual running cost, and the less the benefit of using an outside air economy cycle over operation with year-round minimum outside air. Some of the factors influencing the cost premium of year-round operation with 100% outside air are outlined below.

First the load on the cooling coil is increased in zone 1. The ratio of sensible to latent cooling performed by the coil is reduced, and it must do more dehumidification. Both the chilled water flow temperature and the evaporating temperature of a direct-expansion (DX) cooling coil are ultimately limited by the risk of freezing the evaporator or frosting the DX coil respectively. Hence a deeper cooling coil may be required (with more rows of finned tube) to achieve a similar supply air temperature and moisture content. The capital cost of this is inevitably higher, as is also the pressure drop of the air flowing across it. Fan total pressure is increased, and with it annual fan energy consumption. The fan itself may be more expensive. Since

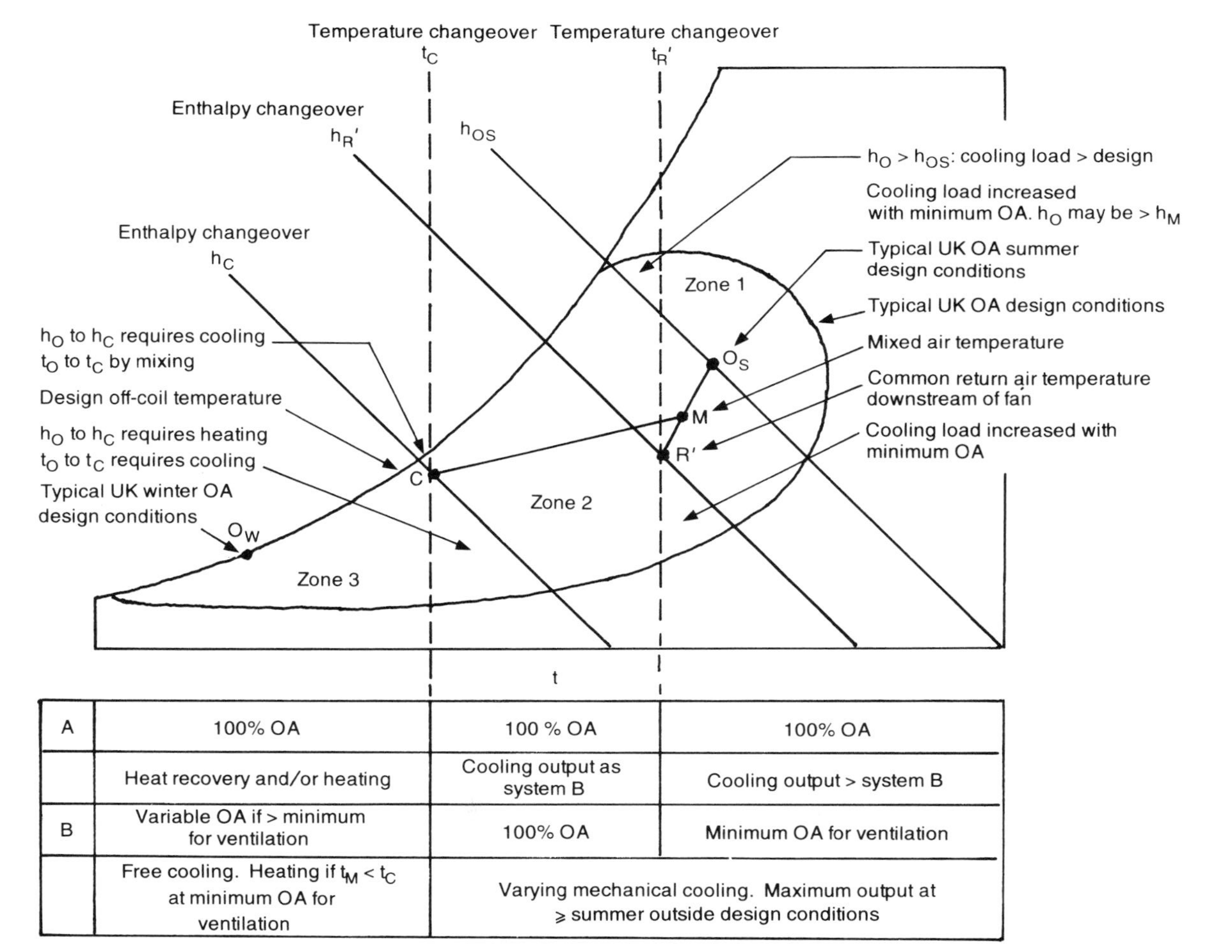

A	100% OA	100 % OA	100% OA
	Heat recovery and/or heating	Cooling output as system B	Cooling output > system B
B	Variable OA if > minimum for ventilation	100% OA	Minimum OA for ventilation
	Free cooling. Heating if $t_M < t_C$ at minimum OA for ventilation	Varying mechanical cooling. Maximum output at ⩾ summer outside design conditions	

Fig. 5.2 Operation with 100% outside air *vs.* use of an outside air economy cycle.

the design cooling load is increased, the size and capital cost of the refrigeration plant are likely to increase, and there may be knock-on effects in terms of the plant space required and the provision of larger electrical supplies to the refrigeration plant, chilled water pumps and VAV fans. Performance and control of the refrigeration plant under part-load is likely to suffer. The larger cooling coil will require the size of the AHU itself to be increased, with its own cost and space implications.

In zone 3 heat recovery from air exhausted from the building is essential to reduce the ventilation heating load in winter. Failure to do so has implications for the sizing and capital cost of the heating plant, its part-load efficiency and annual consumption of fossil fuel. Heat recovery represents an additional capital cost and inevitably increases the system pressure loss for both supply and return fans. Although heat recovery is not required at the system design volume flow rate, the heat recovery devices in HVAC systems are permanently located in the outside and exhaust air streams. Fan size, cost and annual energy consumption are further increased. To reduce the design cooling load and improve utilisation of the heat recovery equipment, it may be practical to transfer heat from the outside air to the exhaust air (recovery of 'coolth') in summer. In the UK, however, unless the exhaust airstream is actively cooled, the low temperature difference which typically prevails between the airstreams (2–7 K for exhaust and outside airstream temperatures in the ranges 23–26°C and 28–30°C respectively) strictly limits the potential benefits of such 'coolth' recovery, even at high heat transfer efficiencies. Such cooling can obviously not be achieved with conventional mechanical or absorption refrigeration, because it must always be more effective to apply this directly to the mixed airstream. The exhaust airstream can, however, be cooled evaporatively using water sprays with suitable atomising nozzles. In various configurations this is the favoured approach for providing cooling of a supply airstream without refrigeration. In any event a device that recovers 50% of the available heat that would otherwise be exhausted from a building is certainly as *effective* as 50% recirculation. However, the application of all common heat recovery devices, such as run-around coils, plate heat exchangers, thermal wheels, heat pipes and air-to-air heat pumps, is *always* associated with some additional expenditure of energy, and this usually in the (most expensive) form of electricity. Hence it is probably true to say that heat recovery is not as *efficient* a process as recirculation. The additional energy input associated with a heat recovery installation may be required both as direct input to a pump circulating a dedicated heat transfer fluid between run-around coils in the exhaust and outside air ducts, or to a refrigerant compressor where the run-around coils are replaced by the evaporator and condenser of an air-to-air heat pump, and indirectly by increasing the air-side system pressure losses. In theory at least, recirculation does not involve any additional expenditure of energy to recover both sensible heat and moisture from the extracted air, although this does, however, ignore the fact that, to achieve acceptable control, the modulating control dampers of the outside-air economy cycle should be sized to minimum *authorities* (and hence pressure drops across the fully-open dampers).

Where only heat is recovered, without any associated moisture recovery, as in the case of run-around coils, plate heat exchangers, heat pipes and air-to-air heat pumps, any requirement for winter humidification will be greatly increased. Thermal wheels are, however, available which incorporate hygroscopic matrices that permit the recovery of both heat and moisture from the exhaust air, and can thus act as a counter to low humidity during winter operation of the OA economy cycle.

In conclusion, while year-round VAV operation with 100% outside air has been utilised, and certainly has an attraction in the ease and simplicity with which ventilation performance can be achieved under all operating conditions, the approach has not found general favour with HVAC engineers. In temperate climates, such as that of the UK, and for applications with moderate ventilation requirements, recirculation is likely to continue to be preferred to heat recovery.

5.8 Volume flow rates of outside air

Since all zones receive outside air in the same proportion, the volume flow rate to any zone in a VAV system may be expressed as:

$$\dot{V}_{OAZ} = \dot{V}_{OAS} \,.\, \dot{V}_{SZ}/\dot{V}_{SYS} \qquad (5.2)$$

where

$\dot{V}_{OAZ}$ = volume flow rate of outside air to a zone ($m^3\ s^{-1}$)
$\dot{V}_{OAS}$ = volume flow rate of outside air drawn into system ($m^3\ s^{-1}$)
$\dot{V}_{SZ}$ = zone supply volume flow rate ($m^3\ s^{-1}$)
$\dot{V}_{SYS}$ = system supply volume flow rate ($m^3\ s^{-1}$).

Equation 5.2 is indeed equally true for any air system, whether of CAV or VAV type. For the VAV system it is useful to consider the zone supply volume flow rate as the product of its design value and a non-dimensional zone load factor (ZLF). Similarly the system supply volume flow rate may be considered as the product of its design value and a System Load Factor (SLF) representing the proportion of the simultaneous design sensible cooling load being experienced at any instant. Hence:

$$\dot{V}_{OAZ} = \dot{V}_{OAS}\,[\dot{V}_{SZ}\,(\text{design})\,.\,ZLF]/[\dot{V}_{SYS}\,(\text{design})\,.\,SLF] \qquad (5.3)$$

where

ZLF = non-dimensional zone load factor, i.e. the proportion of its design sensible cooling load being experienced by the zone at any instant

SLF = non-dimensional system load factor, i.e. the proportion of its design sensible cooling load being experienced by the system at any instant

and all $\dot{V}$ terms are as previously defined.

From Equation 5.3 the volume flow rate of outside air that should be drawn into the system to satisfy the ventilation requirement of any particular zone may be expressed as:

$$\dot{V}_{OAS}\,(\text{required}) = \dot{V}_{OAZ}\,(\text{required})\,[\dot{V}_{SYS}\,(\text{design})\,.\,SLF]/[\dot{V}_{SZ}\,(\text{design})\,.\,ZLF] \quad (5.4)$$

where all terms are as previously defined.

While statutory requirements and methods of controlling a system's intake of outside air must typically be discussed in terms of volume flow rates, it is frequently convenient to consider ventilation rates and requirements in terms of an outside air fraction (OAF). This may be defined as the decimal proportion of outside air in an air stream (although some engineers may prefer to use percentage values), i.e.

$OAF\,(\text{Sys})$ = the system outside air fraction, $\dot{V}_{OAS} \div \dot{V}_{SYS}$

$OAF\,(\text{Zo})$ = the zone outside air fraction, $\dot{V}_{OAZ} \div \dot{V}_{SZ}$

and all $\dot{V}$ terms are as previously defined.

Hence rewriting Equations 5.2 to 5.4 in these terms.

$$\dot{V}_{OAZ} = OAF\,(\text{Sys})\,.\,\dot{V}_{SZ} \quad (5.5)$$

$$\dot{V}_{OAZ} = \dot{V}_{SZ}\,(\text{design})\,.\,OAF\,(\text{Sys,design})\,.\,ZLF/SLF \quad (5.6)$$

$$\dot{V}_{OAS}\,(\text{required}) = [\dot{V}_{SYS}\,(\text{design})\,.\,SLF]\,[OAF\,(\text{Zo,design})/ZLF] \quad (5.7)$$

In the case of VAV systems we are also keenly interested in ventilation performance under part-load conditions. Hence from Equation 5.2:

$$(\dot{V}_{OAZ2}/\dot{V}_{OAZ1}) = (\dot{V}_{OAS2}/\dot{V}_{OAS1})\,(ZLF_2/ZLF_1)\,(\text{SLF}_1/\text{SLF}_2) \quad (5.8)$$

If subscript 1 defines values under zone and system design conditions (i.e. $ZLF = 1.0$, $SLF = 1.0$), while subscript 2 refers to some other part-load condition, then this simplifies to:

$$(\dot{V}_{OAZ2}/\dot{V}_{OAZ1}) = (\dot{V}_{OAS2}/\dot{V}_{OAS1})\,(ZLF_2/SLF_2) \quad (5.9)$$

The presence of the term ZLF in Equations 5.3 and 5.4 immediately suggests the potential problem for VAV ventilation performance, since even under system design conditions not all zones have a zone load factor of unity. This is, after all, one of the principal justifications for using a VAV system.

Let us use a very simple example to compare the ventilation performance of two equivalent all-air systems, one a constant-volume sytem and the other a VAV system. Consider that both approaches serve an identical commercial application of *n* zones, with each utilising a conventional outside air economy cycle (Fig. 5.3). Whichever system is considered, the requirement is to achieve adequate ventilation of all zones at all times when the system is operating. For the purposes of the subsequent discussion, 'adequate ventilation' of any zone will be considered to be achieved if a specified minimum volume flow rate of outside air can be delivered to the zone under all operating conditions. From the preceding sections of this chapter it should be apparent that adequate ventilation performance may not always be achieved so simply. Under (zone) design conditions all values of zone outside air fraction lie within a range from OAF (Zo,min) to OAF (Zo,max).

First let us consider the constant-volume system A. Since system and all zone supply volume flow rates do not (by definition) vary with sensible cooling load, there is no need to differentiate between design conditions and part-load. We can of course simply sum the design ventilation requirements of all zones, in $m^3\ s^{-1}$ of outside air, and ensure that at least this volume flow rate of outside air is continuously drawn into the system, but will this be enough? The approach may be expressed as:

$$\text{Minimum required } \dot{V}_{OAS} = \sum \dot{V}_{OAZ}\ (\text{design}) \tag{5.10}$$

The answer is a qualified 'yes' for the constant-volume system. A look at Equation 5.5 shows that it will be sufficient only if all zones require the same design outside air fraction, which will yield the conditions:

$$(\dot{V}_{OAZ}/\dot{V}_{SZ}) = (\dot{V}_{OAS}/\dot{V}_{SYS})$$

and

$$OAF\ (\text{Zo,min}) = OAF\ (\text{Zo,max}) = OAF\ (\text{Sys})$$

Where the design outside air fraction differs between zones in the constant-volume system, there will naturally be a ventilation deficiency where OAF (Zo) > OAF (Sys) and a surplus over requirements where OAF (Zo) < OAF (Sys). Any deficiency may be simply corrected, and the ventilation requirement of all zones satisfied, if the minimum flow rate of outside air into the system is increased so that the corresponding value of OAF (Sys) in Equation 5.5 matches the maximum design value for any zone, i.e. OAF (Sys) = OAF (Zo,max). In this case all zones with design values of outside air fraction less than OAF (Zo,max) will always receive more outside air than their design requirement. The approach maintains a minimum ventilation rate for the building *and all zones served*, although this may typically be expected to be greater than the sum of the individual zone design requirements.

It is evident that if one or two zones require a significantly greater design outside air fraction, it may be more economical to consider conditioning them independently of the central system. This might be the case, for

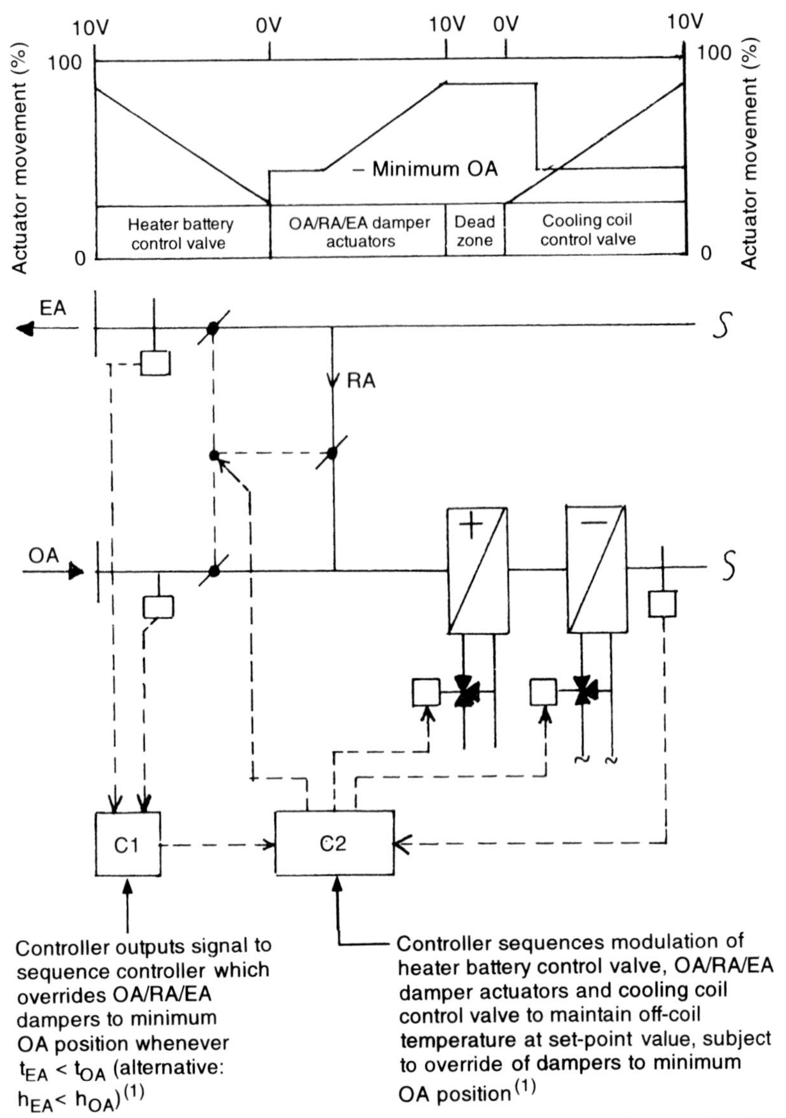

NB: Fans and all other system components and controls have been omitted for clarity

Notes:

(1) The problem of maintaining minimum OA intake rate during VAV operation is not addressed by this figure. Applicable techniques are considered in sections 5.9 to 5.14

Fig. 5.3 Conventional outside air economy cycle.

example, for a meeting room in an office block. Where there are general wide variations in the design outside air fractions of the zones served, an all-air system may not be the most appropriate choice.

As far as the VAV system is concerned, the situation is not quite so straightforward, and the presence of load factor terms in the equations means that we must be very interested in what happens under part-load operation. The same intake rate of outside air into the system, $\sum \dot{V}_{OAZ}$ (design), will in the case of the VAV system represent an increased value of *OAF* (Sys,design), $[\sum \dot{V}_{OAZ}$ (design)]/$\dot{V}_{SYS}$ (design), compared with the constant-volume system – by a factor equal to the *reciprocal* of the VAV system's design load diversity factor. All zones again receive this same proportion of outside air. While the system load factor, or *SLF*, in Equation 5.6 has a value of unity under design conditions, and may thus be disregarded, the same is certainly *not* true of the zone load factor, or *ZLF*. Ignoring any effects resulting from oversizing of the system, at the system design condition a significant proportion of the zones served should be receiving their design supply volume flow rates (i.e. $ZLF = 1$). However, it follows equally, and is in the very nature of VAV operation, that other zones will have lower load factors. The simplest illustration of this would be given by a hypothetical system with n zones, $n/2$ of which are at $ZLF = 1$ while the remaining $n/2$ are at ZLF = [(2 × design load diversity factor) – 1)], which would yield a value of only 0.6 for a design load diversity factor of 0.8. Naturally the zone load diversity characteristics under system design conditions will typically be more complex and varied than this in real applications.

Consideration of Equation 5.6 suggests that, under system design conditions, any zone will receive at least its required outside air volume flow rate if *OAF* (Sys,design) $\geq$ *OAF* (Zo,design)/*ZLF*. Typically the value of *OAF* (Zo,design) will not be constant for all zones, even if the absolute values of $\dot{V}_{OAZ}$ (design) are identical. However, as noted previously for the constant-volume system, if the variation between zones is wide, an all-air design approach may not be the most appropriate.

As for its constant-volume counterpart, under system design conditions all zones can achieve their design ventilation requirements if the system intake of outside air is increased so that *OAF* (Sys,design) equals the greatest value of [*OAF* (Zo,design)/*ZLF*] for any zone.

In typical commercial applications only a very small proportion of the annual operating hours may ever be spent at the system design condition. Part-load operation accounts for the vast majority of the operating time. For much of this operating regime the system in our example will in any event be using 100% outside air. Under these conditions adequate ventilation of any zone, as defined for the purposes of the present discussion, will be achieved as long as the zone load factor does not fall below the required outside air fraction under zone design conditions, i.e $ZLF \geq$ *OAF* (Zo, design). We have seen, in Chapter 4, that considerations of room air distribution performance tend to mitigate against reducing zone supply air volume flow rates much below about 30% of their design value for single-duct throttling VAV systems, and minimum zone supply volume flow rates can typically be established using the local control loops of the VAV terminal units themselves. This would, therefore, be compatible with maintaining the design ventilation rate. Where the use of FATs does not involve the same considerations of room air distribution performance, terminal unit local control loops may still be set to provide the minimum primary air flow rate required for ventilation.

If there is rarely a ventilation problem when a VAV system is operating with full outside air (except through inadequate housekeeping or maintenance), there is still, however, a significant proportion of the potential system operating hours annually during which an all-air system will be operating at part-load with less than full outside air. This occurs either at high ambient temperature with minimum outside air or at moderate-to-low ambient temperature with variable outside air reducing towards minimum. While in the UK's temperate climate the use of an outside air economy cycle invariably makes economic sense, the same is not necessarily true in other more extereme climates, where careful consideration of load and weather profiles and unit energy costs may need to be considered.

Our approach to providing ventilation continues to be based on drawing a set minimum volume flow rate of outside air into the system at all times (minimum building ventilation rate). Under part-load the system outside air fraction, *OAF* (Sys), is inversely proportional to system supply volume flow rate. Again all zones receive a common outside air fraction equal to *OAF* (Sys) or *OAF* (Sys,design) $\div$ *SLF* (Equations 5.5 and 5.6, respectively). At any value of system load factor the worst case from the point of view of ensuring adequate ventilation during part-load operation is the zone with the lowest zone load factor, while zones experiencing load factors equal to or greater than (*ZLF* under system design conditions $\times$ *SLF*) will receive at least the ventilation rate that they would under system design conditions. Since in practice it is very rare for VAV terminal units to be allowed to throttle down to full shutoff, the lowest value of zone load factor may well represent a minimum limit of zone supply volume flow rate set into the terminal unit controller. Table 5.1 shows the solution of Equation 5.6 for a range of combinations of zone and system load factors. The volume flow rate of outside air delivered to any zone, $\dot{V}_{OAZ}$, is *normalised* as a proportion of $\dot{V}_{SZ}$ (design), the supply air volume flow rate to the zone under (zone) design conditions (i.e. *ZLF* = 1). The example is based on providing adequate ventilation to a worst-case zone combining a required zone outside air fraction under (zone) design conditions (*ZLF* = 1) of 0.15, at an associated minimum value of zone load factor under *system* design conditions of 0.6.

This yields a required outside air intake rate into the system (Equation 5.7) of 0.25 times the system design flow rate.

The data in Table 5.1 are plotted in Figure 5.4. For comparison the bracketed values of normalized $\dot{V}_{OAZ}$ show the solution of Equation 5.6 for an intake rate of outside air into the system which is the simple sum of the zone design outside air requirements, $\sum \dot{V}_{OAZ}$ (design), and for the particular case when this is equivalent to a value of OAF (Zo,design) of 0.15 for all zones and the system design load diversity is 0.8. This yields a value of OAF (Sys,design) of 0.15/0.8, or 0.1875, in Equation 5.6. The effect of this reduction in the constant outside air intake rate drawn into the system is indicated in Fig. 5.4.

With a VAV system the provision of a constant minimum *building* ventilation rate inevitably results in variable ventilation rates at the zones. However, the approach can still fully ensure that all zones served by the system achieve at least their design ventilation rates under all operating conditions if the system's minimum intake rate of outside air is suitably increased, above the simple sum of the zone design supply volume flow rates. The criterion for this increase is the volume flow rate of outside air at which the corresponding value of OAF (Sys,design) matches the worst-case value of OAF (Zo,design) $\times$ SLF/ZLF that can realistically be met during *normal* system operation. The effect of oversizing of the VAV system will be to ensure that Zone and System Load Factors are never unity.

It will always be necessary to:

(1) Establish the design ventilation requirements of the zones served, and the corresponding system outside air intake requirement. Providing an initially generous design ventilation allowance, in comparison with minimum statutory requirements, will inevitably make deficiencies in VAV ventilation performance less critical;
(2) estimate by how much actual outside air volume flow rates may fall short of these design requirements for the worst case that may reasonably be expected during normal operation, and for what proportion of the potential annual system operating hours that this condition may reasonably be expected to occur;
(3) establish by how much the outside air intake volume flow rate of the system should be increased to maintain an acceptable minimum level of zone ventilation under all conditions likely to be encountered in normal day-to-day operation of the system;
(4) adopt a design and control strategy to achieve an acceptable level of ventilation in all zones, taking into account the controlling parameters of the particular application.

Having established that maintaining a minimum volume flow rate of outside air into a VAV system can ensure adequate ventilation of all zones at all

Table 5.1 Ventilation performance and requirements for a simple example. Varying values of $\dot{V}_{OAZ}$ as a proportion of $\dot{V}_{SZ}$ (design) for varying zone load factor (*ZLF*)[1].

System load factor (SLF)	*ZLF*								
	1.0	0.9	0.8	0.7	0.6	0.5	0.4	0.3	0.2
1.0	0.25 (0.1875)	0.225 (0.169)	0.2 (0.15)	0.175 (0.131)	0.15 (0.113)	0.125 (0.094)	0.1 (0.075)	0.075 (0.0563)	0.05 (0.038)
0.9	0.278	0.25	0.222	0.194	0.167	0.139	0.111	0.083	0.056
0.8	0.313 (0.235)	0.281 (0.211)	0.25 (0.1875)	0.219 (0.164)	1.188 (0.141)	0.156 (0.117)	0.125 (0.094)	0.094 (0.071)	0.063 (0.047)
0.7	0.357	0.321	0.286	0.25	0.214	0.179	0.143	0.107	0.071
0.6	0.417 (0.313)	0.375 (0.281)	0.333 (0.25)	0.292 (0.219)	0.25 (0.1875)	0.208 (0.156)	0.167 (0.125)	0.125 (0.094)	0.083 (0.062)
0.5	0.5	0.45	0.4	0.35	0.3	0.25	0.2	0.15	0.1
0.4	0.625 (0.469)	0.563 (0.422)	0.5 (0.375)	0.438 (0.329)	0.375 (0.281)	0.313 (0.235)	0.25 (0.1875)	0.188 (0.141)	0.125 (0.094)
0.3	0.833	0.75	0.667	0.583	0.5	0.417	0.333	0.25	0.167

Note

(1) Basic data refer to a constant intake rate of outside air into the system equivalent to 0.25 times its design volume flow rate. For comparison the data in brackets show the effect of a reduction in outside air intake rate to 75% of this, or 0.1875 times the system design flow rate. The basis for these values is described in the text.

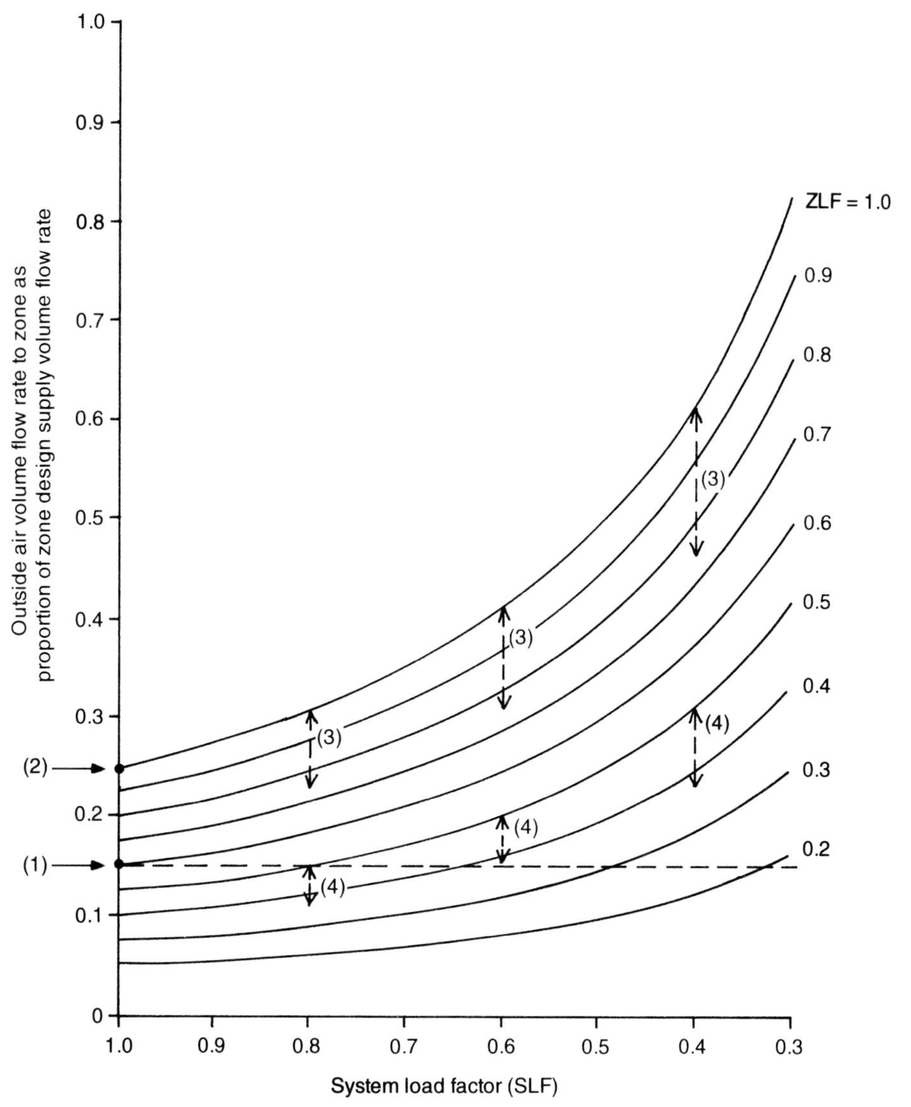

Notes:

(1) Design requirement for this hypothetical example (intended for illustrative purposes only) is an OA supply rate to zone equivalent to an OA fraction of 0.15 under zone design conditions for a worst-case zone at load factor of 0.6 under system design conditions.

(2) Constant OA intake rate into system defined by system OA fraction of 0.25 under design conditions (equation 5.7).

(3) Examples of effect for a zone under design conditions of reducing constant OA intake rate into system to simple sum of zone design OA requirements, based on design OA fraction of 0.15 for all zones and system design load diversity of 0.8.

(4) Ditto for zone at load factor of 0.5.

Fig. 5.4 Ventilation performance and requirements for a simple example.

times, we must now consider how this is achieved in practice, its drawbacks and any alternatives that may exist. Before moving on to this aspect, however, it is worth noting that ASHRAE Standard 62-1989 ('Ventilation for aceptable indoor air quality') gives the following relationship for system outside air fraction:

$$Y = X(1 + X - Z) \tag{5.11}$$

where

Y = the corrected fraction of outside air in the supply system
= corrected total outdoor air flow rate ÷ total system supply volume flow rate

X = the uncorrected fraction of outside air in the supply system
= ($\sum$ outside air flow rates for all system branches) ÷ total system supply volume flow rate

Z = the fraction of outside air in the critical zone (i.e. the zone requiring the greatest outside air fraction in its supply)
= outside air flow rate required in critical zone ÷ supply volume flow rate in critical zone.

5.9 Outside air damper control of minimum building ventilation rate

The basic pressure relationships for an outside air economy cycle are shown in Fig. 5.5. In a constant-volume system the fixed outside air fraction of the system results in a corresponding, and unique, minimum opening position of the outside air damper to ensure adequate ventilation at those times when the system is designed to operate with minimum outside air. Under design conditions a minimum opening position for the outside air damper will also

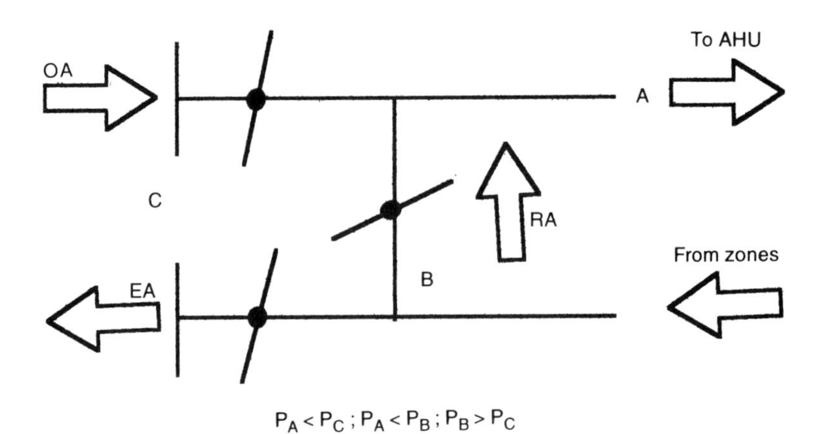

Fig. 5.5 Basic pressure relationships for an outside air economy cycle.

ensure that the required minimum volume flow rate of outside air is drawn into a VAV system. However, if these pressure relationships are not maintained as system volume flow rate reduces under part-load operating conditions, the result may be that either the outside air intake acts as a supplementary exhaust point or the exhaust air discharge acts as a supplementary intake point. The pressure difference between ambient (point C) and the mixed air plenum of the central AHU (point A), ΔP_{AC} controls the rate at which outside air is drawn into the VAV system. Since ambient pressure provides the datum for the duct system, for any position of the outside air damper there will be a corresponding value of negative static pressure in the outside air/return air (OA/RA) mixed air plenum that will ensure that the required minimum volume flow rate of outside air is drawn into the system. As a corollary to this, for any given negative value of static pressure in the mixed air plenum there will be a corresponding position of the outside air damper that will ensure this intake rate of outside air.

The mixed air plenum may be a mixing box section of the AHU itself or simply a section of duct immediately downstream of the junction of the return and outside air ducts, upstream of the connection to the AHU. As the duty of the VAV supply fan reduces under part load conditions, the static pressure in the mixed air plenum will naturally tend to rise (becoming less negative), decreasing ΔP_{AC}, and with it the intake rate of outside air. When operating under design conditions the outside air damper is in its minimum open position, and the required minimum flow rate of outside air is being drawn into the system. As system volume flow rate reduces under part-load conditions, the speed of the VAV supply fan reduces, and with this the static pressure in the mixed air plenum rises, becoming less negative. Since ΔP_{AC} reduces correspondingly, the outside air damper must be progressively opened to maintain the required minimum intake rate of outside air into the system. Although a relatively straightforward procedure in theory, the practical issues associated with control of the outside air economy cycle in VAV systems has exercised the minds of HVAC engineers for a long time. Indeed the VAV outside air economy cycle has been described[7] as 'probably the most difficult control problem facing engineers in the HVAC field'.

There is an opinion that where VAV systems are designed to handle only interior loads due to lighting, people and equipment, cooling load varies in parallel with occupancy. Under these conditions it is suggested that a fixed minimum outdoor air damper position, leading to outside air intake rate decreasing with reducing system volume flow rate, is acceptable. Of course this argument is at best applicable only to deep-plan VAV buildings. Even in deep-plan VAV buildings, the extent of IT-related equipment must render this argument tenuous in most modern commercial environments. It is totally inappropriate for the many shallow-plan buildings where the VAV system copes additionally with varying levels of solar and transmission gains.

Another line of design thought retains a fixed minimum opening position of the outdoor air damper, but adds 'tracking' control of the VAV return fan (Fig. 5.6). In this the duty of the return fan is controlled to maintain a fixed volume flow rate differential below the supply fan. The differential is based on the design building ventilation rate, and is used both to ensure that a minimum of outside air is always drawn into the system and to maintain positive building pressurization. The assumption is that if the VAV supply fan is handling more than the return fan, the 'missing' differential volume flow rate must be coming from somewhere to balance the air flows, and that this 'somewhere' is in fact from outside via the outside air damper. This technique requires the measurement of system supply and extract volume flow rates in the main supply and extract ducts. However, because the volume flow differential to be maintained is small in comparison with the supply and return flow rates being measured, levels of measurement accuracy which are attainable in practice on site can combine to produce a disproportionate level of error in the value of the flow rate differential determined from these measurements (and hence in the ventilation performance of the system). For example, it may be possible to measure a system supply volume flow rate of $20\,m^3\,s^{-1}$ to an accuracy of $\pm 5\%$, i.e. $\pm 1\,m^3\,s^{-1}$. However, this is $\pm 33\%$ of the 15% differential between supply and return flow rates ($3\,m^3\,s^{-1}$). If a similar level of measurement accuracy is applicable to the system return volume flow rate, the combined error in the determi-

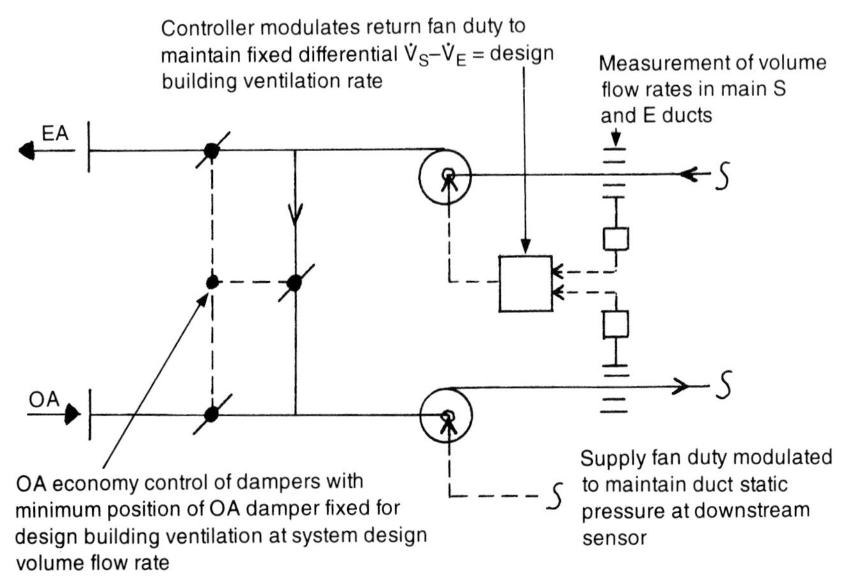

Fig. 5.6 Fixed minimum outside air damper position with tracking control of the return air fan.

nation of the supply–return flow rate differential may be greater or less than this, depending on whether the component errors are additive or self-cancelling.

In the UK it has been conventional to regard tracking control of a VAV return fan purely as a means of controlling the fan itself, in the absence of an extract duct static pressure control loop, and of controlling building pressurisation. Other techniques are then used to maintain a constant minimum level of building ventilation. Conventionally these have relied on overriding normal economiser cycle control of the outside air damper position to maintain a required minimum volume flow rate in the outside air duct (Fig. 5.7). Control of damper position for minimum outside air flow under part-load conditions is based on a comparison of air velocity or velocity pressure measured by a sensor positioned in the outside air duct. The set-point value of this sensor corresponds to the required minimum volume flow rate, and the error between the set-point and measured values is used to derive a modulating control output which adjusts the position of the outside air damper accordingly. Single-point or averaging sensors may be used, but accurate measurement of volume flow rate in the outside air duct typically

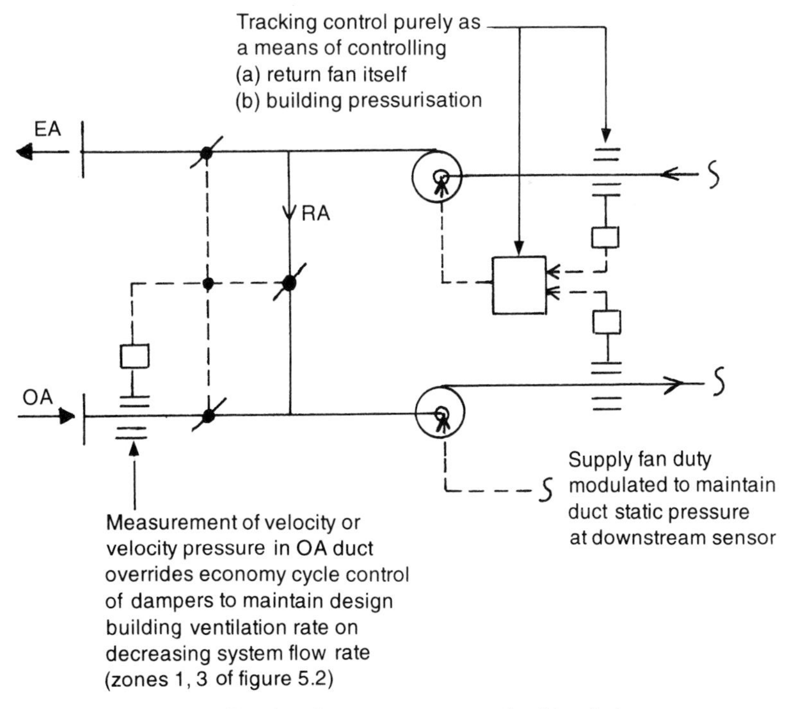

Fig. 5.7 Outside air damper control of minimum building ventilation rate.

requires the availability of sufficient lengths of straight duct both upstream and downstream of the sensor location.

The outside air duct must be sized to handle the full system design volume flow rate when operating on 100% outside air. When handling the minimum outside air volume flow rate, the velocity pressure will be very low, resulting in a lack of sensitivity to changes in volume flow rate at the sensor and leading to poor control. As an example, consider an outside air duct with a duct velocity of $10\,m\,s^{-1}$ under system design conditions. This results in a velocity pressure (based on mean duct velocity) of approximately 60 Pa. Under minimum outside air operation of 20% of system design flow ($2\,m\,s^{-1}$ mean duct velocity), the corresponding velocity pressure is approximately 2.4 Pa. However, the maximum velocity in the duct (typically on the centreline, the usual location for a single pitot-type sensor) will be greater than the mean velocity, by a factor varying from approximately twice for fully developed laminar flow to approximately 1.2–1.25 for fully developed turbulent flow. Hence the centreline velocity pressure could conceivably have a value somewhere between approximately 1.5 times and twice the value based on mean duct velocity. In order to achieve a control accuracy of $\pm 10\%$ of the minimum required flow rate of outside air, a pitot-type device located on the duct centreline would need to be able to accurately sense velocity pressure varying over a range from (at best) approximately 12 Pa at a mean duct velociy of $2.2\,m\,s^{-1}$ (+10%) to approximately 8 Pa at a mean duct velocity of $1.8\,m\,s^{-1}$ (–10%), and (at worst) from approximately 5 to 3 Pa over the same range of mean duct velocity. AHUs are also frequently close-coupled to the outside air intake louvre, with a short outside air duct sized to suit the louvre connection. The louvre itself is sized for a velocity low enough to avoid moisture penetration under adverse weather conditions. Averaging (grid-type) sensors may compensate, to a degree, for non-ideal flow conditions, and devices are available which sense an enhanced velocity pressure, either at single or multiple points (Fig. 5.8).

Following HVCA practice generally, the pressure drop across an orifice plate could be used as a measure of the volume flow rate in the outside air duct. The orifice plate is certainly a more sturdy and reliable device, but will impose a continuous penalty on supply fan energy consumption and system running cost. To obtain an adequate pressure drop signal under conditions of minimum flow would result in a significant pressure loss at the system design volume flow rate. This assumes that maximum pressure recovery is achieved downstream of the orifice plate, which will typically require at least five equivalent diameters of straight duct downstream of the plate.

Velocity sensors working on the 'hot-wire' principle may be used at the velocities that are typically found in outside air ducts during minimum outside air operation, especially if located on the duct centreline. Single-point sensors are most economical, but similarly most susceptible both to the

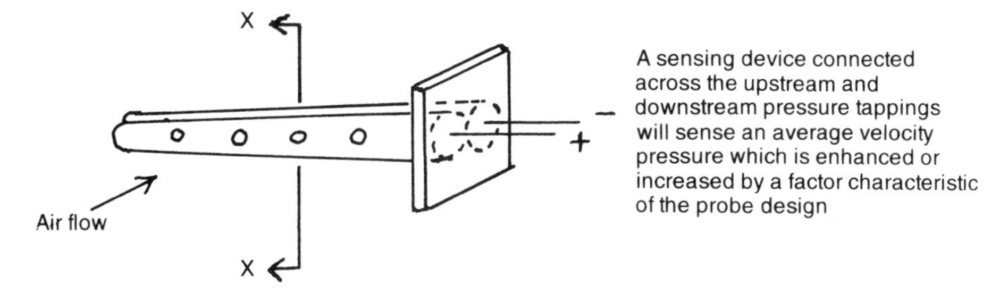

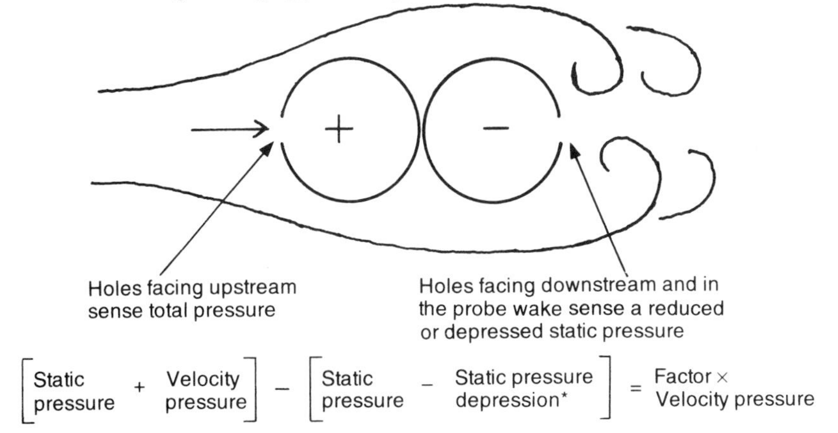

Fig. 5.8 The measurement of enhanced velocity pressures.

effects of unsymmetrical or non-uniform flow and contamination or erosion by the airstream. Multi-point arrays are possible, but more costly. Sensors must be calibrated on site, and accuracy of calibration is fundamental to successful minimum outside air control.

However, if velocity pressure (or velocity) cannot be measured with sufficient accuracy in the outside air duct itself, minimum damper opening may alternatively be linked either to velocity pressure measured in the mixed airstream, or to the supply fan volume control. Both of these latter techniques would seem to require a linear variation of volume flow rate with position of the minimum outside air damper. Indeed, linking of the damper position to the supply fan volume control has been noted as a compromise solution that is only likely to give the correct flow rate of outside air at the maximum and minimum extremes of the fan operating range, at which the volume flow rate passed by the outside air damper is calibrated[8].

In any event, the outside air duct also presents the harshest environment for an airside sensor in an air conditioning system, experiencing both extremes of summer and winter weather conditions. The sensor is also inevitably placed upstream of all filtration, thus being exposed to any dust, dirt or corrosive agents in the outside airstream.

The outside air damper itself is often a large assembly, and is typically fabricated from sheet steel and extruded sections. It is a much less refined device than the modulating control valves used in heating or chilled water systems. Any inherent play in the actuator linkages resulting from manufacturing or installation deficiencies may affect the level of control achieved, as can the cumulative effects of wear and tear. Often neglected in maintenance terms, and again exposed to the harshest environment in terms of the temperature and humidity range, and dust and dirt content of the outside airstream, deterioration in damper control action is not uncommon, and may go unnoticed for long periods in the absence of a proper inspection and maintenance regime. The flow characteristics of control dampers are typically non-linear. Sized to cover the range of system operation, the outside air damper may provide relatively poor control at low volume flow rates. In extreme cases, the damper may simply be sized to fit the outside air duct, although this is rare where the damper is a component part of the central AHU itself. Opposed-blade dampers are typically used in preference to parallel-bladed units, due both to the lower pressure drop required to achieve adequate control authority and the better mixing that they produce.

It is clear from the foregoing discussion that simple outside air damper control is not a satisfactory means of ensuring a minimum building ventilation rate under all operating conditions. The following sections consider alternative techniques that have proved applicable in practice, or may have the potential to do so.

5.10 Minimum outside air dampers

The outside air damper in Fig. 5.7 may be replaced by two dampers in parallel, each actuated independently (Fig. 5.9). The main section exclusively performs mixing of outside and return air for the outside air economy cycle, and as such is allowed to modulate between fully open and fully closed. Control of the 'mixing' outside air damper and the return damper is either to maintain a constant mixed air temperature or a constant off-coil temperature (in sequence with the AHU cooling coil). The second, smaller, damper section is controlled purely to allow a constant volume flow rate of outside air to be drawn into the system. The configuration of the outside air duct must be such that a measurable velocity pressure (or duct velocity) is attained, either upstream or downstream of the minimum outside air

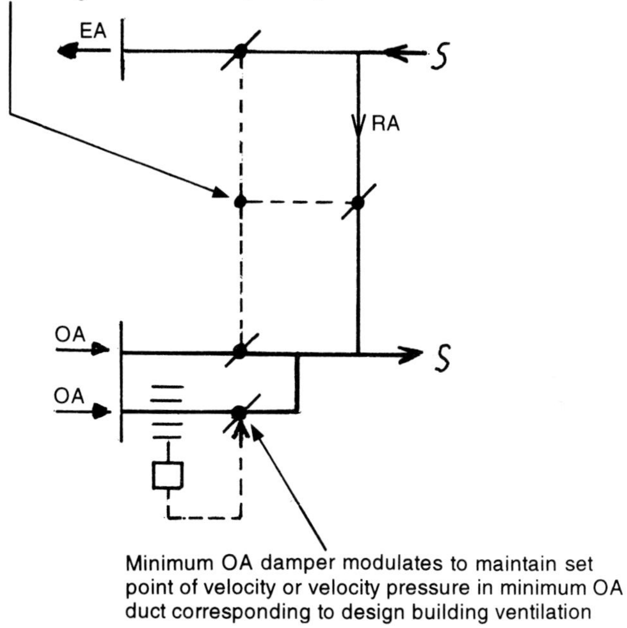

Fig. 5.9 Use of a minimum outside air damper.

damper, under minimum intake conditions. This may be achieved either by partitioning the outside air duct internally to form a separate minimum outside air channel or by means of a separate outside air duct. The minimum outside air damper must be capable of passing the required minimum flow rate when just fully open at the lowest speed (or other duty control setting) of the VAV supply fan.

5.11 Outside air injection fans

A constant minimum rate of outside air injection into a VAV system may be achieved under all operating conditions by means of a dedicated fan which is installed in the minimum outside air duct (Fig. 5.10). The fan is naturally much smaller than the main supply fan, and requires a relatively low fan total pressure, since it has only to offset the pressure losses generated in the minimum outside air duct itself. The fan runs whenever the system requires outside air, and discharges directly into the outside air/return air (OA/RA) mixing section of the central AHU.

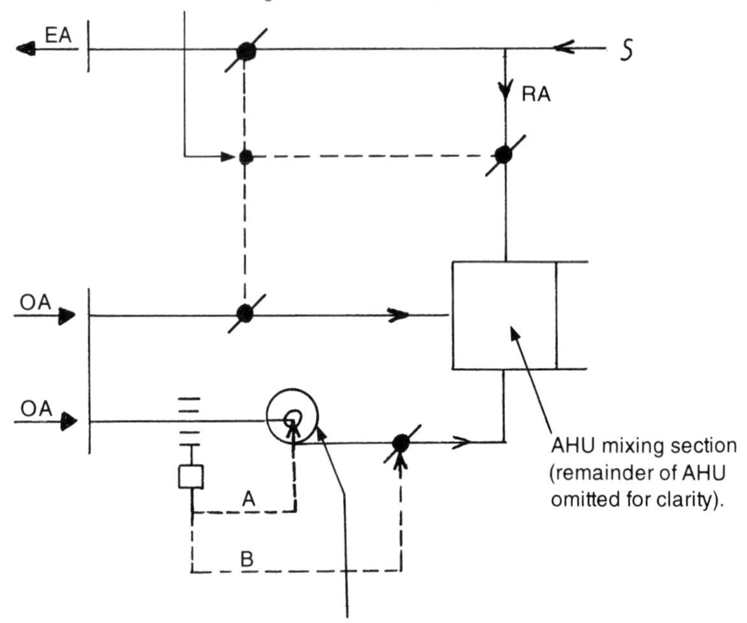

Fig. 5.10 Use of an outside air injection fan.

The outside air injection fan must be capable of providing the required duty under the worst-case condition of low system load (and supply fan speed). At higher system load (increased fan speed) the increased negative static pressure in the mixed air plenum will tend to increase the intake of outside air. Hence unless a fan with a sufficiently steep characteristic can be employed, there will be a need to trim the volume flow rate in the minimum outside air duct. This may be achieved either by combination of a constant-speed fan and a downstream throttling damper or a variable-speed fan. Either damper position or fan speed is modulated to maintain a constant minimum outside air flow rate as sensed by a velocity pressure or air velocity sensor located in the minimum outside air duct.

The outside air injection fan has additional plant and space requirements if an acceptable duct connection is to be achieved. The installation must again be such that the fan duty can be accurately measured during commissioning of the system, and should yield representative measurements at the location of the velocity pressure or air velocity sensor. The additional space requirements of the technique may make it difficult to retrofit into existing installations.

5.12 Plenum pressure control

As proposed by various authors in recent years[9, 10], the concept of *plenum pressure control* is based on the assumption that, with the outside air damper at its fixed minimum position, sufficient outside air will be drawn into the system provided that an adequate negative static pressure is maintained in the OA/RA mixed air plenum (Fig. 5.11).

The mixed air plenum may be a mixing box section of the central AHU itself (most frequently so) or simply a section of duct immediately downstream of the junction of the return and outside air ducts (and upstream of the connection to the AHU). As the duty of the VAV supply fan reduces under part load conditions, the static pressure in the mixed air plenum will naturally tend to rise (becoming less negative). Since the intake end of the duct is open to ambient pressure, the pressure drop across the outside air duct is reduced, and with it the flow rate of outside air drawn into the system.

Plenum pressure control rquires the return air damper to have its own independent actuator. When the economy cycle moves to the minimum outside air/maximum return air setting, the sequence of control may follow one of two variants. In the first, the outside air damper is held in the specified minimum open position on further reduction in system volume flow rate. The combination of outside air inlet louvre and outside air damper is treated as a fixed orifice under these conditions, and the differential pressure is sensed across the combination. Since a reduction in system flow rate results in a rise in static pressure in the mixed air plenum, the intake rate of outside air reduces, and with it the differential pressure sensed across the outside air damper/louvre combination. The return air damper is modulated towards the closed position until the static pressure in the mixed air plenum is reduced to its previous value, and the pressure drop across the outside air duct is again adequate to ensure the correct minimum intake rate of outside air with the outside air damper in its fixed minimum position. With the correct flow rate restored, the differential pressure across the outside air damper/louvre combination also regains its original value. Hence the return air damper may be modulated to maintain this differential pressure as a set-point value.

OPTION A

EA

RA

OA

RA damper overridden to modulate towards closed on drop below set point of differential pressure across OA louvre/damper combination when in fixed minimum open position

Differential pressure sensor across combination of OA inlet louvre and OA damper in fixed minimum open position

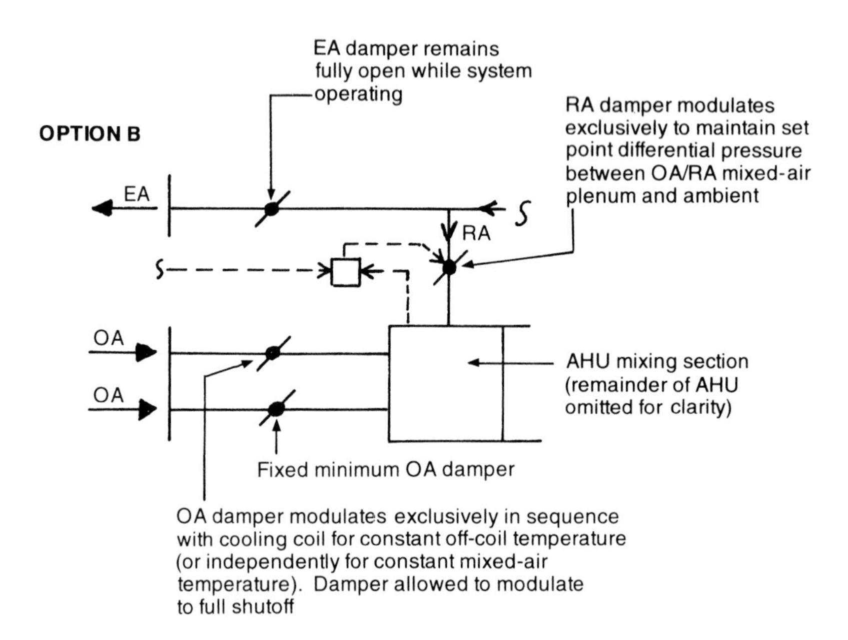

Fig. 5.11 Plenum pressure control.

The technique requires that the minimum position of the outside air damper is accurately repeatable. The differential pressure across the outside air louvre/damper combination must be established at system design volume flow rate and minimum outside air, and the configuration of the outside air duct should again allow accurate measurement of this flow rate. The pressure drop across the damper in its minimum open position should approximate that across the fully open damper when passing the system design volume flow rate.

In an alternative sequence of control the return air damper is modulated accordingly to maintain the pressure differential between static pressure sensors in the OA/RA mixed air plenum itself and in the outside air duct. The location of a static pressure sensor in the mixed air plenum may, however, be problematic due to the degree of air flow turbulence that may be expected there. The technique of plenum pressure control undoubtedly benefits from the capacity of DDC controllers to provide relatively sophisticated control logic.

5.13 Modified economy cycle control

Control of the conventional outside air economy cycle is based on temperature. Figure 5.12 shows in principle a modified economy cycle which is controlled through the outside air intake system[11]. All control is effected by resetting the set-point value of air velocity or velocity pressure sensed in the outside air duct, and the technique probably requires the use of a DDC controller to provide its relatively sophisticated control logic. A requirement for minimum outside air provides the basic value to which the control set-point is reset when outside conditions favour maximising recirculation. To achieve this as system volume flow rate reduces under part load, the outside air damper is modulated towards open. When fully open, the return air damper is modulated towards closed to maintain the negative pressure in the OA/RA mixed air plenum which will maintain the required set-point value of outside air flow rate at the duct sensor.

Reset of the air velocity or velocity pressure set-point to its basic (minimum outside air) value is under the control of either a high-limit temperature sensor in the outside air duct or a comparison between temperature sensors in both outside and exhaust air ducts. When minimum outside air is not signalled, the set-point of air velocity or velocity pressure is reset upwards to maintain either a constant mixed-air temperature or a constant off-coil temperature (controlling in sequence with the AHU heater battery and cooling coil). When operating with minimum outside air, accurate measurement of air velocity or velocity pressure in the outside air duct may remain problematic.

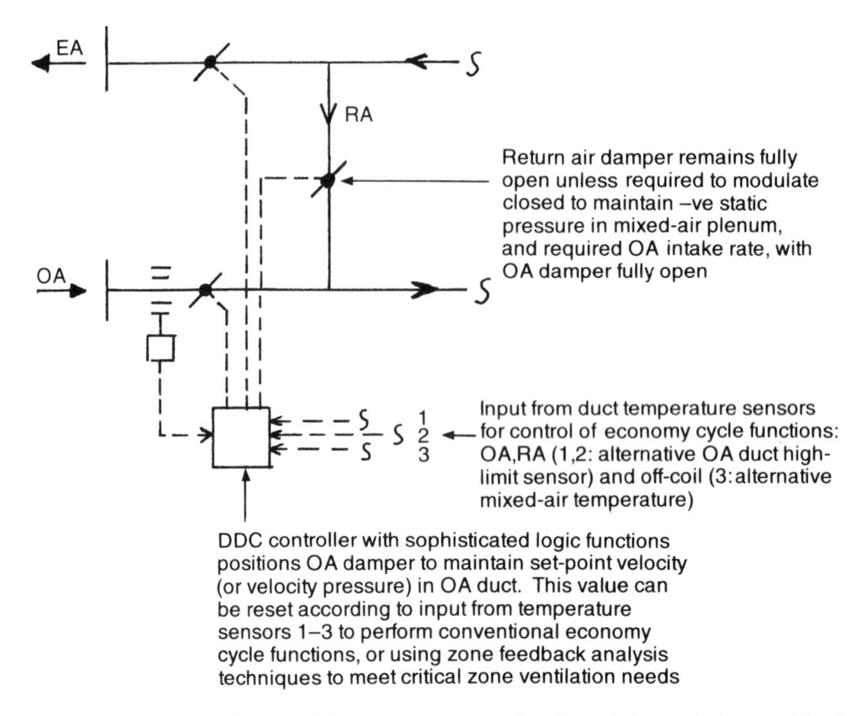

Fig. 5.12 Control of an outside air economy cycle effected through the outside air intake system. Central plant components omitted for clarity.

5.14 Alternatives to fixed minimum building ventilation rate

5.14.1 The use of a DDC-based approach

The commercial availability of DDC controls at a cost comparable to existing analogue control systems offers the potential to utilize both their data processing capacity and their ability to handle sophisticated control logic to improve the ventilation performance of VAV systems. The implementation of sophisticated control strategies should, however, always be viewed against the potential increases in cost and complexity which may be incurred.

5.14.2 Outside air reset according to the needs of the critical zone

First, the DDC approach offers the potential to reset the minimum intake rate of outside air into the system according to the ventilation requirements of the critical zone. Achieving the latter requires feedback of zone air supply rates to identify the zone that is experiencing the lowest load factor, and hence which requires the highest system outside air fraction to achieve its design ventilation rate. Since system supply volume flow rate is itself measured in a VAV system, the required outside air intake rate to satisfy the critical zone may be calculated, together with the velocity pressure or air velocity set-point for the outside air (or minimum outside air) duct that corresponds to this. This approach still assumes a constant ventilation rate for each zone, and will result in all zones other than the critical zone being effectively over-ventilated. It will require either additional cooling or heating at the central AHU, and thus increased energy consumption by the refrigeration or boiler plants. It will be necessary to analyse zone feedback at regular intervals when operating on minimum outside air at high ambient temperature, and if approaching minimum outside air at low ambient temperature. A variable outside air intake rate must always remain within the capacity of the heating and cooling coils in the central AHU to cope with.

5.14.3 Supply air reset at zone level

As an alternative to increasing the outside air fraction for all zones, in some instances it may be possible instead to increase the supply volume flow rate only to those zones which are experiencing a shortfall in design ventilation rate, reheating to maintain control of zone temperature. Feedback and analysis of zone supply flow rates allows identification of zones experiencing a shortfall from their design ventilation rates. Where the DDC approach extends to the terminal unit controllers themselves, the minimum permitted flow rate settings of the VAV terminal units supplying these zones may be reset upwards to provide their ventilation needs, reheating being automatically initiated by the local zone control loop. System supply volume flow rate is increased, and unless the outside air intake rate is increased to compensate, system outside air fraction will actually be reduced to all zones. However, the effect may be slight if only one or two zones are affected. The energy consumption by refrigeration and boiler plants is increased, and an increase in energy consumed by the system fans must also be introduced into any comparison of control strategies. Perhaps more limiting, however, is the consideration that, unless general zone reheat is an inherent feature of the design, its provision for critical zones must be made at the design stage, either on the basis of predicted level of occupancy and pattern of use or simulation of system operation. If the requirement for reheating is very limited, the use of electric zone reheat may even be justified, removing the need for an extended (and pumped) LPHW distribution to serve perhaps just a few terminal units.

5.14.4 Occupancy detection and counting

Figure 5.13 shows a flow chart for ventilation strategies based on zone feedback. Occupancy detection and occupancy counting respectively offer

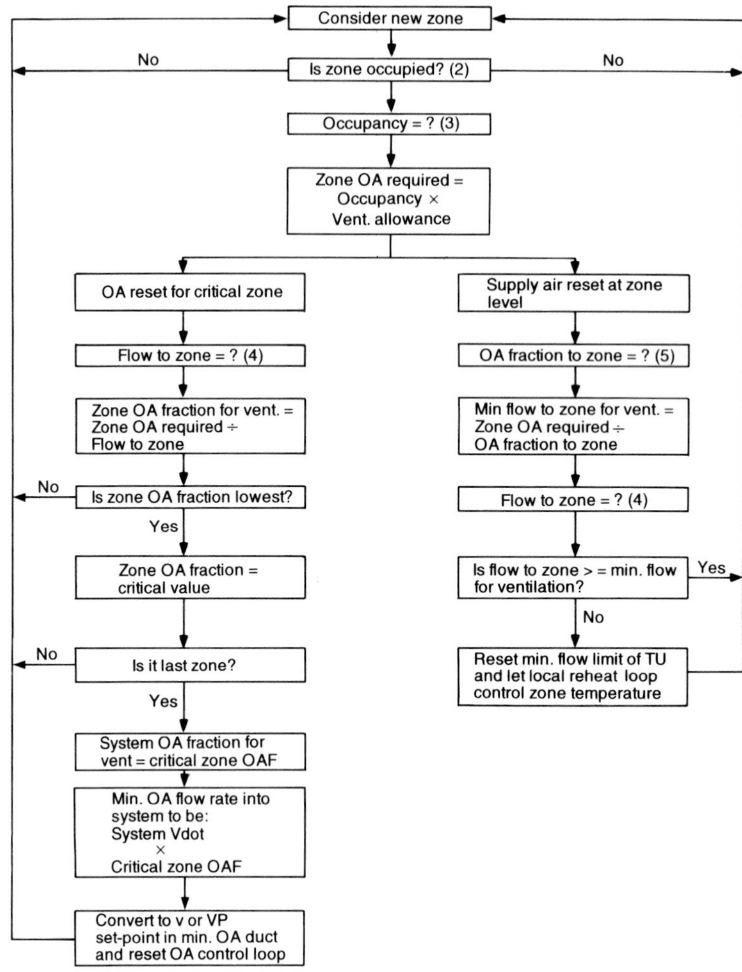

Notes:

(1) Intended to be illustrative of the approach only. It will typically be advantageous to limit the extent of feedback analysis required for any cycle of control resetting, and a rule-based approach may be combined with zone feedback.
(2) 'Yes' if no occupancy sensing and within programmed occupancy period of the space.
(3) Constant at design value if no occupancy counting.
(4) Feedback from terminal unit controller.
(5) Intake rate of OA into system divided by system volume flow rate.

Fig. 5.13 Ventilation control strategies using zone feedback.

the theoretical potential to either eliminate the influence of unoccupied or under-occupied zones when resetting system outside air fraction or to reset terminal unit minimum air flow rate according to occupancy level. However, devices need to have low cost and simple installation if each zone is to be equipped.

5.14.5 CO_2 level control

It is important to realize that the instantaneous outside air flow rate to a zone may not significantly affect its indoor air quality (IAQ). Since in mixing air distribution outside air dilutes the concentration of airborne contaminants generated within the zone, it is the *average* delivery rate of outside air over time that determines the results. Substantial changes in outdside air supply may therefore be tolerated, so long as they are only transient. This is important, since any form of control to ensure minimum zone (or building) ventilation requirements will not act instantaneously, especially if it is based on the analysis of feedback from multiple zones.

As far as achieving acceptable IAQ is concerned, an alternative to specifying fixed rates of outside air supply is to measure the dilution effect of the outside air (which is, after all, what it is supplied for), adjusting the supply rate in response to its effect. With this approach errors in measuring absolute volume flow rates hold less significance. The volumetric carbon dioxide content of air can readily be measured to fairly low concentrations using commercially available sensors, although these remain expensive. Futhermore, significant amounts of CO_2 are injected into the return air stream by occupancy. However, an adequate CO_2 level is not necessarily a direct reflection of good IAQ. Contaminants given off by photocopiers, printers and outgassing from the building fabric and finishes do not contribute to CO_2 level, and are to a greater or lesser extent independent of occupancy. These sources of contaminants may dominate at low occupancy levels. Therefore, even where CO_2 sensing is employed, the outside air supply rate should always consist of two components – one a constant minimum supply rate, and the other CO_2-related. The use of CO_2 concentration level as a controlled variable suffers a further drawback in that there is an inherent lag in the dilution process. Increasing the supply rate of outside air will not instantaneously check a rise in CO_2 level, and there will therefore be a tendency for any such control loop to cycle between providing full and minimum outside air. To some extent this may be overcome by providing both maximum and minimum outside air supply settings, or by resetting the minimum level according to CO_2 level. System outside air fraction may be determined from measured CO_2 levels according to the following expression, which is derived from a CO_2 balance of the outside, supply and return airstreams:

$$OAF(\text{Sys}) = (CO_{2\text{RA}} - CO_{2\text{SA}}) \div (CO_{2\text{RA}} - CO_{2\text{OA}}) \qquad (5.12)$$

where suffixes OA, SA and RA signify values of CO_2 level (ppm) for outside air, supply air and return air respectively.

This expression has the value of zero for 100% return air and unity for 100% outside air. Typical CO_2 levels found in practice are 350–450 ppm for outside air and 500–1000 ppm for return air where occupancy is predominantly sedentary. The level in the supply air will be intermediate between the outside and return air levels, according to their mixing proportions. CO_2 sensors are typically more suited to mounting within a zone itself, although measurement of concentration levels in ducts may be made using a sampling pump.

The characteristics of CO_2 level control may thus make it more suited to a high-limit function. In this the velocity pressure or air velocity set-point for control in the outside air intake duct may be reset upwards, from a constant minimum level based on design occupancy, according to CO_2 level measured either in the return air duct or in a critical zone (or zones).

A further proposal uses measurements of CO_2 level in the outside, supply and return air ducts to determine system outside air fraction (Equation 5.12), and hence outside air intake rate (from measured system volume flow rate)[12]. This is then compared with a minimum outside air intake rate set-point chosen as the higher of alternative values based either on building pressurisation requirements, programmed occupancy or CO_2 level in the return air duct or a critical zone (Fig. 5.14). The error in outside intake rate is used to generate a control signal that modulates whichever method of minimum air intake rate control is employed in a corrective manner. Multiplexing is proposed as a practical means of obtaining the required measurement accuracy of ± 10 ppm with commercially available sensors having much lower accuracy, and of eliminating drift and temperature errors. The technique employs a single CO_2 sensor, in combination with a sampling pump, to measure concentration levels in the outside, supply and return ducts in sequence. Each cycle of measurement takes several minutes, since the equipment must be thoroughly purged between each measurement, and this delay must be taken into account in the design of any control system employing the technique. Sampling of outside air from outside the building is preferred, with sampling of supply air from well downstream of the fan, where the airstream is well mixed.

5.14.6 Air quality sensing vs. CO_2 level control

As an alternative to CO_2 level control, multiple-contaminant IAQ sensors for room or duct mounting are now commercially available. Based on a preheated semiconductor that reacts to a number of gases, they are much cheaper than CO_2 sensors. Their lower cost enables them to be more widely used, perhaps ultimately on a one-per-zone basis to allow realistic feedback of measured ventilation performance at zone level. However, such multicontaminant IAQ sensors presently have no defined calibration, being set up using empirical values and repeated tests on site under varying occupancy conditions.

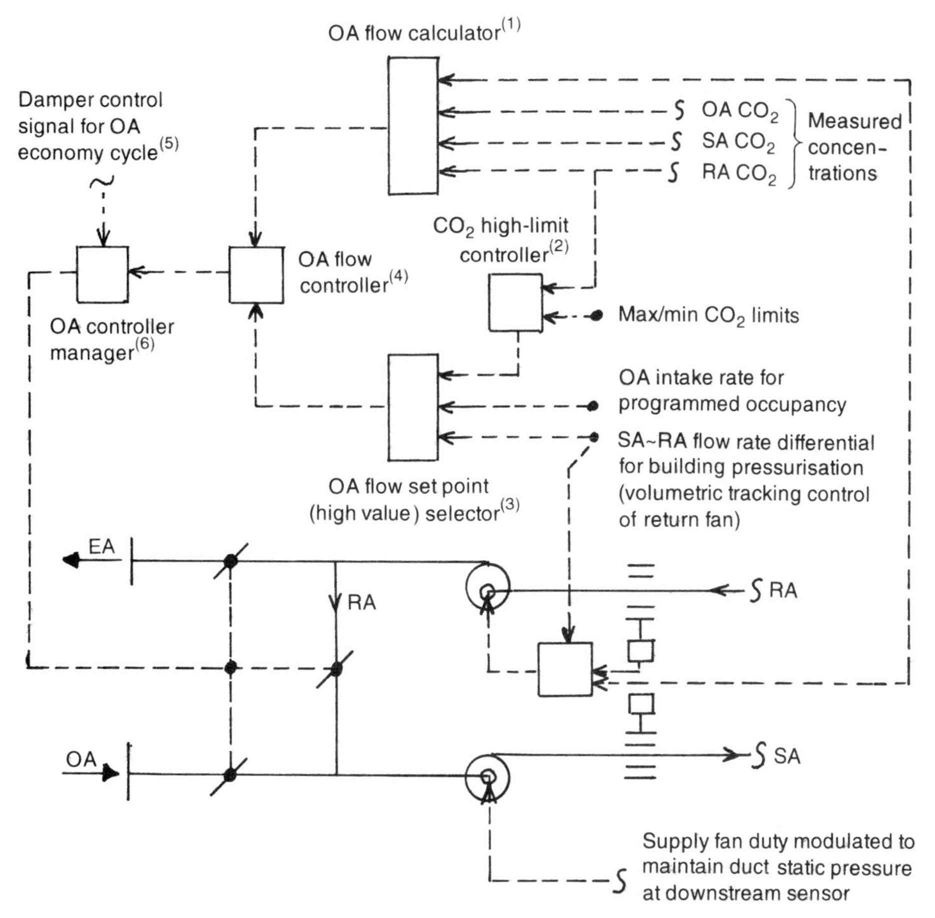

(1) Uses CO_2 concentration balance for measured values to calculate actual system OA fraction (equation 5.12), then combines with measured SA flow rate to generate OA flow rate as a controlled variable.
(2) Generates proportional OA flow rate for CO_2 high-limit control, based on measured value of RA CO_2 concentration (measurement preferably in a critical zone or the RA duct from it).
(3) Selects highest of 3 inputs as OA flow rate control set point.
(4) Outputs a control signal as a function of the error between (calculated) actual OA flow rate and set point value for ventilation control.
(5) Derived from control logic and components for sequence control of central plant heating/cooling (not shown for clarity) and OA/RA/EA dampers, with minimum/full/variable OA operating changeovers according to figure 5.2.
(6) Selects either the ventilation (OA flow) or economy (temperature) based control inputs according to whichever will provide the greatest OA flow rate, and outputs the signal to position the OA/RA/EA dampers.

Fig. 5.14 CO_2-based high-limit ventilation control strategy.

5.14.7 Fixed scheduling of outside air intake rate

In the absence of zone feedback, Fig. 5.4 and Table 5.1 suggest that thorough part-load analysis of VAV system performance may allow the development of a simple schedule of outside air intake rate against system supply volume flow rate for a particular system. In addition to the actual ventilation requirement, factors which may be expected to influence this include the expected range and pattern of zone load variation, the performance of the VAV terminal unit itself and the room air distribution performance.

5.15 Bypass filtration

Where VAV operation is achieved by bypassing unwanted supply air around a constant-volume fan, a high-efficiency filter/air cleaner (typically based on activated charcoal) may be installed in the bypass (Fig. 5.15). The effect of reduced outside air supply rates is offset by supplying cleaner air with a lower concentration of VOCs. An IAQ sensor in either the return duct or a critical zone independently resets the supply air temperature – up during cooling operation, down during heating. In response the zone dampers throttle the

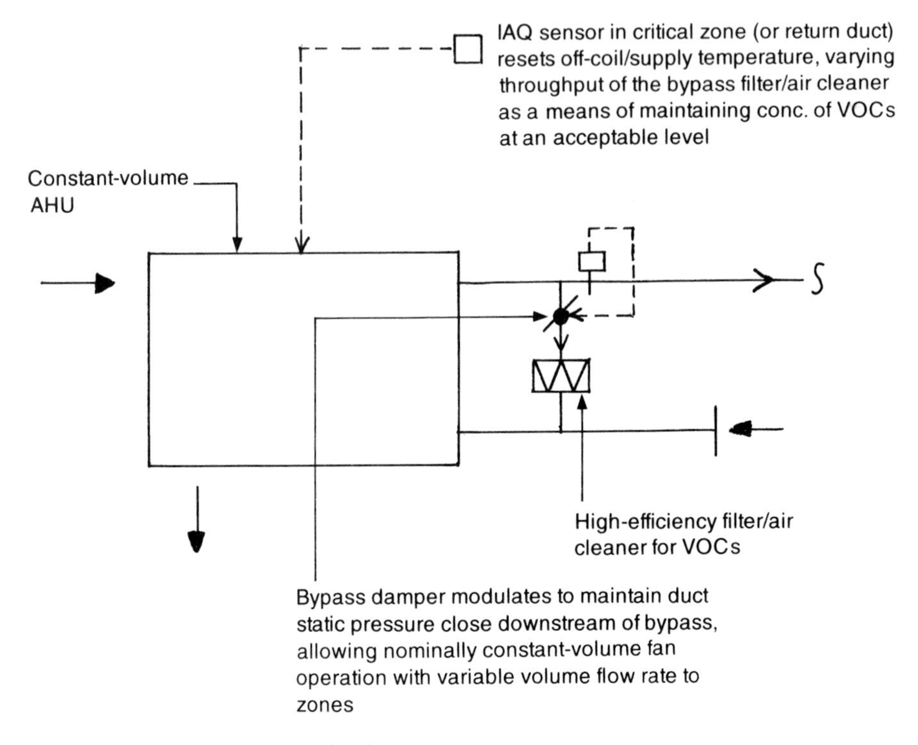

Fig. 5.15 VAV with bypass filtration.

air supply, driving the duct static pressure up and driving more air through the high-efficiency filter/air cleaner in the bypass duct[13].

5.16 The influence of building layout on ventilation strategy

We have seen in earlier sections that, without any feedback on actual zone occupancy or IAQ levels, the absolute minimum intake volume flow rate of outside air into a VAV system should be at least equal to the sum of the zone design ventilation requirements. We have also seen that this action will not *in itself* be sufficient to ensure that all zones actually receive their individual design ventilation requirements, even under system design conditions. Since whatever measures are adopted to improve on this performance will inevitably carry with them either a capital or running cost penalty (or both), the question must inevitably be asked, does it really matter?

The answer depends not only, as we have seen, on the amount of time likely to be spent operating with minimum outside air, and on how generous is the design ventilation allowance, but on the type and layout of the accommodation served by the system.

As far as the accommodation served is concerned, consider the simple example shown in Fig. 5.16. This represents an office block of several identical storeys with principal facades facing east and west. It is air condi-

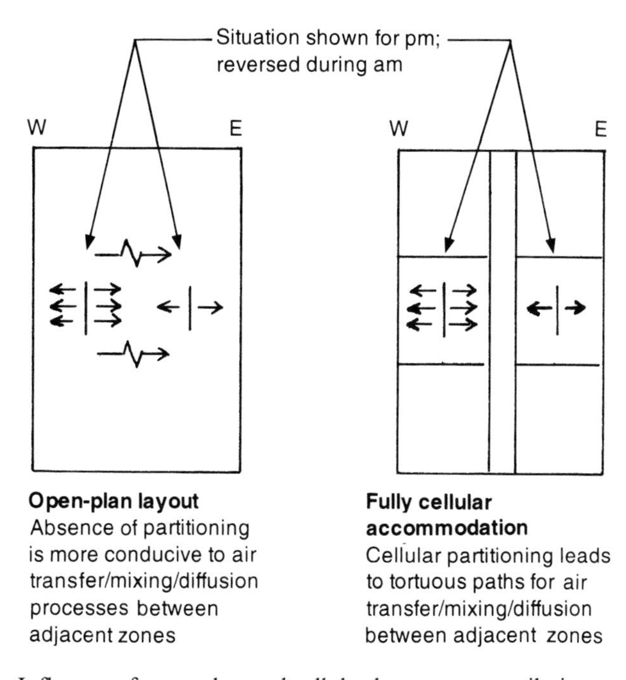

Fig. 15.16 Influence of open-plan and cellular layouts on ventilation.

tioned by means of a VAV system, which draws in a minimum volume flow rate of outside air equal to the sum of the individual zone ventilation requirements. In case (a) the main floor areas are occupied by open-plan accommodation. At each floor level whatever the volume flow rates of outside air that are delivered through individual diffusers, ultimately the air all ends up in the same single open-plan area. For this to be fully useful in ventilation terms, however, does require that all supply air, from whichever diffuser delivers it to the space, ends up more-or-less uniformly mixed throughout the whole open-plan volume, or at least the occupied part of it. Assuming that this will be fully achieved in practice may prove rather optimistic. Any deficiencies in this respect will inevitably be less critical if the design ventilation allowance per occupant or unit of treated area is generous, rather than simply satisfying the minimum statutory requirement for the application. However, while the open-plan approach is generally more favourable to VAV system ventilation performance at any particular floor level, there is still the floor-to-floor aspect to consider. Unless the system serves a single storey building, or a single floor of a large multi-storey building, any variation in the load factor of each floor (floor sensible cooling load expressed as a decimal proportion of its design value) will result in a floor taking either a greater or lesser proportion of the available system intake of outside air. The extent of this will need to be considered. Unless the uppermost floor of a multi-storey building is used exclusively for plant and utility accommodation, it may generally be expected (because of the presence of the roof) to experience greater heat gains during summer and greater heat losses during winter – all of which will affect the supply volume flow rate that it receives, and with this the amount of outside air.

In case (b) the accommodation at each floor level is fully cellular. Surplus outside air delivered to any one zone will be extracted from that zone, with no capacity to directly supplement zones deficient in the outside air received. Differences in zone pressurisation associated with varying supply volume flow rates and VAV air extract strategies, coupled with movement of occupants between zones, may indeed allow some limited sharing between zones via an indirect (and potentially tortuous) path. Naturally the floor layouts of many real buildings fall between the two extremes of cases (a) and (b), but the implications for VAV ventilation performance need to be considered in all cases.

5.17 Building pressurisation

It is not typical in UK design practice to control the volume flow rates of air extracted from individual zones. The cost alone of an equal number of extract VAV terminal units usually precludes this. From a review of the subject literature design practice in the USA would appear to follow a similar philosophy to that in the UK. However, the literature is less clear about design practice in continental Europe. Without the facility for individual control of extract rate at zone level, only the total system extract air volume flow rate can be controlled to match the total system supply air volume flow rate, less an allowance for the requirements of building pressurisation. This allowance is typically further increased to allow for air exhausted from the building via any independent local mechanical extract ventilation systems drawing their make-up air, either directly or indirectly, from the VAV supply.

In responding to a change in system supply air volume flow rate, the extract system will therefore try to change the volume flow rate extracted from every zone by an equal proportion. A balance must always ultimately be established between the volume flow rates of air supplied to a zone, extracted from it, and infiltrating to it (or exfiltrating from it). This balance will be established at a particular level of pressurisation of the zone relative to adjacent areas – manifest as static pressure differences between them. Looked at rather crudely, 'what air goes in must come out somewhere' (Fig. 5.17).

Buildings do not act like balloons because air can leak out of them (or into them) via gaps and cracks around doors and windows, in walls, ceilings and floors, and even to some degree through seemingly quite solid building materials themselves. Any imbalance between VAV supply and extract rates for a zone can only exhaust from it (or enter it) via these leakage paths. Since air flowing through any one of these leakage paths experiences frictional and dynamic pressure losses in the same way that the airstreams flowing through the VAV supply and extract ducts do, air will only move along a path if a static pressure difference exists between its ends. An imbalance of VAV supply over extract for a zone will result in an increase in zone static pressure until the static pressure difference between it and adjacent areas is of sufficient magnitude that enough air leaks out of the zone to restore the overall balance between the supply to it and the extract/exhaust from it – in this case the latter representing the sum of the VAV extract (via grilles, diffusers or air-handling luminaires) and the leakage to adjacent zones or 'free' air (outside the building). This is case (b) in Fig. 5.17. The opposite will be true for an imbalance of extract over supply, zone static pressure falling until air drawn in from adjacent zones or outside the building restores the overall balance. This is case (c) in Fig. 5.17.

The 'leakier' a zone is, the lower will be the change in its static pressure for a given supply–extract imbalance, and vice versa. Conversely, the greater must be the imbalance of supply over extract to maintain a given level of zone pressurisation in a 'leaky' zone. Zone 'leakiness' is a function not only of the number of leakage paths from the zone, but also their nature. A large

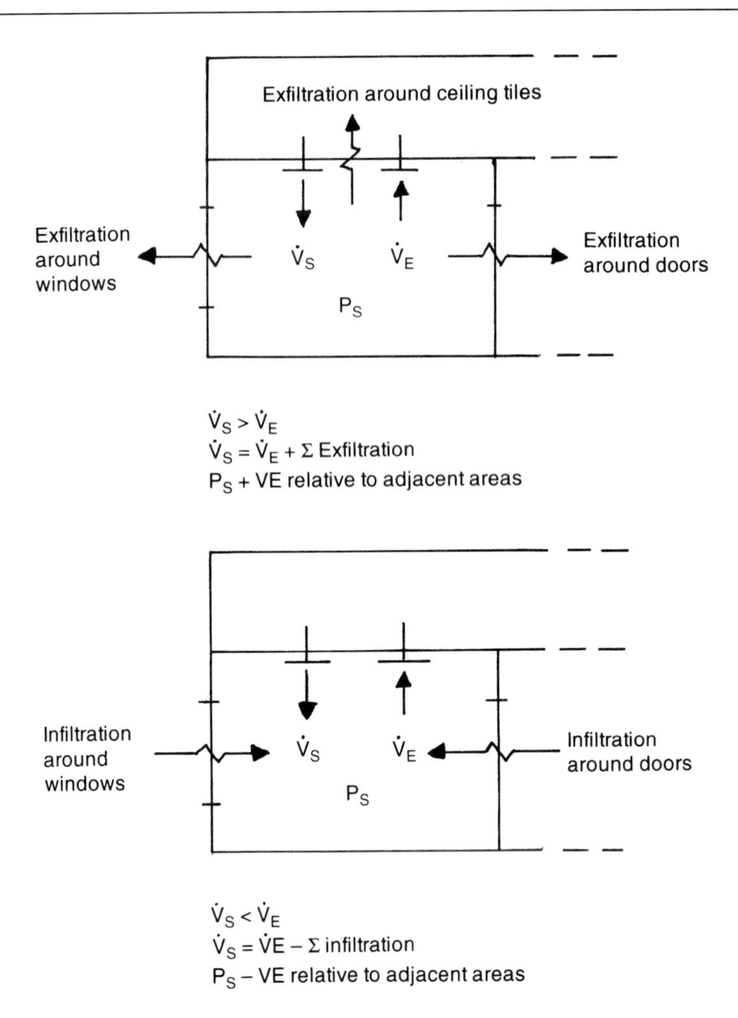

Fig. 5.17 The effects of building pressurisation.

opening of short length will allow a high rate of airflow to or from adjacent areas at a modest static pressure difference between them, whereas a small and tortuous leakage path will show opposite characteristics. As zone supply rates vary with changing zone sensible cooling load, and zone extract rates vary with changing system load, the imbalance between VAV supply and extract may be expected to vary, resulting in zone static pressures fluctuating to maintain the overall balance between their supply and extract/exhaust.

Close control of space pressurisation is not practical without individual control of zone extract volume flow rates. While this may be necessary in some industrial air conditioning applications, it is typically not necessary in commercial applications where the provision of occupant thermal comfort is the primary design aim. In these applications it is indeed preferable to maintain a slightly positive, or at least neutral, space pressurisation to minimise the cooling or heating load which results when outside air infiltrates a space. While zone pressures are therefore allowed to fluctuate, the changes will typically go unnoticed (and be tolerated) by occupants as long as these do not become of a sufficient magnitude to affect the convenient opening and closing of doors, or to result either in draughts from cracked doors or the intrusion of odours from adjacent spaces. The possibility of lighweight ceiling tiles lifting under VAV pressurisation effects is another consideration, since if tiles must be clipped in place access for commissioning and routine maintenance is made more difficult, and the risk of tile damage is increased. In any event the implications for fire protection and limiting the spread of smoke must be at the forefront of a designer's mind if significant pressurisation effects are possible. This probably puts an upper limit in the range 25–50 Pa on the acceptable maximum level of pressure differences between adjacent areas resulting from operation of the VAV system. It must be remembered that seemingly low pressure differentials can generate significant forces when acting across sizeable surface areas. As far as operation of the VAV system itself is concerned, a change in zone pressure may be expected to have a similar effect to an equal and opposite change in static pressure at inlet to the VAV terminal unit. However, such changes will ultimately be compensated by the throttling action of the terminal unit itself, generated by the zone control loop.

Pressurisation effects in VAV systems are naturally most significant, and potentially complex, when the control zones represent individual cellular spaces. Many modern office buildings are of open-plan design, and use the void above a suspended ceiling as a common return air plenum for an open-plan area. There are thus no barriers to prevent equalisation of space pressure between zones served by different terminal units. The rate of air extraction from this is representative of the overall diversified supply volume flow rate to the floor area served, and it may be expected that pressurisation effects will be minimised in such a layout. In any event the effects generated by the VAV system will be combined with the natural influences of wind pressure and stack effect, which are themselves changing, and may be expected to generate complex scenarios where a high proportion of cellular accommodation is combined with high-rise construction.

5.18 Method of air extract

The method employed to extract air from the zones and return it to the central AHU is significant to the operation and performance of VAV systems (Fig. 5.18). When using the ceiling void as a return air plenum, a large proportion of the sensible heat gain from lighting is removed directly to the central AHU with the return air. With ducted extract from individual zones

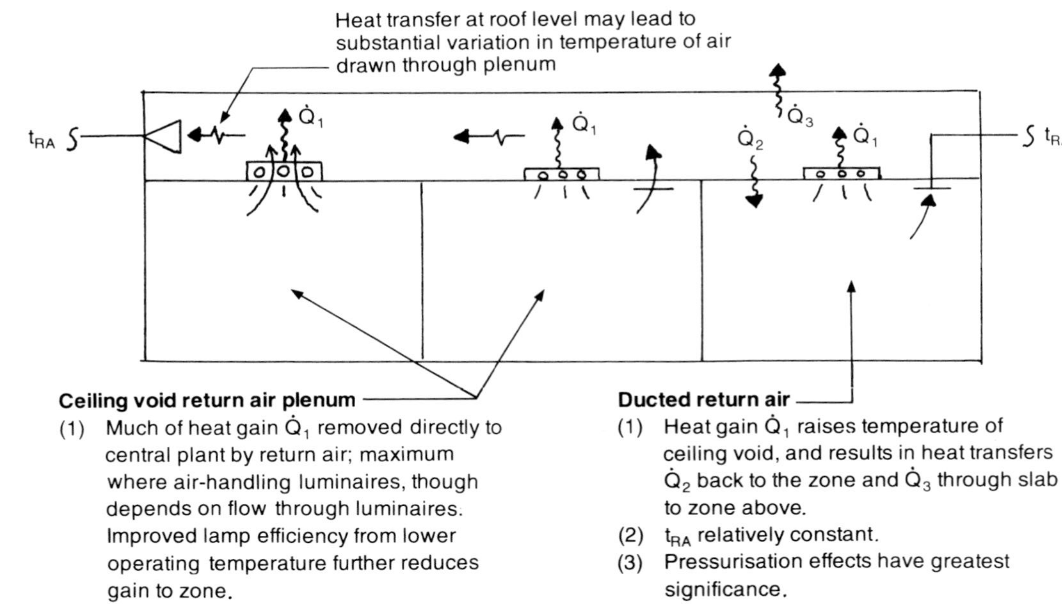

Ceiling void return air plenum

(1) Much of heat gain $\dot{Q}_1$ removed directly to central plant by return air; maximum where air-handling luminaires, though depends on flow through luminaires. Improved lamp efficiency from lower operating temperature further reduces gain to zone.

(2) Ceiling plenum temperature and t_{RA} vary with flow rate of RA; temperature of air extracted through air-handling luminaires not representative of mean zone temperature, and ∴ not suitable for control purposes.

(3) Pressurisation effects least significant.

(4) Potential for short-circuiting of air in vicinity of stub extract duct.

Ducted return air

(1) Heat gain $\dot{Q}_1$ raises temperature of ceiling void, and results in heat transfers $\dot{Q}_2$ back to the zone and $\dot{Q}_3$ through slab to zone above.

(2) t_{RA} relatively constant.

(3) Pressurisation effects have greatest significance.

Fig. 5.18 Ceiling void return air plenum vs. full ducted extract.

any sensible heat gain from lighting passing into the ceiling void increases the air temperature in the void, and results in heat transfer back to the zones both below (through the false ceiling) and above (through the floor slab), increasing zone supply volume flow rates. While the return-air temperature is relatively constant with a full ducted extract system, the temperature of the return-air plenum (and the air drawn through it) depends upon the volume flow rate of the return air itself. On the uppermost floor of a building the energy balance on a return-air plenum influences the sensible heat gain or loss to the space through the roof.

Extracting air into the ceiling void return-air plenum through the light fittings themselves may be expected both to increase the efficiency of the lamps (at the lower operating temperature that results) and to reduce further the heat gain to the space from lighting. The heat extraction from air-handling luminaires is dependent on the volume flow rate through the luminaire, and may be expected to reduce steeply with increasing turndown in zone supply volume flow rate. If all extract is through the luminaires, the effect of an inevitable leakage through ceiling panels, doors, etc. becomes more significant. With a variable volume flow rate, the temperature of air extracted through air-handling luminaires is not representative of mean zone temperature, and must therefore not be used for terminal unit control.

Increasing the return air temperature by using the ceiling void as a return air plenum affects operation of the outside-air economiser cycle. The range of operation with 100% outside air is extended a little. The mixed air temperature is increased slightly with minimum outside air, and the mixing proportion of outside air (ignoring considerations of ventilation) is increased a little without penalty during 'free cooling'. Where heat can be gained by, or lost from, a return air plenum through a roof, the temperature of air drawn through the plenum may vary substantially.

It has been proposed[14] as a potential advantage of VAV systems that thermal insulation may be omitted from much of the supply ductwork in return air ceilings (to yield a capital cost saving) with less concern about the consequences of an increase in the supply air temperature to the zones, as this will automatically be compensated by a corresponding increase in the supply air volume flow rate, assuming that this has been allowed for in the design. It is not known whether this practice is common in the USA. However, it is not accepted design practice in the UK, and would appear not to take into account the risk of condensation forming on the surfaces of the supply ducts in the ceiling void. This generates both a risk to building hygiene and a potential for damage to the building fabric. Since the greatest temperature change due to heat transfer to/from the supply ductwork occurs for small-diameter ducts, some application of thermal insulation will in any event be essential both towards the end of long runs of main duct and in branch ducts

– indeed ASHRAE suggests all ducts with a design air handling capacity of 700 l s^{-1} or less.

The capital and running cost advantages of restricting the extent of VAV extract ductwork by employing a ceiling return air plenum are readily apparent. However, a word of caution has been sounded[8] that this practice may lead to difficulty in achieving the correct balance between the supply and extract systems due to the potential for short-circuiting of air close to the stub extract duct which draws air from the plenum (Fig. 5.18).

References

1. Yu, H.H.S. and Raber, R.R. (1992) Air-cleaning strategies for equivalent indoor air quality. *ASHRAE Transactions*, **98**, Part 1, 173–181.
2. Crandell, M.S. (1987) NIOSH indoor air quality investigations: 1971 thro' 1987. In *Proceedings, Air Pollution Control Association Annual Meeting*, Dallas.
3. Kirkbride, J., Lee, H.K. and Moore, C. (1990) Health and welfare Canada's experience in indoor air quality investigation. In *Indoor Air '90*, Proceedings of the Fifth International Conference on Indoor Air Quality, Toronto (D.S. Walkinshaw, ed.) **V5**, 99–106.
4. Tamblyn, R.T. (1983) Beating the blahs for VAV. *ASHRAE Journal*, **25**, September, 42–45.
5. Int-Hout, D. and Berger, P. (1984) What's really wrong with VAV systems. *ASHRAE Journal*, **26**, December, 36–38.
6. Kelley, R. (1995) Room air circulation: the missing link to good indoor air quality. *Heating, Piping and Air Conditioning*, **67**, September, 67–69.
7. Avery, G. (1986) VAV: designing and controlling an outside air economiser cycle. *ASHRAE Journal*, **28**, December, 26–30.
8. Chartered Institution of Building Services Engineers (CIBSE) (1985) *CIBSE Applications Manual: Automatic Controls and their Implications for Systems Design*, CIBSE, London.
9. Kettler, J.P. (1995) Minimum ventilation control for VAV systems: fan tracking vs workable solutions. *ASHRAE Transactions*, **101**, 625–630.
10. Graves, L.R. (1995) VAV mixed air plenum pressure control. *Heating, Piping and Air Conditioning*, **67**, August, 53–55.
11. Haines, R.W. (1994) Ventilation air, the economy cycle and VAV. *Heating, Piping and Air Conditioning*, **66**, October, 71–73.
12. Janu, G.J., Wenger, J.D. and Nesler, C.G. (1995) Strategies for outdoor airflow control from a systems perspective. *ASHRAE Transactions*, **101**, Part 2, 631–643.
13. Meckler, M. (1994) VAV/bypass filtration system controls VOC's, particulates. *Heating, Piping and Air Conditioning*, **66**, March, 57–61.
14. American Society of Heating, Refrigerating and Air Conditioning Engineers (ASHRAE) (1987) All-air systems. Chapter 2 in *ASHRAE 1987 Handbook: HVAC Systems and Applications*, ASHRAE, Atlanta.

Chapter 6
The Loads on VAV Systems

6.1 The sensible cooling loads on VAV systems

The sensible cooling load in typical commercial applications may be considered to be composed of two component elements. The first is due to the *climate* of the location, and includes solar heat gains through glazing and the opaque fabric, air-to-air transmission heat gains, and the effects of air infiltration through gaps and cracks in the building envelope. As noted in Chapter 1, a need for air conditioning is frequently associated with a need to seal the building envelope (non-opening windows) against the ingress of noise and dirt from city environments.

In temperate climates the 'climatic' component of the sensible cooling load is typically dominated by solar gain through glazing. The general trend in architectural design since the 1960s has been towards the use of extensively glazed facades in commercial buildings. In partial compensation for this, the treatment of glazing to reduce the intensity of direct solar gain has become increasingly sophisticated (Fig. 6.1). The need for double glazing is primarily based on the reduction of perimeter heat loss in winter. At low outside air temperatures large areas of single glazing can lead to 'cold radiation' effects, and there is a benefit to thermal comfort from the increased room-side surface temperature afforded by double glazing. Associated with this is a reduced risk of surface condensation. In contrast, however, it has seldom been considered economically viable to use double glazing solely on the basis of cooling load reduction in air conditioned buildings. Especially in noisy city-centre locations, double glazing may be required on acoustic grounds, although the design criteria for thermal and acoustic double glazing are fundamentally distinct. Internal blinds have generally been considered essential on any glazing receiving direct sunlight, both to reduce cooling loads and since sitting or standing in the direct glare of the sun is incompatible with a feeling of thermal comfort. Placing the blinds between the glass panes in double glazing, rather than inside the room, significantly reduces the component of solar radiant heat ultimately transferred to the space from the blind (Fig. 6.1). Potentially such blinds are less likely to be harmed by careless treatment or mechanical damage, but are more difficult to access for

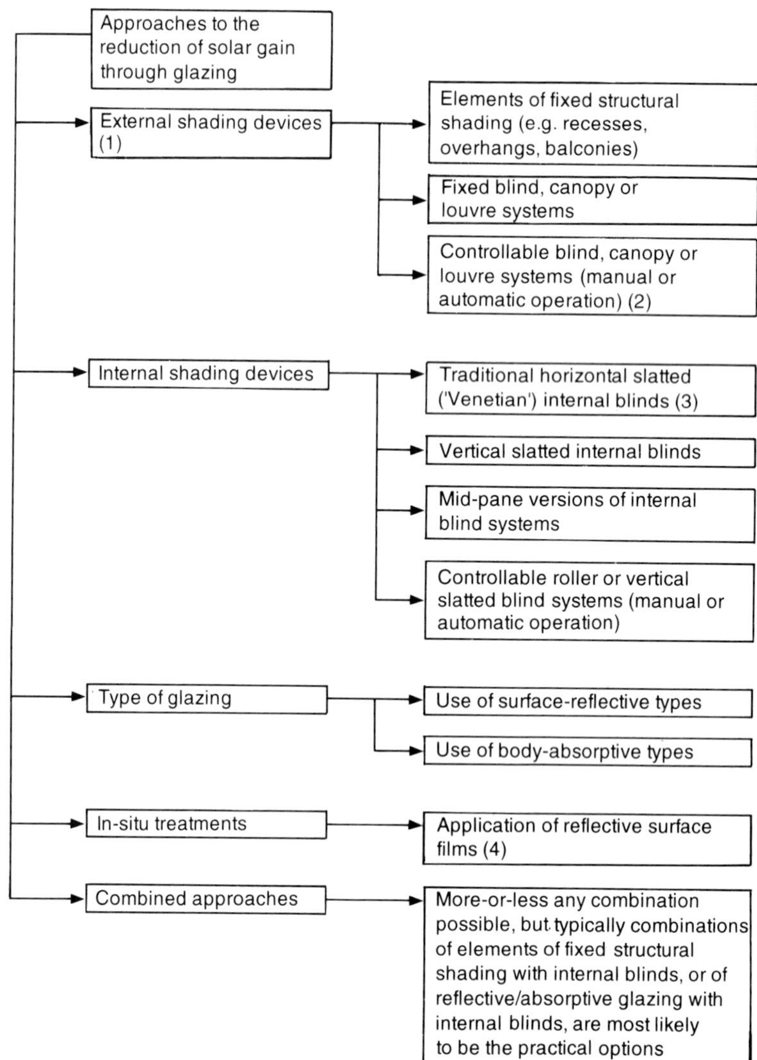

Notes:

(1) All external shading devices may present aesthetic and architectural difficulties.

(2) Likely to be most effective; inevitably the most expensive; a potential maintenance headache and recurring cost.

(3) Simplest to apply; the cheapest solution; the slats are dust traps, and operating cords are prone to damage from heavy-handed treatment in the event of sticking or jamming.

(4) May be difficult to apply successfully in practice, and quality of installation work is all-important; may be combined with other functions (safety, security).

Fig. 6.1 The treatment of glazing to reduce the intensity of solar gain.

repair or replacement in the event of such damage occurring, or of an inherent fault in the operating mechanism. In recent years the use of solar reflecting and absorbing glasses has increased. There has similarly been an increased interest in the use both of external blinds and of structural shading features, since direct solar gain is eliminated in the areas shaded by such devices.

The second component of the sensible cooling load is that generated *internally* within the building through its occupation and use. This internal component includes the sensible heat gains from people, lights and the whole range of business machines now typically found in commercial applications. Indeed with the improvements in the efficiency of luminaires and lighting controls that have come to the market since the early 1970s, business machines may have superseded lighting as the most significant source of casual heat gain in many applications. Under this heading are included the ubiquitous desktop computer or PC, workstations, visual display units, printers and modems, as well as the more general items of office equipment, such as photocopiers and facsimile machines.

The rise in the significance of business machines as a principal source of casual heat gain in air conditioned environments has been an inevitable result of the revolution in information technology (IT) that has swept through the office working environment in the UK, as elsewhere in the world. It has indeed proved problematic for HVAC designers, due in large part to the difficulty in correlating manufacturers' ratings for items of the new office technology with the actual levels of sensible heat gain which they may realistically be expected to impose on the system, particularly where a number of similar items of equipment are involved. A typical office desktop personal computer with associated visual display unit (VDU) may have an average power demand of 160 W.[1] This is almost twice the sensible heat gain from an occupant doing sedentary work at a nominal room temperature of 22°C, and it is not uncommon for the ratio of personal computers to occupants to approach unity in modern office buildings. Similarly the average power demand for a printer can vary from less than 50 W for a simple dot matrix type to over 250 W for sophisticated laser printers, although the ratio of these to occupants may typically be expected to be rather less than is the case for personal computers, and an inverse function of the sophistication (and cost) of the printer. Best predictions suggest that the trend of an increasing density of computers in office buildings is expected to continue towards a general 'one-per-capita' level by around the year 2000, associated with larger VDU screens consuming more power (*ibid.*). The intensity of use of desktop machines is also expected to rise, reducing the diversity of this element of the small-power load in offices. Numbers of facsimile machines (average power consumption 30–40 W) may also reach a level of one per person in offices over the same timescale. In the longer term, however, it may be expected that advances in equipment technology will reduce the power consumption of office equipment.

As far as lighting is concerned, however, the situation is rather different. Since the introduction of VAV systems there has typically been an overall reduction in the sensible cooling load from lighting in commercial buildings. Indeed a reduction of around 50% has been achieved in the electrical power required per square metre, relative to typical 1980s values, to achieve illuminance levels in the range 300–500 lux. The reasons for this include the introduction of krypton-filled fluorescent lamps with improved 'triphosphor' coatings, the use of luminaires with high-performance louvred and prismatic lens diffusers, and the introduction of low-loss high frequency electronic control gear. Associated with the increasing amount of work involving the use of VDUs in offices, there has been an increased interest in the use of task-related lighting in conjunction with reduced levels of general illumination. Recognition of the wastefulness of the all-too-common practice of leaving artificial lighting installations on at full output even during periods of peak solar gain has focused attention on lighting control. Automatic switching may be time- or daylight-related, and high frequency fluorescent fittings are suited to photocell-controlled 'topping up' of daylight levels. As far as VAV systems are concerned, the overall effect has been to diminish the level of base cooling load formerly afforded by lighting. Where daylight-related switching or automatic 'topping up' of daylight levels is employed, lighting may have to be regarded as a variable load, either in part or in total.

The potential savings from the refurbishment of lighting in existing buildings can be sizeable, and payback periods are generally favourable. Furthermore such refurbishment can often be carried out without the major disruption involved in general HVAC or structural refurbishment. In combination these factors make the updating of lighting installations attractive to building owners and tenants. The effect on an existing VAV system may be to remove a large part of the base sensible cooling load. System control and stability may suffer, and some of the theoretical reduction in the energy consumption of the VAV system that might reasonably be expected to accompany lower lighting gains may not be achieved in reality. There are two potential reasons for this. First, under conditions where VAV terminal units were already operating close to (or at) minimum flow rate limits prior to introduction of the new lighting installation, full (or any) adjustment for a reduced level of gain from the new installation may not be possible, and in turn will not be reflected in the system load. (Simply resetting terminal unit minimum flow limits downwards is unlikely to be feasible without reconsidering both the existing room air distribution design and zone ventilation requirements.) Secondly, increased operation at low system load factor is likely to be accompanied by reductions in fan and drive efficiency, particularly where variable-speed drives do not employ modern solid-state inverter technology.

6.2 Variations in the sensible cooling load

Outside air temperature is variable, and the intensity of solar radiation highly so. Random changes from minute to minute and hour to hour, caused by changing patterns of cloud cover and the vagaries of the local microclimate, are superimposed on daily and seasonal cycles. The general influence of daily and seasonal climatic effects may of course be anticipated using published climatic data. The effects of the microclimate are typically more difficult to quantify. Established and recognised patterns may be modified by the addition of a new building itself, and there is seldom the opportunity to take meaningful local measurements for a particular site prior to the commencement of design work. Some aspects are, however, amenable to analysis and quantitative assessment, for example the degree of exposure and the influence of an adajcent large area of water.

Figure 6.2 shows the effect of the daily pattern of solar radiation for a building with principal axis aligned north–south. Values of cooling load due to solar gain through vertical glazing were added for one square metre each of east-facing and west-facing glazing at hourly intervals from 0800 to 1800

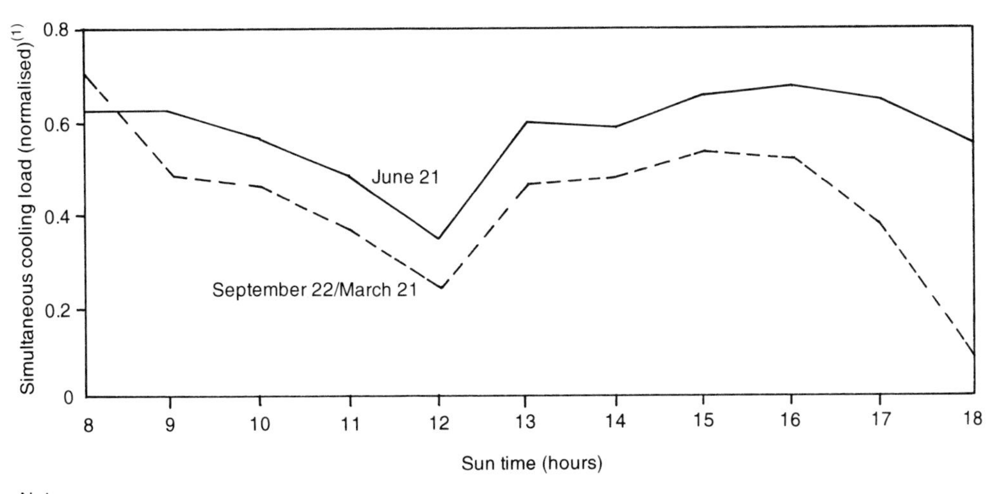

Note:
(1) Simultaneous total for equal areas of east- and west-facing vertical glazing, divided by the sum of the peak values for June 21

Fig. 6.2 Example of the importance of cooling load diversity to VAV system design.

hours sun time on 21 June. These simultaneous cooling loads for each hour were normalised by dividing each by the sum of the peak values for both aspects on that day. These occurred at 0800 hours sun time for east-facing glazing, and at 1600 hours sun time for west-facing glazing. Simultaneous loads were also determined using hourly data for 22 September/21 March, and these were again normalised using the same summed peak values for both aspects on 21 June. Both sets of data are plotted in Fig. 6.2, which is based on hourly values of cooling load due to solar gain through vertical glazing as taken from Table A9.15 of the CIBSE Guide for a building of lightweight construction at latitude 51.7°N, which is representative of UK design criteria for London and the Southeast of England[2]. Glazing is a single pane of 6-mm clear glass with internal blinds. These are assumed to be under occupant control, and used intermittently according to need. Plant operating period is ten hours daily.

The simple assessment of Fig. 6.2 illustrates an important consideration in VAV system design, which is that it is advantageous for design load diversity to serve zones facing in different directions from the same system. This is fundamentally different from the situation with a constant-volume all-air system, where it is advantageous in reducing the amount of reheating that occurs for control purposes, to zone systems according to the orientation of the building facades. Many commercial buildings have relatively shallow-plan layouts. Such buildings tend to have two long opposing glazed facades parallel to the major longitudinal axis of the building. The greater the variation in solar cooling loads experienced by each of these opposing facades, and the more out of phase (the further apart in time) are the peaks, the greater is the benefit to design load diversity of serving them from the same VAV system. Thus the ideal application for a VAV system is a building with longitudinal axis north–south, i.e. major glazed facades facing east and west. The east facade experiences its peak load in the early part of the working day, the west during the late part. If separate VAV systems were zoned to serve each facade, benefits would still accrue to both systems from VAV operation. However, there would be no benefit to system sizing. Zones on the same facade would experience peak cooling loads at the same time, and each individual system would have to meet the simple sum of its zone peak design loads (except only for any diversity possible in internal heat gains). Indeed, there would be a further disadvantage to system sizing, namely because the ratio of wetted perimeter to cross-sectional area decreases with increasing duct size. Hence, two ducts each of cross-sectional area $A\,\text{m}^2$ carrying identical volume flow rates of air do not need a duct of cross-sectional area $2A\,\text{m}^2$ to carry the same total volume flow rate at the same frictional pressure loss rate.

6.3 Annual operating profiles for VAV systems

If the daily and seasonal climatic load variations are taken into account, an operating profile may be drawn up summarising the way in which the system supply volume flow rate varies over a typical year's operation. Figure 6.3 shows a hypothetical operating profile in generalised form for VAV systems serving the type of commercial application most frequently encountered in the UK. This would be a multi-storey office block of relatively shallow planform, well insulated, with a moderate-to-high level of glazing (up to approximately 40% of the area of the principal facades). Floor layouts are typical, mainly open-plan, and represent general-purpose offices with an occupant density of the order of 10 m^2 net area per person. Although strictly-speaking hypothetical, and kept deliberately simple, the operating profile of Fig. 6.3 is firmly based on data presented by a number of sources. These are discussed in a later section. In particular it owes much to a study carried out in the UK by Jones[3]. This considered a notional office building of 14 storeys situated in London. Each storey provided 36 modular office pairs astride a north–south longitudinal axis. The north–south direction of the building's major axis, coupled with 50% glazing of the long facades, would be expected to give a good result as a VAV application, allowing maximum advantage to be taken of the variation in solar cooling load through the vertical glazing. The benefit from this would be less for any other orientation of the building, and least so for an east–west alignment of the major axis. The effect would also be decreased by a reduction in the extent of glazing on the long facades.

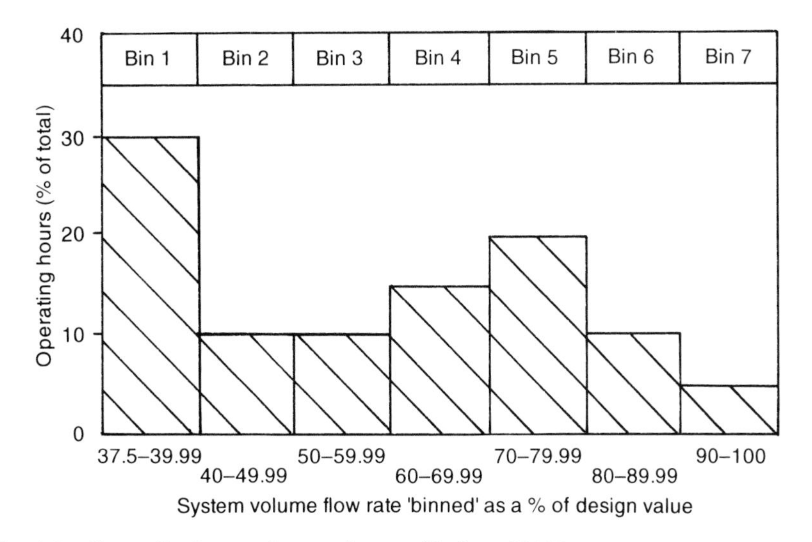

Fig. 6.3 Generalised annual operating profile for a VAV system.

Despite its relative simplicity, Fig. 6.3 will allow the illustration of several important points concerning VAV design and operation, and it is instructive to consider it in some detail.

The operating profile itself is a histogram showing the number of operating hours annually during which the system volume flow rate lies within defined ranges. In Fig. 6.3 both operating hours and volume flow rate are normalised as percentages of the total annual operating hours and the system design volume flow rate respectively. In drawing up an annual operating profile, the system volume flow rate is ideally calculated at hourly intervals throughout the annual operating period. In the UK the climate can vary quite considerably from year to year, particularly in the amount and duration of periods of hot, sunny weather. While strictly speaking the true operating profile of any VAV system will probably not be exactly the same for any two years, it is usual (excepting for the effects of system modifications and changes in the type or pattern of building use) to regard the operating profile of a particular system as constant from year to year over its life-cycle. Thus for a system operating 10 hours a day, 5 days a week, and 52 weeks a year, there are potentially 2600 different hourly values of volume flow rate each year. This amount of data presents a considerable processing load, and one that is certainly beyond the scope of manual analysis techniques. A less calculation-intensive alternative is to determine the system volume flow rates for each operating hour of a 'typical' day in each month, subsequently factoring each condition for the number of operating days in the month. In both cases many of the volume flow rates will be quite close to others, and it is generally considered acceptable, and certainly advantageous, to simplify it using a 'bin method'. In this the individual volume flow rates are sorted into ranges or bands. The mean of the upper and lower limits of each range is subsequently taken as typical for all of the hours (one hour per volume flow rate) in that range. The ranges are termed 'bins', and the data is described as 'binned'. The bin ranges must be chosen to be narrow enough so that the mean of a bin's limits is reasonably representative of the mean of all the individual volume flow rates it contains, whilst still being wide enough to simplify the required data processing and analysis to a worthwhile extent.

Figure 6.3 is consistent with typical practice in using a number of bins each having a range equal to 10% of the system design volume flow rate. An exception is made to accommodate the lowest bin. As shown, the profile is for a system with a minimum limit on volume flow rate. This is typically either for the purpose of ensuring a minimum level of ventilation to all zones, or to maintain acceptable room air distribution. However low the system cooling load may fall, the system volume flow rate will not be reduced below this minimum value – 37.5% of the design value in Fig. 6.3. It is necessary to introduce some variation to the bin ranges at the lowest volume flow rates. However, in the example shown there is a negligible loss of accuracy from regarding the contents of the lowest bin as all at 37.5% of the design value.

The operating profile of Fig. 6.3 has three features of particular significance for VAV design and operation. The first is the minimal amount of system operating time spent at or near its design volume flow rate. This results from the inherently 'worst-case' nature of design cooling loads, which may never be encountered in operation. It is quite possible that many VAV systems never actually operate at full design volume flow rate. In any event it cannot be proven, as it is not the general practice in the HVAC industry to monitor the actual operation of an installation once the building itself has been handed over to a client. As a rider to this generalisation, there is always the possible exception of a change in building use resulting in an increase in internal sensible heat gains. A move from general office use to computer-intensive financial services operations would be an example of this. Furthermore, oversizing is common in HVAC systems[4]. By oversizing is meant sizing plant and equipment for cooling loads that are never likely to be met in practice. In a speculative development the exact nature of the building occupation and use may be unknown, and there will be a strong desire on the part of the developer to cover all possible eventualities, potentially resulting in an excessively stringent design brief. Faced with imprecise design data, or uncertainty of the final accuracy of load calculations, the results of which may vary significantly according to the calculation procedure adopted, system designers are frequently tempted to 'allow a little bit more – to be on the safe side'. The result is an oversized system. Whatever the basis of the oversizing, it has strong implications for the operation, energy performance, and control of VAV systems.

The second significant feature of Fig. 6.3 is the peak of operating time in the lowest bin. The profile may indeed be somewhat conservative in this respect. The proportion of operating time spent in this region has important implications if the potential minimum energy consumption of VAV supply fans is to be achieved in practice. The build-up of operating time in the lowest bin is a result of the minimum limit on system volume flow rate. Modern commercial buildings may generate a cooling requirement throughout the year. However, minimum volume flow rates typically in the range 20–40% of the system design value still provide a significant cooling capacity, which is available whether required or not. Much of the operating time in the lowest bin arises during the period late autumn to early spring in the UK.

The third significant feature of the operating profile in Fig. 6.3 is the second, subsidiary, peaking of operating time between 70 and 80% of system design volume flow rate. According to the particular system, this may be displaced downwards towards lower volume flow rates or upwards towards higher volume flow rates. The build-up of operating time in this range represents operation during the most frequently occurring warm weather between late spring and early autumn. As a simple guide, this may be thought of in terms of the morning and afternoon peaks of the curve for September and March in Fig. 6.2.

The shape of the operating profile may be defined in more detail by using more volume flow rate bins. This may generally be expected to improve accuracy in estimating supply fan energy consumption, and in assessing the energy implications of fan control strategies. However, a larger number of bins does not automatically guarantee a significant improvement in the usefulness of the profile, and inherently carries a penalty in terms of the data processing load. Bin ranges of 5 or 10% of the system design volume flow rate will typically be appropriate.

Variations in the level and pattern of the internal sensible heat gains are largely unpredictable, and these cannot typically be taken into account by an operating profile based on design cooling load calculations. Such an operating profile thus only takes into account the daily and seasonal variations in sensible cooling load due to typical annual weather. Surveys of a number of offices over a period of days of normal use found that the power demands of small equipment loads were fairly uniform across the day[1].

Data on the sensible cooling load profiles of VAV systems serving modern office (or office-type) buildings exist from a number of studies of VAV systems. Some simple idealised system volume flow rate distributions are possible (Fig. 6.4). One such approach assumes a minimum volume flow rate occurring *N* times more than the maximum, the occurrence of intermediate values varying linearly between these two extremes[5]. In another distribution the peak frequency of occurrence is at a volume flow rate intermediate between identified maximum and minimum values. The frequency of occurrence of volume flow rates on each side of the most-frequent value varies linearly between the peak frequency and those of the maximum and minimum volume flow rates.

The bulk of the available data on the annual operating profiles of VAV systems derives from studies undertaken in the USA. Principal among these are those due to Brothers and Warren[6], Spitler *et al.*[7] and Englander and Norford[8]. While the first two of these employed computer simulations, the last-mentioned was based on measured data. Brothers and Warren used the DOE-2.1 software package to derive an hour-by-hour simulation of the thermal performance of what they described as a well-designed, low-energy-use commercial building of 929 m^2. In a study partly funded by the United States Department of the Army, Spitler *et al.* used the BLAST (Building Loads Analysis and System Thermodynamics) software package to simulate the energy performance of two prototype buildings, one a single-storey military headquarters building (1115 m^2 of offices and classrooms) and the other a 20-storey office building (2090 m^2 per floor). In both cases the simulations were repeated for city locations covering the principal climatic

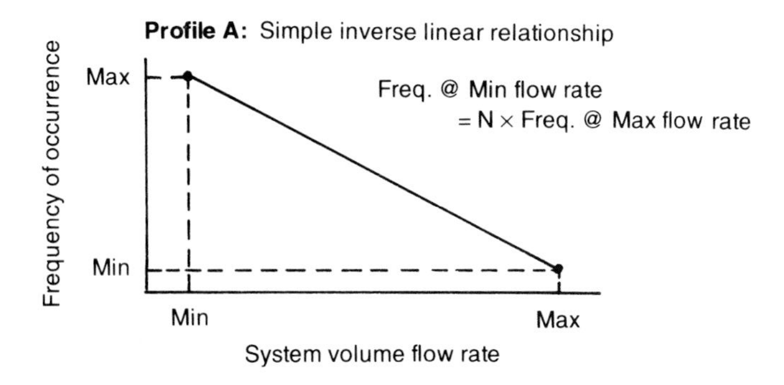

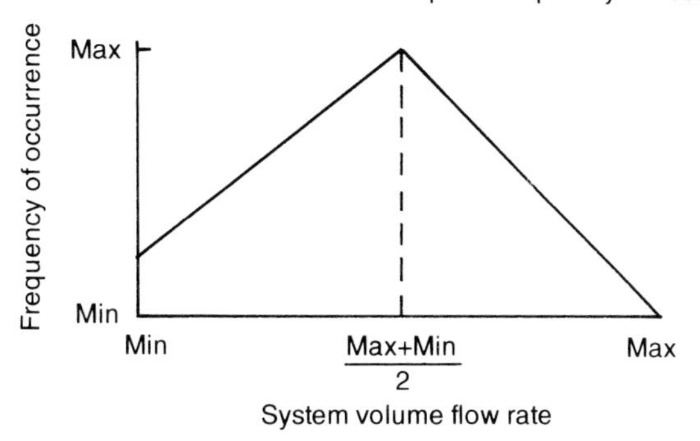

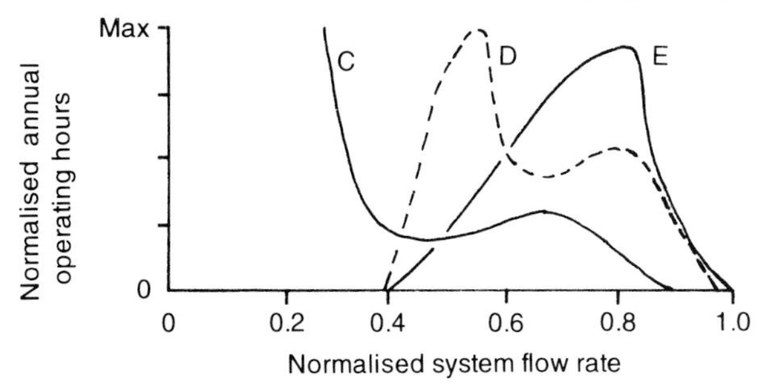

Fig. 6.4 Summary of VAV system annual operating profiles.

regions of the USA, and all cases featured a minimum volume flow rate limitation at the VAV terminal units. In a case study of retrofitting variable-speed drives to VAV system fans in a large office building, Englander and Norford showed volume flow rate histograms for both supply and return fans derived from available measured data sampled hourly over a base year.

Figure 6.4 shows the three generalised characteristic types of VAV system annual operating profile that can be identified from all these studies. In all cases system volume flow rates are normalised as a decimal proportion of the design value. In reality the three types of operating profile are not totally distinct and independent from each other, sharing a common dependency on the form and design of the building, on the climate of its locality and on the composition of the sensible cooling load served. As far as the last-mentioned item is concerned, the relative proportion of the sensible cooling load generated internally by people, lights and equipment is naturally significant. Since internal sensible heat gains are generated across the plan area of a building (A_{Plan}), while solar and conduction gains are generated across its external surface area (A_{Ext}), the ratio A_{Ext}/A_{Plan} may be a useful guide to this relative proportion, although of course it will not in itself allow for high or low absolute levels of either class of heat gain. Use of the ratio of peak internal sensible heat gains to the external surface area of the building has also been suggested[7].

6.4 The effects of oversizing

System design is not a precise affair. The Fielden Report on engineering design suggests that scientific principles, technical information and imagination are all involved[9]. The objectives are to define the system to *perform the specified functions with the maximum of economy and efficiency*. The emphasis of the design is not independent of the client, and the market for commercial buildings ranges from the owner-as-end-user to the developer/speculator. There will be a significant variation in economic perspective across this range, and in the willingness to define the end-use of the building in terms of allowances for internal sensible heat gains.

Since in any event it is doubtful that there is such a thing as the correctly sized system, oversizing is probably an inevitable consequence of the avoidance of undersizing. What is in question is the degree of oversizing. Among HVAC engineers there will be very many different opinions on this. Procel was quite categoric in stating that VAV systems were 'totally unforgiving', and should not be overdesigned[10]. In any event the system designer should be aware that an oversized system is more expensive in capital cost, and has greater space requirements for plant, equipment and distribution ducts than one that is sized appropriately for the application. Both control and stability may suffer, energy performance may be compromised, and the service life of plant and equipment shortened. Overcooling of zones may occur at minimum specified flow rates. While system volume flow rate is reduced, so also are the efficiencies both of fans and their means of duty control, and there is an excessive level of wasteful throttling of duct static pressure at the terminal units. Reduced sensor accuracy, and valve and damper authority, can lead to poorer control of all plant and equipment, system instability and accelerated wear of plant and equipment.

Let us briefly examine the effect of oversizing of a VAV system by reference to Fig. 6.3. For the purposes of this the simplifying assumption will be made that there exists a reasonably uniform spread of volume flow rates across the range of each bin. A typical scenario will be supposed in which the actual level of internal heat gains experienced during normal operation of the system is significantly lower than estimated during design. As noted, this is in any case a difficult area for HVAC system designers. Coupled to this are the demands of a speculative development where the ultimate end user of the facilities is unknown at the design stage. There is an understandable desire on the part of the developer to cover all potential users, whether these may require minimal small power use or an intensive IT coverage, especially when he or she has heard something (but not enough) about VAV systems. It would, in any event, be a brave person who would turn away a potentially lucrative lease or sale of their development on the grounds that the potential lessor or purchaser did not need the full capacity of the building's air conditioning system. At design stage oversizing can also result from the use of cooling load calculation programs which overestimate climatic effects, from the addition of allowances to compensate for missing or imprecise design information, from the addition of unspecified safety margins, and from the limitations of standard product ranges. A system may also be rendered oversized by subsequent changes to the building and its pattern of use.

The effect of the overdesign may be simulated by adjusting all the system volume flow rates binned in the operating profile of Fig. 6.3 for appropriately reduced values of internal sensible heat gain. The system design volume flow rate used to normalise these values remains constant, however. This is appropriate because the intended design cooling capacity of the system is not changed by lower-than-expected actual cooling requirements. All hourly-calculated system volume flow rates will be reduced from their previous values (Fig. 6.5). The operating time in bin number 7 will decrease, as some of the 'hours' move into the adjacent, lower bin (number 6). A 10% reduction in volume flow rates will be sufficient to almost empty bin 7. Bin number 6 receives some hours from the adjacent higher bin, but loses more than it gains (again on the assumption of uniform distribution within each bin, as it started with more hours) to the adjacent, lower bin (number 5). The process continues through the bin ranges of the histogram, from high to low volume

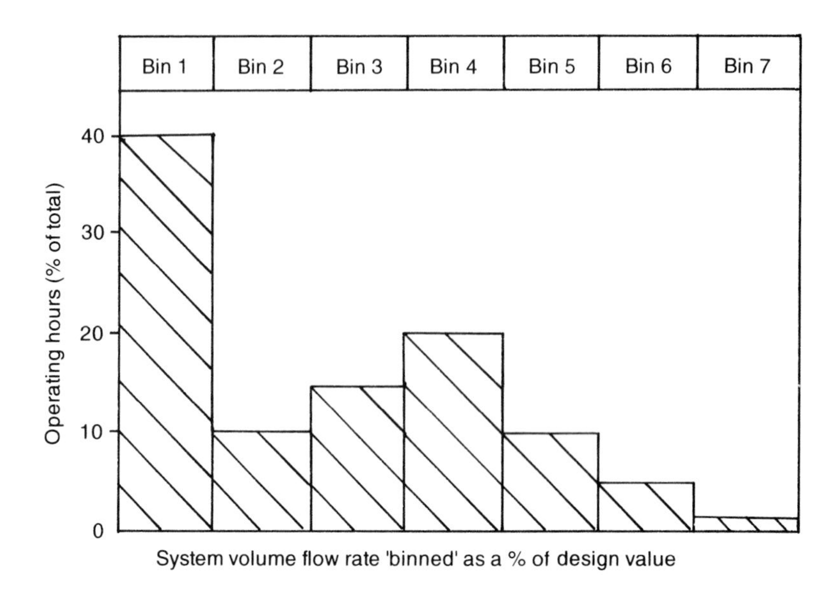

Fig. 6.5 The effect of oversizing on VAV system annual operating profiles.

flow rates. Owing to the minimum limit of system volume flow rate, at the bottom end the lowest bin only accepts hours, losing none.

Each $W\,m^{-2}$ by which design internal sensible heat gains are overestimated will cause progressively greater reductions in the hourly values of system volume flow rate under conditions of low solar gain. This may be expected to exaggerate the shift of operating time in moving down the bin ranges from 7 to 1. The overall effects relate to the three main features of the generalised profile without overdesign. The amount of operating time in the higher bins is reduced. If the overdesign is great enough (or the bin ranges are narrow enough), the profile will shift away from the system design volume flow rate, condensing the operating time into fewer bins. The operating time at minimum volume flow rate builds up. The secondary peak of operating time is shifted towards lower volume flow rates, and is reduced.

6.5 The nature and diversity of the loads served

The nature of the sensible cooling loads served by a VAV system is of significance to both its design and operation. The climatic component of the sensible cooling load may reasonably be expected to affect all zones with a particular orientation simultaneously, although in some city-centre locations the effects of shading from adjacent high-rise buildings may be significant. Ignoring this, however, in general if the sun is shining on one east-facing zone, it will typically be shining on them all, and the effects of air and sol-air temperature on heat transfer through the building fabric will also be common to zones with the same orientation.

Whilst the climatic element of the load is highly variable, the internal load may reasonably be regarded as generally 'statistical' in nature. This is not to imply a random variation, but rather to recognise that statistical techniques can usefully be applied to predict likely levels of internal heat gains. For example, in its published guidance on small power loads in offices, the Energy Efficiency Office notes that if each PC in a population of 100 is used on average for 20% of the time, probability theory suggests that up to 30% of the machines may be in use *simultaneously*[1]. Using the data in this source (summarised for PCs in Fig. 6.6), if the population is reduced to ten, the probability of simultaneous use increases to *c*.55% of the population. On the other hand, for an increase in population to 160 PCs, the probability of simultaneous use is only marginally less than that of the original population of 100. If the building end-user's type of working is totally computer-based, the only diversity allowable in simultaneous PC use may be due to holidays, sickness, etc. Many occasional end users of PCs automatically switch on their machines at the start of each working day/session, and only switch them off prior to leaving at the end of the day/session. Add to this the possibility of automatic standby ('sleeping') modes of operation for PCs and VDUs featuring reduced energy use, and the overall picture is seen to be potentially complex in assessing realistic levels for this type of internal sensible cooling load. In any event, it is typically true that the larger the population con-

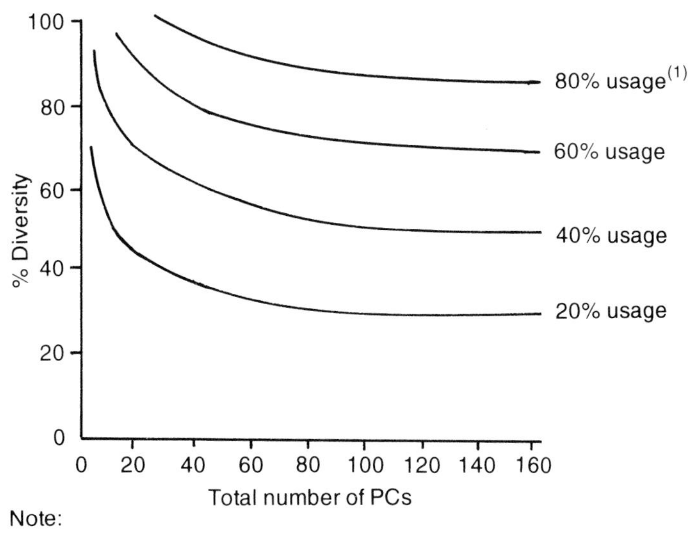

Fig. 6.6 A statistical approach to simultaneous sensible heat gains from office equipment.

sidered (which is ultimately a function of the area served), the greater may be the confidence with which diversity factors based on probability theory can be applied. This is most true for a system as a whole, and least so for individual zones. As a result the overall sensible heat gains from all internal sources can vary markedly between adjacent zones, and over time within individual zones. However, in many instances it may be practical to divide them between a constant, base-load element (including, for example, the cooling load from general office lighting) and a variable element which can be assessed statistically for combinations of zones and groups of zones working backwards from the index terminal unit to the system supply fan.

The make-up of the design cooling load, in terms of the proportions of each component, has implications for design diversity and system control. Design load diversity allows the economic sizing of VAV system plant, equipment, and distribution ducts. Main duct sizes are based on the design volume flow rates at peak *simultaneous* system load. Branch ducts serving individual zones must of course be sized to handle the peak design volume flow rate for the zone, and ducts serving a number of zones with a single orientation must be similarly treated (although wherever possible this latter case is to be avoided for this very reason). In principle the sizing of any duct serving a large enough group of zones may benefit from some applied diversity in their internal sensible heat gains, although this may not be so straightforward to achieve in practice. Figure 6.2 suggests that the maximum volume flow rate to satisfy the cooling load due to solar gain through the vertical glazing of an identical pair of east- and west-facing zones will never exceed approximately 70% of the sum of the peak volume flow rate for this element. The same will, therefore, be true for any duct section serving equal numbers of identical east- and west-facing zones, and the principle is taken into account by calculation of peak simultaneous loads. Naturally many real buildings do not fall into such a neat and simple category as regards orientation. The more highly variable the zone sensible heat gains are (conventionally the lower the proportion of the total sensible heat gain accounted for by internal sources), the greater the diversity of the connected zone loads, and the smaller the duct size can be for a given design rate of frictional pressure loss, and the lower is the potential minimum annual fan energy consumption of the system.

6.6 Implications for energy use

The daily, seasonal, and annual variations of system supply volume flow rate give rise to the third principal justification for the use of VAV systems – that of reducing fan energy consumption. It is probably fair to say that this is often regarded as the justification for a VAV application by non-HVAC specialists. It is probably equally fair to say that reducing fan energy consumption is the most important contribution of a VAV system to energy-conscious design, even among HVAC specialists. However, a truly energy-conscious approach to systems design ought to focus on the concept of a *life cycle energy use*, which includes the total energy consumption involved in the extraction of all raw materials, the design, manufacture, delivery to site and installation of all the component plant and equipment items, and the operation of all energy-consuming components (fans, pumps, compressors, humidifiers, boilers, electric heating elements, etc.) over the projected operating life of the system. In this respect the contribution of the reduced level of embodied energy (which is analagous to capital cost in life cycle costing) of the smaller air handling and refrigeration plants and ductwork systems permitted by VAV design load diversity is also relevant. However, there is presently little practical data available to HVAC designers to enable them to compare the embodied energy of alternative system designs, or even of different sizing approaches to the same system or plant components.

6.7 Implications for fan performance

Fan power depends on volume flow rate, total pressure and efficiency. Hence for a VAV supply fan:

$$\dot{W}_S = \dot{V}_{SYS} \times FTP_S \times 100 \div \eta_S \tag{6.1}$$

where

$\dot{W}_S$ = supply fan power (W)
$\dot{V}_{SYS}$ = system volume flow rate ($m^3\,s^{-1}$)
FTP_S = supply fan total pressure (Pa)
η_S = overall efficiency of supply fan, motor and drive components (%).

The overall efficiency of the fan, motor and drive components is conventionally determined as the product of the individual efficiencies of these components (i.e. $\eta_S = \eta_{S,FAN} \times \eta_{S,MOTOR} \times \eta_{S,DRIVE} \times 10^{-6}$, where all terms are percentage values), but must take into account the efficiency either of the variable-speed drive or of whichever of the alternative forms of fan duty modulation is employed.

Fan energy consumption is the integral with respect to time of this expression. As the terms of the fan power equation are not simple functions of time, the annual energy consumption of VAV fans is typically approximated by a numerical integration using the 'binned' volume flow rate data of a system annual operating profile, such as that of Fig. 6.3. This concept is considered further in Chapter 8.

Use of the peak simultaneous sensible cooling load means that in ideal VAV applications (principal building axis north–south) system volume flow rate may be of the order of 70–80% of that of a constant-volume all-air system. Fan power consumption of both VAV supply and extract fans at the design condition is a similar proportion of their constant-volume counterparts. This assumes that fan total pressure can be maintained at the greater volume flow rates of the constant-volume system by simply increasing duct sizes from their VAV values. However, the effect must also be considered that, since the annual operating time spent at volume flow rates in the highest bin range is so small, VAV supply fans are typically selected to have maximum efficiency at a volume flow rate rather lower than the system design value. For the operating profile of Fig. 6.3 this would be in the bin range of the secondary peak of operating time (70–80% of system design volume flow rate). The fan should then exhibit good efficiency characteristics over a range of volume flow rates representing some 45% of the annual operating period. A similar consideration also applies to VAV extract fans.

The cumulative effect of independent throttling action by the various VAV terminal units is to control system supply volume flow rate during part-load operation. When combined with suitably controlled supply and extract fans, this results in a reduction in fan power and annual energy consumption. Best performance is typically achieved with variable-speed centrifugal or variable-pitch-in-motion vane-axial fans. Other combinations of fan and duty control are possible, although they are unlikely to realise the energy saving potential of the former combinations. In particular, centrifugal fans with inlet guide vanes have been widely used.

The VAV supply fan must also maintain sufficient fan total pressure, or else the VAV terminal units will be unable to satisfy the volume flow rate requirements of their zones. Except for the case of a VAV system serving a single control zone, the minimum fan total pressure that qualifies as 'being sufficient' is typically not simply a function of system volume flow rate. While 'air-side' pressure losses across the main cooling coil, etc. are a function of system volume flow rate alone, the loss of total pressure in the ductwork system depends additionally on the volume flow rate distribution between the zones served.

In the constant-volume design approach system volume flow rate, its distribution between the control zones, and the fan total pressure all remain nominally constant. Fan total pressure is determined by the pressure losses across plant items and the minimum requirements of the index duct run, the path through the ductwork system which has the greatest loss of total pressure for the specific volume flow rate distribution. In constant-volume systems there is only one volume flow rate distribution, with every zone nominally receiving its design volume flow rate. In the absence of any particularly tortuous or limiting routes through the ductwork system (in the sense of changes of direction and cross-sectional area, or restrictions on duct size), the initial prime candidate for the index run is typically the path to the supply outlet furthest from the fan).

Whenever a fan is operating in a ductwork system a balance will always exist between what leaves the fan and what is either discharged from the supply outlets or leaks from the ductwork along the way to them. An unconstrained or 'natural' volume flow rate distribution will result which ensures that the component of fan total pressure available for duct losses is fully absorbed along each path through the ductwork system. Without deliberate action during design and/or commissioning, this flow distribution is unlikely to match that needed to supply each zone with its design volume flow rate. Air-side *balancing*, as part of the system commissioning procedure, involves increasing the loss of total pressure along each non-index path through the ductwork system, based on having the required volume flow rates in all sections of the path, so that the component of fan total pressure available for duct losses is just fully absorbed. Conventionally *proportional* balancing brings all measured volume flow rates to (ideally) the same proportion of their design values. This proportion can then be brought to unity by adjusting the performance of the system fan. In theory only one notional condition of balance exists that will allow the final volume flow rate distribution throughout the ductwork system to match the design requirements. However, in practice it is rarely either practical or necessary to achieve an exact system balance, but rather to achieve a balance within an acceptable tolerance, taking into account accuracy of air flow rate measurement, duct leakage and the nature of the spaces served (whether cellular or open plan). Hence there is indeed a range of approximate system balances that will typically be acceptable, and generally indistinguishable in terms of the system performance that will be achieved.

Increasing the losses of total pressure along non-index paths through ductwork systems, so that the loss along each path matches the component of fan total pressure available for duct losses, is typically achieved by throttling off excess pressure across balancing dampers. The positions of these are set manually and, once set, their position remains fixed.

It is theoretically possible to design an inherently balanced ducted air distribution system by adjusting duct sizes to absorb the predicted excess pressures in non-index runs through increased frictional and velocity pressure losses, rather than by throttling across a partly-closed damper. Potential reductions in the capital cost of the smaller ducts may make it attractive to do so, or at least to reduce the amount of throttling carried out by (smaller) balancing dampers. Fan total pressure, and hence energy consumption, is not affected by the presence of balancing dampers as long as no throttling of the flow takes place along the index run.

In the VAV design approach both system volume flow rate and its distribution are continuously adjusted by the throttling control action of the VAV terminal units. In effect the system is continually re-balanced to suit the pattern of zone cooling loads prevailing at any given time. While similar considerations apply to establishing the index run under system design conditions, it does not remain constant under part-load conditions. All of the pressure losses along any path through the ductwork system occur either in the ducts and fittings themselves, the supply diffuser and any duct-mounted equipment, or across the VAV terminal unit, and the losses along all paths are again equal to the component of fan total pressure available for duct losses. The path through the ductwork system with the highest loss of total pressure in the ducts and fittings, diffuser, etc. naturally has least need of throttling action by its terminal unit to absorb the total pressure available to the path. The throttling action of a VAV terminal unit derives from a damper which moves to progressively shut off the air flow passage to the zone and increase the pressure drop across the unit at any given volume flow rate. Since the same total pressure is available to all paths through the ductwork system, the index run in a VAV system is identified by the terminal unit in which the throttling damper is nearest to the fully open position, and which is the index terminal unit.

The losses of total pressure in ducts and fittings (or plant items) cannot be reduced as they are an inherent consequence of the required volume flow rates of air. However, fan total pressure can be minimised by avoiding the throttling of any excess pressure by the VAV terminal unit on the index run (the index terminal unit). Hence the fan total pressure that is 'just sufficient' is that when the throttling damper in the index terminal unit is 'just fully open'. All other terminal units have some excess pressure, and will automatically adjust their level of throttling to balance the air distribution system to the required volume flow rates at the available fan total pressure.

As in a constant-volume system, the index run in a VAV system can be predicted under design conditions and its pressure drop calculated. Under part-load conditions *feedback* is required from individual VAV terminal units to identify the index unit, and whether or not the available fan total pressure is just sufficient. Without such feedback the only way in which to ensure that sufficient fan total pressure is always available is to use a worst-case criterion. This is in fact what has traditionally happened in VAV systems, where throughout its annual operation duct static pressure has been maintained constant at a nominated point in the air distribution system, and at a value based on the system's requirements under design conditions. This inevitably results in fan total pressure being maintained at a higher value than is strictly required for much of the system operating time. This aspect of the control of VAV supply fans is discussed in Chapter 8, and the fan energy consumption and operating cost penalty associated with it is considered further.

Sufficient fan total pressure must also be provided by VAV extract fans. However, this is typically a simpler case. Without independent control of zone extract volume flow rates, the extract system essentially behaves as a fixed resistance to air flow – with a total pressure loss that is a function of system volume flow rate alone.

Because of design load diversity, the minimum system volume flow rate is always greater than the minimum zone volume flow rate, as a percentage of its design value. A simple example helps to illustrate this. The minimum system volume flow rate, expressed as a percentage of its design value, is given by the expression:

$$\text{Minimum system flow rate (\% Design)} = \sum \textit{Values of } VFZ\%_{\text{(Limit)}} \text{ for all zones} \div (n \times DLDF) \qquad (6.2)$$

where

$VFZ\%_{\text{(Limit)}}$ = minimum limit on zone supply volume flow rate, expressed as a percentage of its design value
n = number of zones
DLDF = design load diversity factor (or *simultaneous peak cooling load* ÷ $\sum$ *sum of zone peak loads*).

If DLDF has a value of 0.8, and no zone is to receive less than 30% of its design volume flow rate, the minimum system volume flow rate is 37.5% of its design value. It is at low system volume flow rates, represented by bins 1 to 3 in Fig. 6.3, that traditional control techniques for VAV supply fans have maintained the greatest excess of fan total pressure over the minimum actually required by the system. This is significant because it is an area which may represent up to 50% of annual operating time.

References

1. Building Research Energy Conservation support Unit (BRECSU). (1993) Energy Efficiency in Offices – Small Power Loads, BRECSU, Garston, Energy Efficiency Office Best Practice Programme, Energy Consumption Guide 35.
2. Chartered Institution of Building Services Engineers (CIBSE) (1986) *Estimation of Plant Capacity*, CIBSE Guide, Section A9, CIBSE, London.
3. Jones, W.P. (1980) Energy consumption by variable air volume systems. *Building Services and Environmental Engineer*, **2**, July, 16–19.
4. Brittain, J.R.J. (1997) *Oversized Air Handling Plant – A Guide to Reduce the Energy Consumption of Oversized Constant or Variable Air Volume Air Handling Plant*, BSRIA, Bracknell, BSRIA Guidance Note 11/97.
5. Holmes, M.J. (1976) *Backward Curved Centrifugal Fans in VAV Systems – Selection and Energy Consumption*, Building Services Research and Information Association, Bracknell, Technical Note 2/76.

6. Brothers, P.W. and Warren, M.L. (1986) Fan energy use in variable air volume systems. *ASHRAE Transactions*, **92**, Part 2B, 19–29.
7. Spitler, J.D., Hittle, D.C., Pedersen, C.O. and Johnson, D.L. (1986) Fan electricity consumption for variable air volume. *ASHRAE Transactions*, **92**, Part 2B, 5–18.
8. Englander, S.L. and Norford, L.K. (1992) Saving fan energy in VAV systems – Part 1: Analysis of a variable-speed drive retrofit. *ASHRAE Transactions*, **98**, Part 1, 3–18.
9. Fielden, G.B.R. (1963) *Engineering Design*, HMSO, London.
10. Procel, C.J. (1974) Variable air volume systems: loads and psychrometrics. *ASHRAE Transactions*, **80**, Part 1, 473–479.

Chapter 7
VAV Terminal Units

7.1 The role of the terminal unit

In general terms a VAV terminal unit may simply be described as a *factory-made component of the air distribution network* in a VAV system. The primary characteristic of a VAV terminal unit is that it controls volume flow rate in the branch of the distribution network in which it is installed. On the supply side of a VAV system this is the volume flow rate of conditioned air supplied to a zone from the central air handling plant. In its basic form, this directly determines:

(1) the rate of removal of sensible heat from the space;
(2) the rate of removal of moisture from the space;
(3) the rate of supply of outside air to the space.

With the addition of a suitable form of air heating at the terminal unit, or by mixing of airstreams at two different temperatures, item (1) may be extended to include the rate of removal or addition of sensible heat to the space.

The terminal unit thus controls the supply of energy to the space for cooling, heating and dehumidification (Fig. 7.1). It directly influences the dry-bulb temperature and relative humidity within the space, and its ventilation rate. It inherently also influences the level of air movement and the air pressure within the space, although the nature and degree of this influence depends on both the type of terminal unit and the particular system design. In commercial air conditioning applications feedback control of the volume flow rates of conditioned air supplied by the VAV terminal units is typically provided for dry-bulb temperature only. As far as control of the terminal units is concerned, the other factors inherently influenced – space humidity, ventilation, room air movement, and air pressure in the space – are typically allowed to 'float'. This is the main drawback of the VAV system. Its acceptability or otherwise needs to be considered and fully established for each particular application, and its implications taken into account in the design and control of the system as a whole.

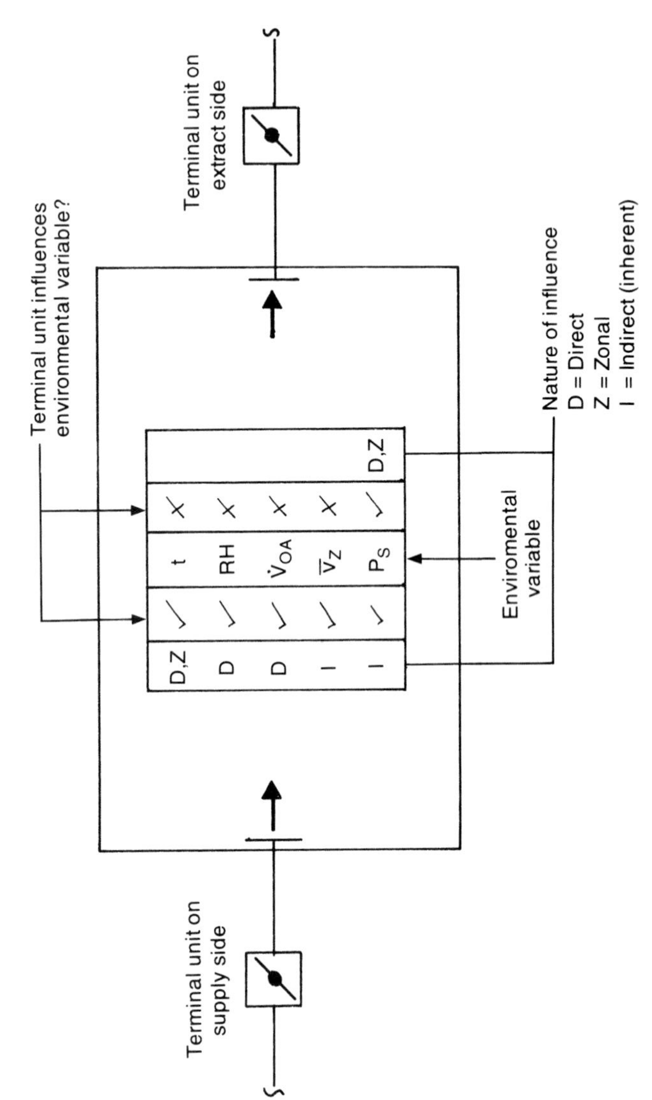

Fig. 7.1 The role of the terminal unit in VAV systems.

When installed on the extract side of a VAV system, a terminal unit controls the volume flow rate extracted from the zone and returned to the central plant. This directly influences air pressure within the space.

7.2 The variety of terminal unit design

A wide variety exists in the design of VAV terminal units (TUs). Some differ fundamentally from each other, while between others the differences are confined to details of configuration, operation, or performance. Some idea of the potential range of equipment, configuration and detail design options available can be seen by referring back to Fig. 2.1. Important differences exist between pressure-independent and pressure-dependent units, between electrically or pneumatically actuated and system-powered units, between units which incorporate a supply diffuser and discharge air directly into the space served and those serving remote supply diffusers, between throttling, fan-assisted and induction terminal units, between series-fan and parallel-fan types of fan-assisted terminal (FAT), and between single-duct and dual-duct units. These characteristics are illustrated by the typical configurations shown in Figs 7.2–7.30, and a general comparison of terminal unit characteristics is presented in Table 7.1.

7.3 An overview of terminal unit control

Terminal unit control in a VAV system covers a wide range of functions. The more common functions that are likely to be encountered are listed below (not necessarily in order of importance). For this purpose primary air is the supply reaching the terminal unit from the central AHU.

(1) Measurement of zone air temperature.
(2) Adjustment of zone temperature set-point, either within the terminal unit controller or within the zone sensor itself.
(3) Measurement of primary air velocity pressure or velocity.
(4) Set-point adjustment of primary air volume flow rate, including upward reset for minimum ventilation rate.
(5) Adjustment of high- and low-limit set-points of primary volume flow rate.
(6) Control of the primary air damper.
(7) Control of three-port motorised control valve on LPHW air heater battery.
(8) Step control of electric air heater battery.
(9) High-limit safety cutout of an electric air heater battery.

Table 7.1 General comparison of terminal unit characteristics.

Characteristic	Single-duct VAV box	Dual-duct VAV box	Fan-assisted terminals	Induction VAV terminal	System-powered VAV terminal	Pressure-dependent VAV terminal	Thermally-powered diffuser
Location	Ceiling/floor void	Ceiling void	Ceiling void	Ceiling void	Ceiling void	Ceiling void	Ceiling void
Supply air inlet	Ducted	Ducted hot, cold High space, cost	Ducted	Ducted	Ducted	Ducted	Ducted
Recirculation	None	None	CV, unducted Zone, ducted	CV, unducted	None	Depends on design	None
Zone discharge	Ducted	Ducted	Ducted	Ducted	Typically direct	Typically direct	Direct
Operation	Throttling	Throttling	Primary supply throttled; mixed + recirculated by fan (fan power significant in running costs)	Primary supply throttled; mixed + recirculated by induction (needs high P_S at nozzles)	Throttling	Throttling	Throttling
Flow to zone	Variable cooling, may be constant (min) or variable heating	Variable cooling, may be constant or variable heating	Constant 'series' fan; variable cooling/constant heating 'parallel'	Nominally constant	Variable	Variable	Variable
Control	VCV[1] or 'pressure-independent' typical	VCV typical	VCV primary air. CAV to zone (only + fan for 'parallel')	VCV primary air, with linked throttling of induced flow	VCV	Direct control of damper position by zone controller	Integral direct-acting control of damper position
Branch pressure control	Not required	Not required	Not required	Not required	Not required	Required (PRDs[2])	Required (PRDs[2])
Damper type (terminal unit)	Single/opposed-blade, 'butterfly'	Single-blade	Single-blade	Single-blade	Neoprene bellows	Single-blade, curved, 'specials'/proprietary	Circular disc/baffle

Damper actuation	Electric, pneumatic	Electric, pneumatic	Electric	Electric, pneumatic	Duct pressure. Higher min. P_S than electric, etc. (100–150 Pa)	Electric, pneumatic	Direct acting (thermal expansion of wax amplified mechanically)
Room air distribution	Remote diffuser(s)	Remote diffuser(s)	Remote diffuser(s)	Remote diffuser(s)	Integral diffuser (lin slot typical)	Integral diffuser (lin slot typical)	Integral square-faced diffuser
Space heating capability	Available if LPHW, electric heater battery	Provided by hot duct supply	Available if LPHW, electric heater battery	Limited. Electric heater possible	Very limited. Warm-up using central plant	Available if LPHW, electric heater battery fitted	Available if electric heater battery fitted upstream diffuser neck
Cold air distribution	Not typically suitable	Not typically suitable	Suitable (especially 'series' fan)	Suitable	Not suitable	Not suitable	Not suitable
Room-side acoustic attenuation	Secondary silencer may be fitted	Secondary silencer may be fitted	Secondary silencer may be fitted	ΔP loss is against; less suited low *NR*	Limited to treatment of terminal casing	Limited to treatment of terminal casing	None possible
Electrical connections	Power 24 VAC; control 0–10 VDC, etc.	Power 24 VAC; control 0–10 VDC, etc.	Power 24 VAC; 240 VAC to fan; control 0–10 VDC, etc.	Power 24 VAC; control 0–10 VDC, etc.	None typical; 0–20 V phase-cut control possible	Power 24 VAC; control 0–10 VDC, etc.	None
Typical size range[3]	150–1500 l sU–1 1.5–15 kW	150–750 l s^{-1} 1.5–7.5 kW	100–1000 l s^{-1} 1–10 kW	Not available	25–200 l s^{-1} 0.25–2 kW	40–180 l s^{-1} (100–250 'swirl' type); 0.4–1.8 kW (1–2.5 'swirl').	40–400 l s^{-1} 0.4–4 kW
Control sensor (temperature)	Remote in zone	Remote in zone	Remote in zone	Remote in zone	Integral bimetal (induced room air); interface for remote electric, etc. possible	Remote in zone	Integral wax-filled element for induced room air

Notes
(1) Variable-constant-volume
(2) Pressure regulating damper
(3) Based on nominal maximum ratings of smallest to largest units in typical commercially available ranges.

(10) Air proving switch, interlocked to cut out the electric air heater battery on low volume flow rate.
(11) On/off control of terminal unit fan (parallel configuration).
(12) Manual override of primary air damper to either fully open or to full shut-off.
(13) Warm-up cycle.
(14) Upload of data to/download of control settings from a central control and management computer.

Naturally not all of these functions will be required at any particular time. The HVAC designer has three general options regarding an approach to control system design – namely pneumatics, analogue electronic, and direct digital control (DDC). Pneumatic controls are now typically less common, since the underlying demand for more control functions cannot be met without a substantial cost penalty. Analogue electronic controls are still the norm for many applications, while the microprocessors of the DDC approach can provide all the functions required of a terminal unit. With its wider adoption, the cost premium of adopting the DDC approach has now reduced, bringing it in line with the cost of a conventional analogue electronic system.

7.4 VAV box: single-duct throttling terminal unit with variable-constant-volume action

7.4.1 Design and construction

Figure 7.2 shows a terminal unit of the type commonly referred to as a 'VAV box'. The unit is of the so-called pressure-independent type. 'Pressure independence' is the characteristic that the volume flow rate of conditioned air supplied by the terminal unit to the zone served remains nominally constant, at any set-point value between specified maximum and minimum settings, despite fluctuations in the static pressure at the inlet to the terminal unit – as long as this remains within maximum and minimum limits specified by the unit's manufacturer. The set-point volume flow rate for this constant-volume action is itself varied (reset) according to the output of the zone temperature controller serving the terminal unit. For this reason the characteristic of pressure independence is more correctly termed variable-constant-volume action. This type will be described in some detail, since it can in many respects act as a general model for VAV terminal units as a whole, is a type frequently encountered in existing systems, and remains a common approach to VAV system design.

The basic terminal unit comprises a galvanised sheet steel casing of rectangular cross-section. At its upstream end the casing is blanked off, save

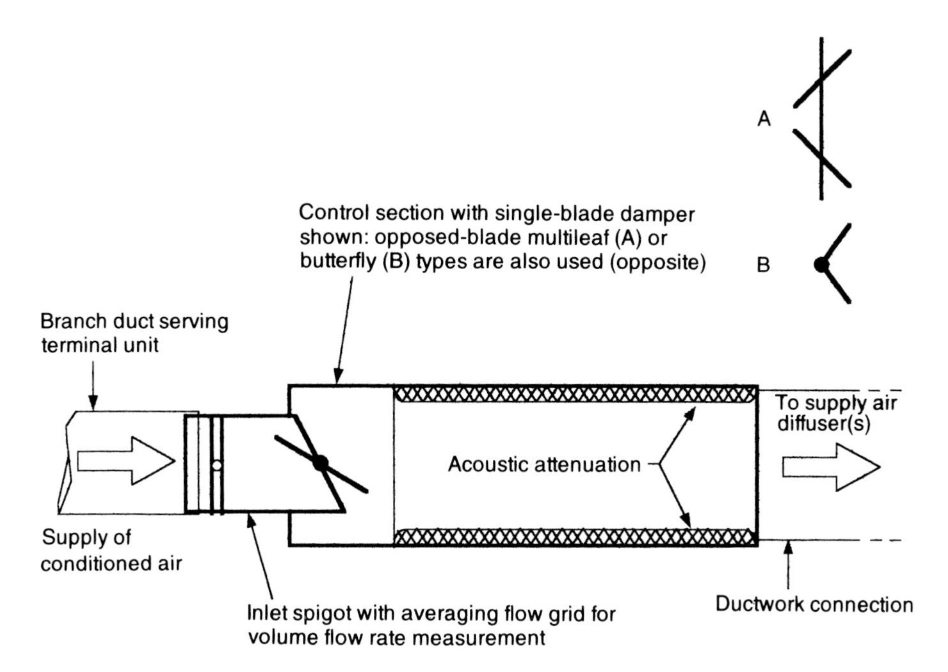

Fig. 7.2 Confirguration of a 'VAV box': single-duct throttling terminal unit with variable-constant-volume ('pressure-independent') action and remote supply diffuser(s).

for a stub inlet duct (or spigot) of circular cross-section. In an installation this provides the connection to the supply branch duct serving a zone. The unit is fabricated using traditional sheet metalworking techniques (lockforming, welding, etc.), and all joints are typically sealed for airtightness. The upstream end of the casing is occupied by the control section. This contains a volume (strictly-speaking, volume flow rate) control damper. This is most commonly of single-blade type, but may also be of opposed-blade multileaf or butterfly design (as inset in Fig. 7.2), although one UK manufacturer has used the twisting of a cylinder of flexible material to create a throttling action. Damper blades are again typically fabricated from sheet metal. Single-blade and *butterfly* types are usually of single-skin design, while some multileaf types may have stainless steel aerofoil blades. Damper blades are typically fitted with edge seals to minimise leakage past the damper in the fully closed position – although it is not usually the practice in the UK to allow VAV systems to go to full shutoff during normal operation. An opposed-blade damper may occupy almost the whole cross-section of the control section, whilst single-blade and butterfly types are typically mounted in an extension of the inlet spigot which extends into or through the control section. The damper controls the volume flow rate of air with a throttling action as it closes off the cross-sectional area available for air flow through the unit. The closing action has the conventional throttling effect of either increasing the pressure drop across the damper at constant volume flow rate, or of decreasing volume flow rate at constant pressure drop. As is typical of damper action generally, the relationship between damper position and volume flow rate or pressure drop is non-linear. Damper position is measured, according to standard convention, in terms of degrees from either the fully open or fully closed positions. The damper typically moves through rather less than 90° – perhaps as little as 45° – between fully open and fully closed. In some designs of terminal unit the control section, complete with inlet spigot, may be withdrawn or detached from the main body of the casing.

Downstream of the control section the casing functions as an acoustic attenuator for noise generated by the damper unit. To achieve this it is typically lined with mineral wool, which is given a surface finish (bonded glassfibre scrim or perforated steel mesh is typically used) to provide resistance to erosion by the air stream at air velocities up to *c*.20 m s^{-1}. The airway within this attenuating section may additionally be profiled to reduce the propagation of damper-generated noise directly downstream.

The discharge from the downstream end of the casing is ducted to remote supply diffusers. VAV terminal units have traditionally been installed in ceiling voids, and mounting lugs for threaded rod or wire hangers are provided near each corner of the casing. Terminal units have increasingly also been installed in floor voids. In ceiling void installations (Fig. 7.3) the downstream end of the terminal unit is frequently fitted with a purpose-

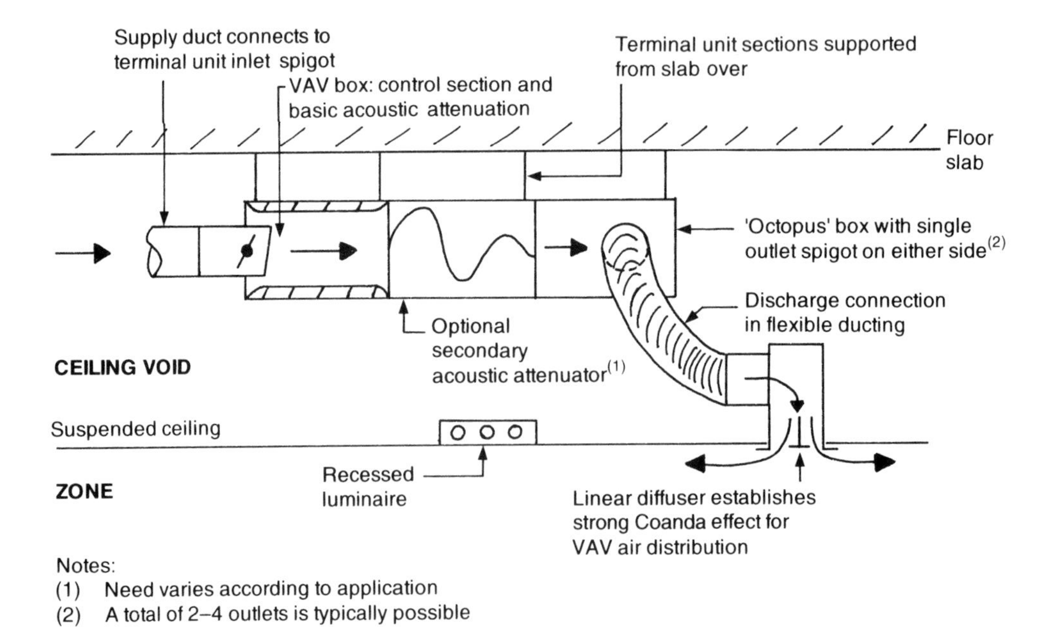

Fig. 7.3 Ceiling void installation for VAV box.

designed multiple-outlet header box or plenum (commonly termed an 'octopus box'). This typically has between two and four stub outlet ducts of circular cross-section, and may be acoustically lined if the application has a low design noise rating. The supply from each outlet can conveniently be ducted to a different supply air diffuser using small-diameter spirally-wound ductwork or flexible ducts if the diffusers are close to the terminal unit. Unwanted discharge outlets are simply capped off if not required. Some designs of 'octopus box' make provision for manual balancing dampers to be fitted in the outlet spigots. Where terminal units are mounted in floor voids (Fig. 7.4), the void itself may be sealed and used as a supply air plenum, removing the need for direct ducted connections to the floor-mounted supply diffusers.

Manufacturers typically offer additional (or secondary) acoustic attenuators that can simply be bolted to the basic terminal unit, if required, using mild steel angle or proprietary flanged cross-joints. Most fundamental of the accessories to the basic cooling-only terminal unit offered by most manufacturers, LPHW heater batteries of usually between one and four rows of aluminium-finned copper tubes can be bolted to the downstream end of the unit.

The damper actuator and terminal unit controller are mounted on the side of the unit casing. The extended damper spindle is typically driven by an electric motor through suitable gearing, although pneumatic actuation remains possible, and continues to be offered by some manufacturers.

Achieving the characteristic of pressure independence requires the sensing of volume flow rate at the terminal unit. Figure 7.5 shows a generic approach to this which is widely used in the UK, in which the inlet spigot to the terminal unit contains an *averaging flow grid* for sensing either air velocity or an enhanced velocity pressure. The grid is sometimes termed a *flow-cross*, although a variety of designs differing markedly from a cruciform shape are used by various terminal unit manufacturers. As shown in Fig. 7.5, the averaging grid consists of two sets of small-diameter aluminium tubes which radiate out across the inlet spigot from a plastic hub on its centreline. These aluminium tubes are sealed off at the inside surface of the inlet spigot. The wall of each arm of the upstream cross is perforated in three places facing directly into the oncoming airstream (parallel to the centreline of the inlet spigot). These holes sense the total pressure of the airstream in a similar manner to a pitot tube. Each arm of the cross is open at its inner end to a small chamber within the hub. As is the general case for the flow of an incompressible fluid in a closed conduit running full, the total pressure is not constant across the inlet spigot. The pressure in the hub chamber is the average of the total pressure sensed at each of the 12 holes in the arms of the upstream cross. Depending on the actual configuration of the averaging flow grid, the number of sampling points for total pressure is typically between 6

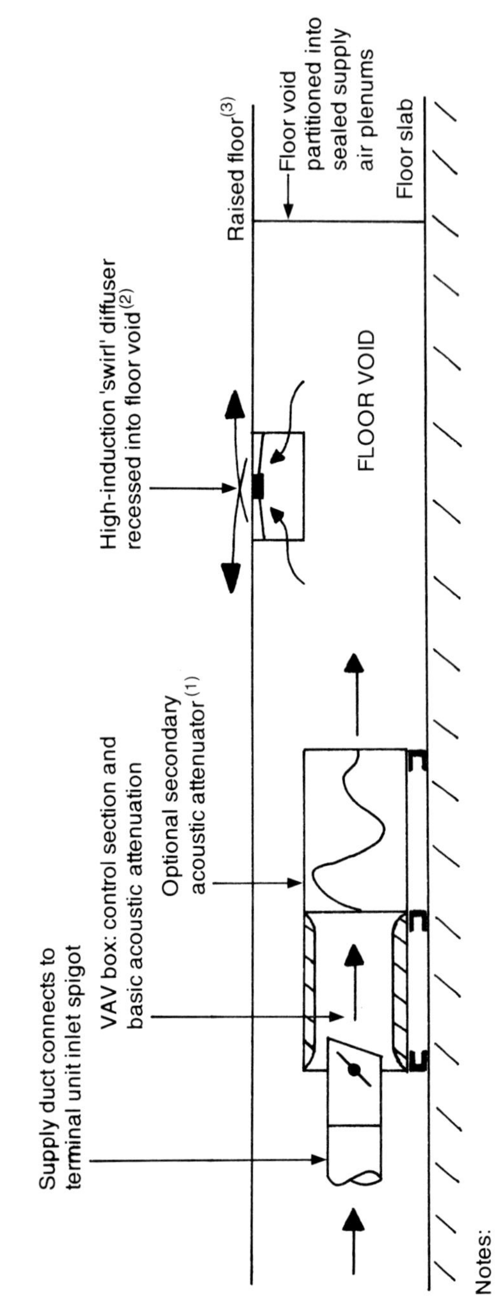

Notes:

(1) Need varies according to application.
(2) Requires provision of dirt collector within diffuser body, and regular cleaning as part of routine maintenance.
(3) Where the floor void is used as a supply air plenum, the floor design must both avoid lifting of tiles under fluctuating plenum pressure and minimize air leakage between adjacent tiles.

Fig. 7.4 Floor void installation for VAV box.

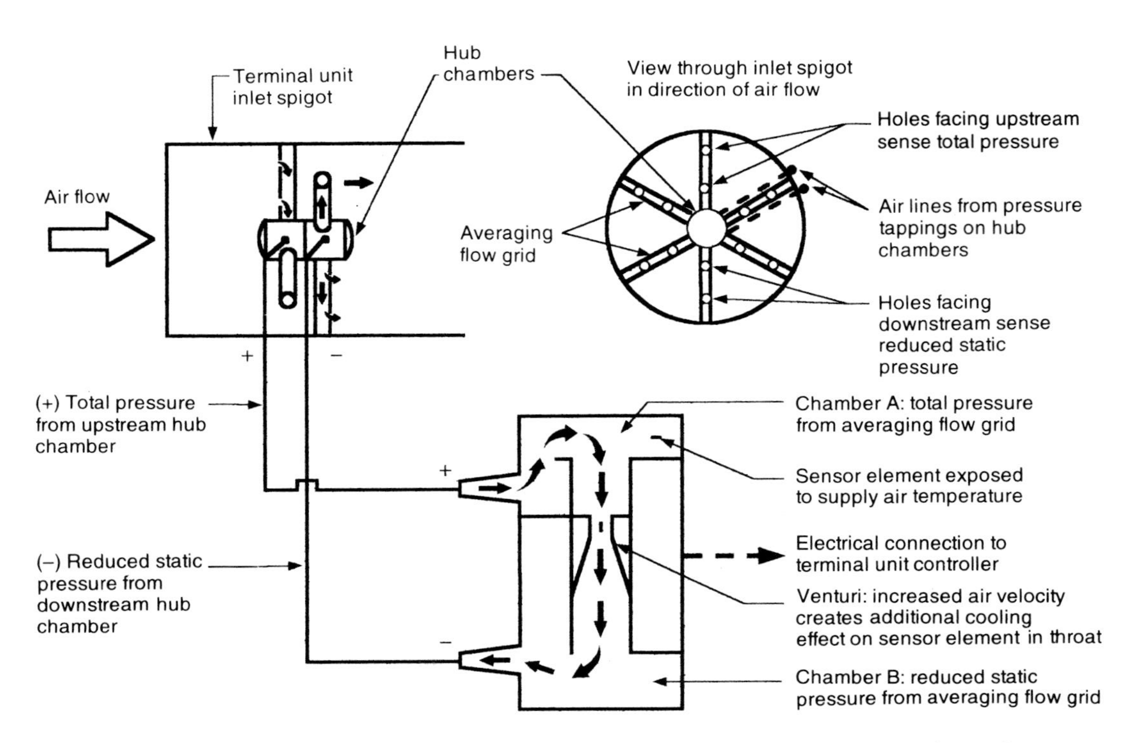

Fig. 7.5 Dynamic differential flow sensing for VAV box with variable-constant-volume ('pressure-independent') control action.

and 12. The wall of each radial arm of the downstream cross also has perforations. Identical in size to those in the upstream cross, and disposed the same distances from the hub along each radial arm, these face directly downstream (again parallel to the centreline of the inlet spigot). Situated in the turbulent wake of the arms of the cross, these holes sense a reduced static pressure. The arms of the downstream cross are again open to another small chamber in the central hub, and the pressure in this is again the average of that sensed at each hole in the downstream cross. In some designs the arms of the upstream and downstream crosses are staggered. Both hub chambers carry a pressure tapping. Flexible pneumatic tubing connects each to the flow sensing unit.

The flow sensing unit generally takes the form of a small plastic box mounted locally to the flow grid on the terminal unit casing. Inside (Fig. 7.5), a typical arrangement divides the unit into two chambers. These are connected by a small channel, moulded integrally with the box, which contracts abruptly to a narrow throat section and subsequently expands gradually back to the channel width. The pneumatic tubing from the upstream total pressure tapping on the hub of the averaging grid is connected to the chamber (A) at the upstream end of the sensing unit, and that from the tapping sensing the downstream reduced static pressure is connected to chamber (B) at the unit's downstream end. The difference in pressure between the two chambers in the hub of the averaging flow grid causes air to flow from the upstream to the downstream chamber via the pneumatic tubing and the inter-chamber venturi in the flow sensing unit – before rejoining the airstream in the terminal unit inlet spigot.

In the example considered the flow sensing unit contains two small copper sensor elements. One (reference) element is positioned in the upstream chamber (A), where it is exposed to the supply air temperature in the inlet spigot of the terminal unit. The second element is located in the throat section connecting chambers A and B, and a current is applied to maintain it at a constant temperature difference above the reference element, in the presence of the air flow through the throat. Measurement is carried out by electronic circuitry. A heat balance will exist between the heat generated by the element in the throat and the air flowing, via the throat, from the upstream to the downstream hubs of the averaging flow grid. If this volume flow rate increases, with heat generation remaining constant, the temperature of the element will fall, and with it the temperature differential relative to the reference element (and vice versa). This is restored by an increase in current and heat generation within the throat element. The current required to maintain the temperature differential between the elements is therefore a function of the pressure differential between the hubs of the averaging flow grid, and hence of the volume flow rate of air in the terminal unit inlet spigot – the higher the flow rate, the higher the required current. Maintaining a temperature difference between the elements allows sensing of the volume flow rate to be independent of the actual value of the supply air temperature. It is also claimed to make the technique relatively resistant to contamination by dust or particulate matter in the supply airstream, since both elements may be expected to be affected to roughly the same degree. While this should not be a significant factor in comfort air-conditioning systems, it may be so in an industrial VAV application.

Although conceptually similar, the averaging flow grid differs from a conventional pitot-static traverse (assuming the same measuring points) in important respects. A pitot-static traverse yields simple velocity pressures. Air velocities are determined from the individual velocity pressures at each traverse point, which are then averaged to give a mean air velocity over the duct cross-sectional area. The averaging grid yields the difference between the two average pressures in the hubs of the grid. The first is the mean of the total pressures sensed by the front grid, the second the mean of the *reduced* static pressures sensed by the rear grid. This difference is several times greater than the average velocity pressure. One manufacturer quotes a factor of between 2.8 and 3 times greater, according to the diameter of the terminal unit inlet spigot. This is an advantage at low volume flow rates, where velocity pressures are very low due to their proportionality to volume flow rate squared. As may be expected, the relationship between volume flow rate in the terminal unit inlet spigot and the *enhanced or amplified* velocity pressure signal from the averaging flow grid is non-linear. The output signal provided by the flow sensing unit to the terminal unit controller must be linearised to the volume flow rate in the inlet spigot to give the basis of the unit's pressure independence characteristic.

This approach to sensing of volume flow rate at the terminal unit is commonly termed *dynamic differential pressure sensing*. It can give good accuracy of flow sensing over a wide range of volume flow rate, and the averaging flow grid makes it at least relatively independent of the characteristics of the flow profile in the inlet spigot of any particular terminal unit. On the negative side there is inevitably some penalty to fan total pressure resulting from the pressure loss caused by the averaging flow grid itself, although this should typically be minimal.

Nevertheless, alternative techniques of sensing volume flow rate are possible. First among these alternatives is the technique of *static differential pressure sensing*. This retains the averaging flow grid, but bases the flow sensing unit instead on a static pressure cell that senses the magnitude of the pressure signal from the flow grid by its effect on a flexible diaphragm. Whilst in theory this offers a more accurate technique, it is typically a more expensive option due to the higher cost of the static pressure cell itself, and one that is rather more susceptible to the effects of cell misalignment or careless handling of the sensing unit. However, it may have an advantage in an industrial VAV application where the effects of a dusty or polluted airstream would be detrimental to other sensing techniques.

A further technique deletes the downstream cross of the averaging grid and substitutes a *velocity sensor* in the remaining hub chamber (Fig. 7.6). The openings in the upstream cross channel a characteristic flow of air across a sensor which is typically based on the 'hot-wire' principle, prior to discharging it back into the airstream in the terminal unit inlet spigot. Although not

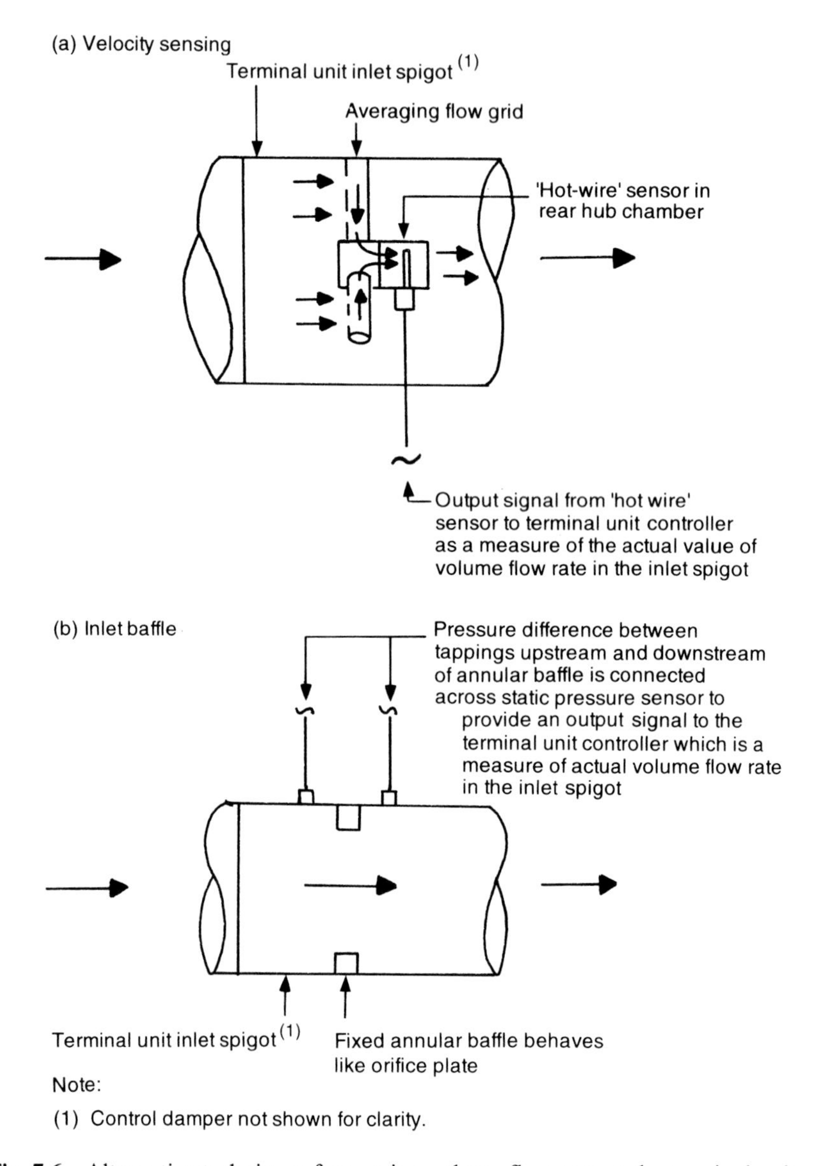

Fig. 7.6 Alternative techniques for sensing volume flow rate at the terminal unit.

necessarily the mean axial velocity, the velocity of this flow is certainly a function of the volume flow rate in the inlet spigot, and provides a sensor output that can be calibrated and linearised against this. A hot-wire sensor can also be used to measure air velocity at a single point without an averaging flow grid. In their favour, hot-wire velocity sensors can provide good accuracy over a wide range of volume flow rate, but are more susceptible to erosion and contamination by any particulate matter in the airstream. The single-point configuration without averaging grid gives the lowest pressure loss, but its accuracy is reduced for disturbed or non-ideal flow profiles in the inlet spigot.

Sensing of volume flow rate is also possible using the pressure signal across a *baffle* in the inlet spigot as a measure of volume flow rate (Fig. 7.6). The baffle effectively performs as a type of orifice plate. The 'orifice' size is made close to the internal diameter of the inlet spigot itself, in order to minimise the penalty to fan total pressure associated with the device. This would be significant if a conventional orifice plate were used in the inlet spigot. The resulting diameter ratio of orifice to conduit is therefore greater for the baffle than the maximum value of approximately 0.75 that would typically be allowed by international standards for an orifice plate used as a primary flow rate measuring device. The baffle technique can therefore not be expected to achieve the accuracy of volume flow rate measurement typically associated with conventional orifice plates.

The electrical power requirements of the terminal unit are typically met by the provision of a single 24 VAC supply, both the terminal unit controller and damper motor deriving their needs from this. Additionally the controller typically accepts a single control input in the range 0–10 VDC (0–20 V phasecut is widely offered as an alternative) from a zone temperature controller. Individual VAV boxes are available to handle zone design supply air volume flow rates ranging from less than $150\,\mathrm{l\,s^{-1}}$ to more than $1500\,\mathrm{l\,s^{-1}}$. Associating a supply air temperature differential of 8 K with these volume flow rates gives a range of zone design sensible cooling load from less than 1.5 to 15 kW. Unit dimensions vary considerably according to the particular accessory 'fit' chosen. However, for the range of duty quoted, inlet spigot diameter may reasonably be expected to vary from less than 150 mm to approximately 400 mm. Similarly, across this duty range the dimensions of a basic cooling-only unit typical of designs used in the UK may be expected to vary between approximately 300 W × 250 H × 1200 L (respectively casing width, height and length over the inlet spigot, in mm) and approximately 1000 W × 450 H × 2200 L. A two-row heater battery might add less than 100 mm in length, and an 'octopus box' of the order of 500 mm. If required, a secondary attenuator might add a further 900 mm (or more or less, in increments of 300 mm, according to need).

7.4.2 Operation and control

Figure 7.7 shows basic cooling-only operation of a 'pressure-independent' single-duct throttling VAV terminal unit of the 'VAV box' type shown in Fig. 7.2, when employed in a comfort conditioning application. A generalised schematic of the control arrangement is also shown. The terminal unit controller positions the volume flow rate control damper in response to the control signal from the zone temperature controller. Maximum and minimum volume flow rates are typically specified by the system designer for each terminal unit. To prevent the risk of downstream terminal units being starved of air under extreme conditions (or, for example, if blinds are left open in a

(a) Generalised schematic for 'pressure-independent' control

Zone temperature sensor: may be integral with zone controller

VAV terminal unit

Supply of conditioned air from central AHU

Velocity sensor or differential pressure sensor in inlet spigot

Terminal unit controller

Zone air temperature controller

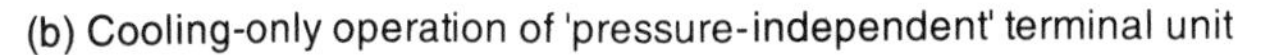

(b) Cooling-only operation of 'pressure-independent' terminal unit

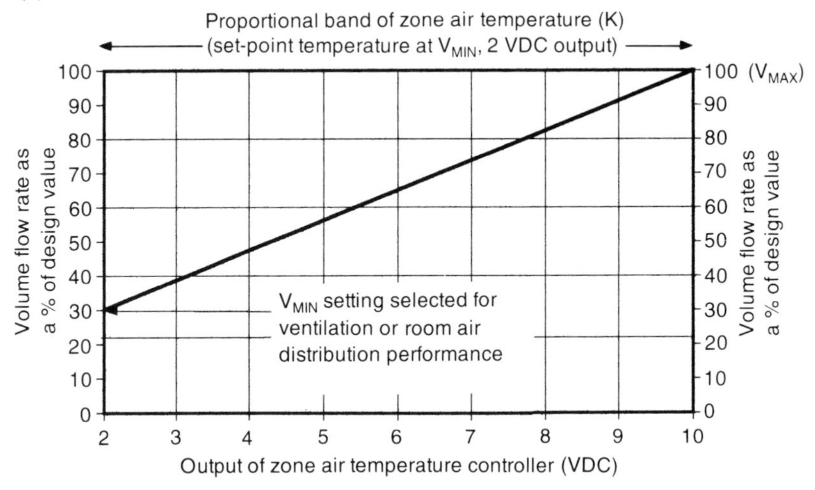

Fig. 7.7 Operation and control schematic for variable-constant-volume ('pressure-independent') single-duct throttling terminal unit: cooling-only operation.

zone) the high limit of volume flow rate is the design flow rate to the zone. The low limit is based either on considerations of ventilation or room air distribution (or a combination of both). Conventionally both limits are preset into the terminal unit controller by the manufacturer at works. However, should the need arise, they may subsequently be reset on site, subject only to the ultimate high limit of the manufacturer's maximum rating of the unit ($\dot{V}_{NOM}$, the upper limit of calibration of its flow sensor) and any minimum required differential between $\dot{V}_{MAX}$ to $\dot{V}_{MIN}$.

Within these high and low limits the volume flow rate of conditioned air supplied to the zone is proportional to zone cooling load and bears a linear relationship to the output of the zone controller. This is a function of the error between the set-point air temperature for the zone and a value sensed at a location chosen to be representative of the zone as a whole, or at least of the occupied region of the space. With analogue electronic terminal unit and zone controls, an input range of 2–10 VDC from the zone controller corresponds to the range of volume flow rate from $\dot{V}_{MAX}$ to $\dot{V}_{MIN}$. Hence the relationship between the volume flow rate set-point of the terminal unit controller ($\dot{V}_{SPV}$) and the output of the zone temperature controller is:

$$\dot{V}_{SPV} = \dot{V}_{MIN} + [(\dot{V}_{MAX} - \dot{V}_{MIN})\,(\textit{Zone controller output} - 2)/8] \quad (7.1)$$

Volume flow rates may be in $l\,s^{-1}$, $m^3\,s^{-1}$ or any other consistent units. Zone controller output is in VDC. The latter typically employs a control action that is either proportional or proportional-plus-integral (P + I). Simple proportional control is generally preferable for stability, ease of commissioning and economy of operation, although P + I may be necessary where the terminal unit performs both zone cooling and heating functions. Ease of commissioning is an important consideration, since any sizeable VAV installation will have a large number of zones, each with their own individual temperature controllers. Simple proportional control, with the set-point temperature of the conditioned space only achieved when the zone cooling load is ($\dot{V}_{MIN} \div \dot{V}_{MAX}$) × 100% of its design value, inherently allows an upward drift in zone temperature across the proportional band as load increases to its design value. This will result in the consumption of less energy for cooling than the P + I case, where an approximately constant zone temperature set-point is maintained. In comfort terms this upward drift in zone temperature is probably acceptable, if the proportional band stretches from approximately 2–4 K above a nominal set-point value of 21–22°C, since it will typically be accompanied by increasing outside air temperature. Under these conditions the differential between zone and outside air temperature may remain more-or-less constant. The higher the outside air temperature rises, the more acceptable a small increase in zone temperature becomes. The relationship between the output of a proportional controller and operation of a VAV box is shown in Fig. 7.7.

Where DDC systems are employed, a range of volume flow rate set-point from 0 to $\dot{V}_{MAX}$ corresponds to a continuous analogue output of 0–10 VDC from the zone controller, with the $\dot{V}_{MIN}$ low-limit and any supplementary override functions being performed within the software control algorithms of the zone controller itself.

The control signal from the zone temperature controller does not directly position the volume flow rate control damper in the VAV terminal unit. It must first be *conditioned* within the terminal unit controller to an *internal* set-point value which can be directly compared with the calibrated and linearised signal derived by dynamic differential pressure sensing using the averaging flow grid in the terminal unit inlet spigot (or one of the alternative techniques discussed in Section 7.3.1). This signal is also representative of a volume flow rate – in this case the instantaneous actual value of volume flow rate to the zone. Hence it is commonly referred to as the 'actual-value' signal. Conditioning of the external set-point signal from the zone temperature controller is necessary to ensure that both this and the calibrated and linearised output from the terminal unit's flow sensor have the same value at identical flow rates, for zero error and a stationary damper under these conditions. Without conditioning, the two signals cannot be directly compared, since each has a different span of volume flow rate corresponding to 2–10 VDC. With analogue electronic controllers calibration and linearisation of the output signal from the terminal unit's flow sensor yields an output varying linearly from 2-10 VDC across the full span of volume flow rate from zero flow to the manufacturer's maximum rating of the terminal unit, $\dot{V}_{NOM}$. This relationship may be expressed as:

$$\textit{Actual value output} = 2 + 8\,(\textit{Actual volume flow rate}/\dot{V}_{NOM}) \qquad (7.2)$$

Where DDC controls are employed the calibration and linearisation process is typically carried out from 0 to 10 VDC.

In any event the conditioned input from the zone controller and the calibrated and linearised output from the terminal unit's flow sensor provide the set-point and measured values required for feedback control of the volume flow rate to the zone. The deviation between them provides the error signal that is used to produce a suitable proportional control output to drive the damper motor and position the damper in a corrective sense. The speed of the damper motor may be a function of this error, to achieve an integral action that removes any offset between the flow rate demanded by the zone controller and that measured in the inlet spigot to the terminal unit.

With DDC zone controllers the damper motor may be synchronous and pulsed with a DC signal output from the controller itself. Two-speed operation of damper motors is possible, turning down from full speed to half speed when the volume flow rate error lies below a threshold value. This both lessens overshoot of the controlled variable and enables the use of longer sampling intervals. Pneumatic controls are still an option for VAV terminal units, and here damper angle is a function of the controller's error signal.

Any change in flow rate in the inlet spigot caused by system pressure fluctuations will be sensed by the terminal unit controller, and automatically compared with the set-point value demanded by the zone controller. Since any error will again generate a damper movement in the corrective sense, the terminal unit controller also acts to maintain a constant volume flow rate until a demand to change (reset) is received from the zone controller, in response to a change in the load on the zone. This gives rise to the correct description of the 'pressure-independent' type of terminal unit as a variable-constant-volume type. A general block diagram for this type of control is shown in Fig. 7.8. Where a velocity sensor is used in the terminal unit inlet spigot, the zone temperature controller resets the air velocity set-point of the terminal unit controller, and the variable-constant-volume/pressure-independent characteristic may also simply be referred to as 'velocity reset'.

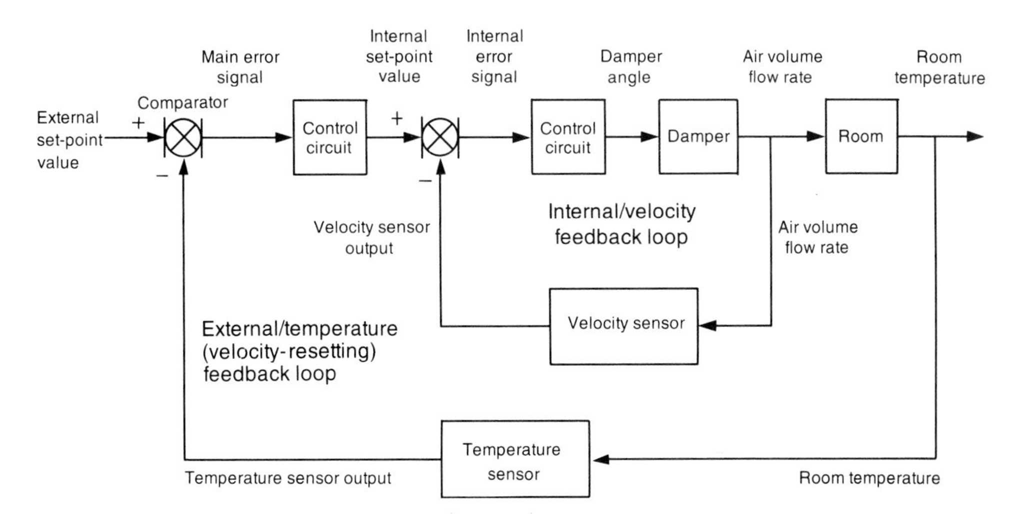

Fig. 7.8 Block diagram for variable-constant-volume ('pressure-independent') control action.

Each terminal unit and zone controller combination may function independently, or all may be linked into a centralised control system using *direct digital control* (DDC) techniques. Several terminal unit controllers may be slaved to a single master unit, in which case only the master unit receives a direct input from a zone air temperature controller. The 'actual-value' signal derived from the master terminal unit controller's sensing of volume flow rate is passed to the slave unit(s) as their control signal (Fig. 7.9). This may be of particular benefit where DDC zone temperature controllers are used, since the number of continuous 0–10 VDC outputs is typically limited, their

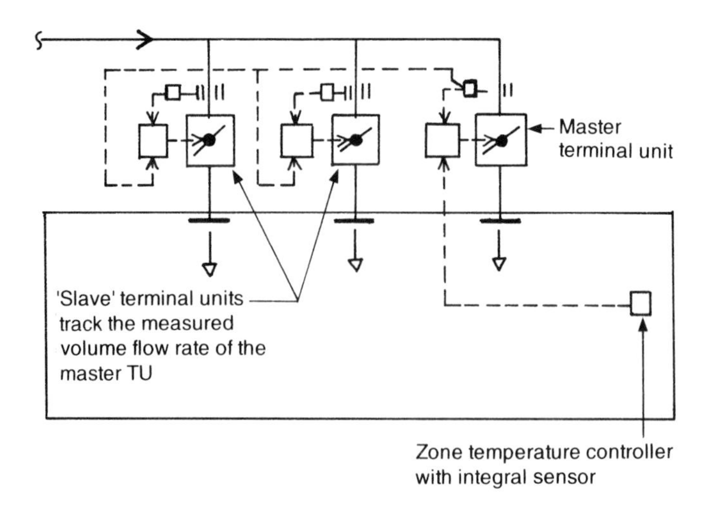

Fig. 7.9 Master–slave control of VAV boxes.

provision being considerably more expensive than straightforward digital outputs.

With DDC control of terminal units two arrangements are possible. In the first (Fig. 7.10a) the zone DDC controller receives input from the zone temperature sensor, and provides a continuous 0–10 VDC analogue output to the terminal unit controller, which conditions this input, derives the appropriate error and generates a suitable motor control signal to drive the damper. The 'actual-value' calibrated output signal is fed back to the zone controller. The latter may condition this feedback signal for control of other, 'slave' terminal units or process it for other system control functions (e.g. a 'low-flow' alarm for feedback control of the system supply fan). In the second option (Fig. 7.10b) the zone DDC controller contains the actual control algorithms for driving the terminal unit damper motor using a three-point (open/close) output to the terminal unit controller. In this case, the latter retains only the flow sensing and actuation functions. The 'actual-value' output signal from the terminal unit controller is again available for feedback to the zone controller, or as a control input to 'slave' terminal units.

Some care must be taken in both specifying and interpreting the volume flow rate limits of a VAV terminal unit. The maximum limit ($\dot{V}_{MAX}$) is invariably specified as a percentage of the manufacturer's nominal maximum volume flow rating for the unit ($\dot{V}_{NOM}$). However, the minimum limit ($\dot{V}_{MIN}$) may be specified as a percentage of the maximum volume flow rate limit ($\dot{V}_{MAX}$, rather than $\dot{V}_{NOM}$). The 'actual-value' signal derived from the sensing of volume flow rate in the inlet spigot may also be provided as an output from the terminal unit controller, for use in checking the performance of the terminal unit during commissioning or troubleshooting procedures. It may

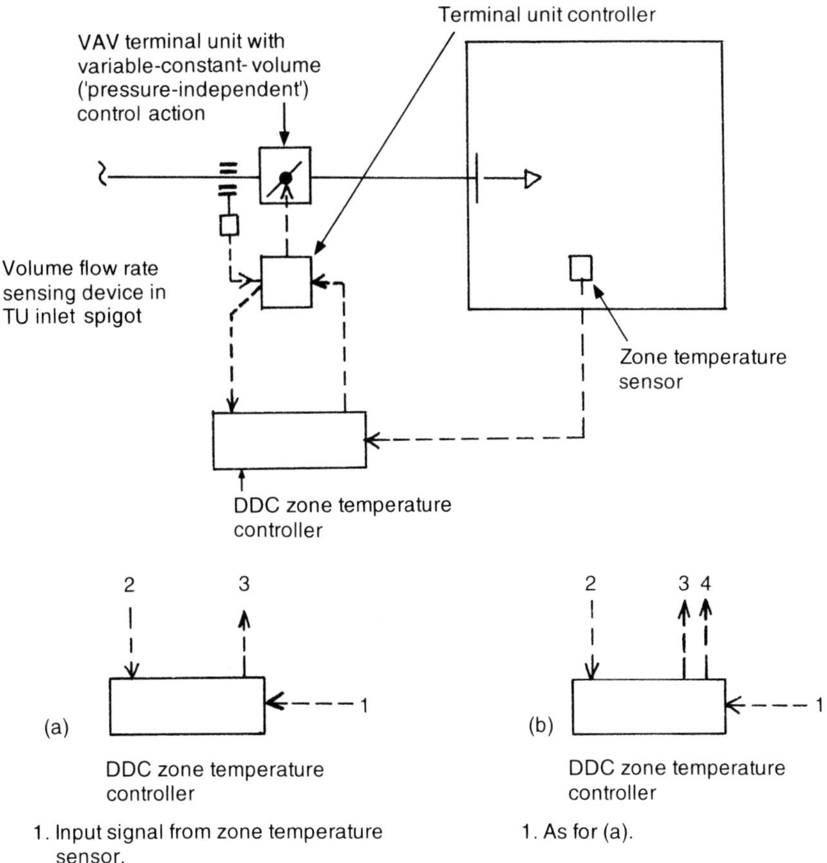

(a)

1. Input signal from zone temperature sensor.
2. 0–10 VDC measured volume flow rate feedback signal from terminal unit controller.
3. 0–10 VDC set-point volume flow rate output signal to terminal unit controller.

(b)

1. As for (a).
2. As for (a).
3. Damper opening control output to terminal unit controller.
4. Damper closing control output to terminal unit controller.

Note:

DDC zone controller applies zone sensor and feedback inputs to control algorithms which generate 'open'/'close' outputs to position damper. Functions of terminal unit controller are limited to volume flow rate sensing and damper operation.

Fig. 7.10 DDC control options for VAV boxes.

also be used to provide a feedback of measured zone supply air volume flow rate to a building management system (BMS) for control or monitoring purposes.

In conventional analogue electronic control systems the zone temperature controller is typically unobtrusive enough to be installed in the occupied space, using an integral air temperature sensor. According to the application and client preference on the subject, control set-points and overrides may either be preset and fixed or allow some degree of occupant adjustment. The present generation of DDC zone controllers are considerably bulkier than their analogue counterparts, and are thus typically located at the terminal unit itself in the ceiling (or floor) void. This conveniently puts these expensive items out of sight and mind of the occupants, removing any temptation towards inquisitiveness, interference or the venting of tempers. However, the remote temperature sensor which feeds the controller may still be provided with local set-point adjustment. Each zone controller should respond to its own dedicated zone temperature sensor. Averaging of sensors in adjacent zones should only be employed if there is a risk of interaction between the terminal units. In any event, the customary considerations regarding positioning of temperature sensors apply, as for any HVAC system. For example, position the zone controller out of direct sunlight, away from the direct influence of radiators, etc. used for perimeter heating, and away from sources of local casual heat gain. Alternatively the temperature of the air extracted from the zone may be sensed. The temperature of air extracted through light fittings should not be used, as this is subject to variable heat pick-up with the changing volume flow rate in a VAV system.

Where the terminal unit also provides individual space heating, this may be achieved with a zone controller having two 0–10 VDC outputs. As before, the cooling output is fed to the terminal unit controller to position the air-side damper, while the heating output is fed either to the control valve of the terminal unit's LPHW heater battery or to the control unit for its staged electric reheat, as appropriate. In any event, cooling and heating should always be sequenced, with an intervening 'dead band' or 'zero energy band' (Fig. 7.11).

Various override and fail-safe options (to fully open or fully closed) are typically available for terminal unit controllers. The nature of the control modes that may be available depends on the sophistication of the zone controller itself, the degree of central control that is exercised over the VAV system, and the extent of interlinking that is possible with other elements of the HVAC control system. However, typical possibilities include heating/cooling changeover, night setback for heating and cooling (down and up respectively), and compensation of the zone temperature set-point against outside air temperature during cooling operation. Where DDC controls are employed, zone controllers may be linked to a central supervisory PC,

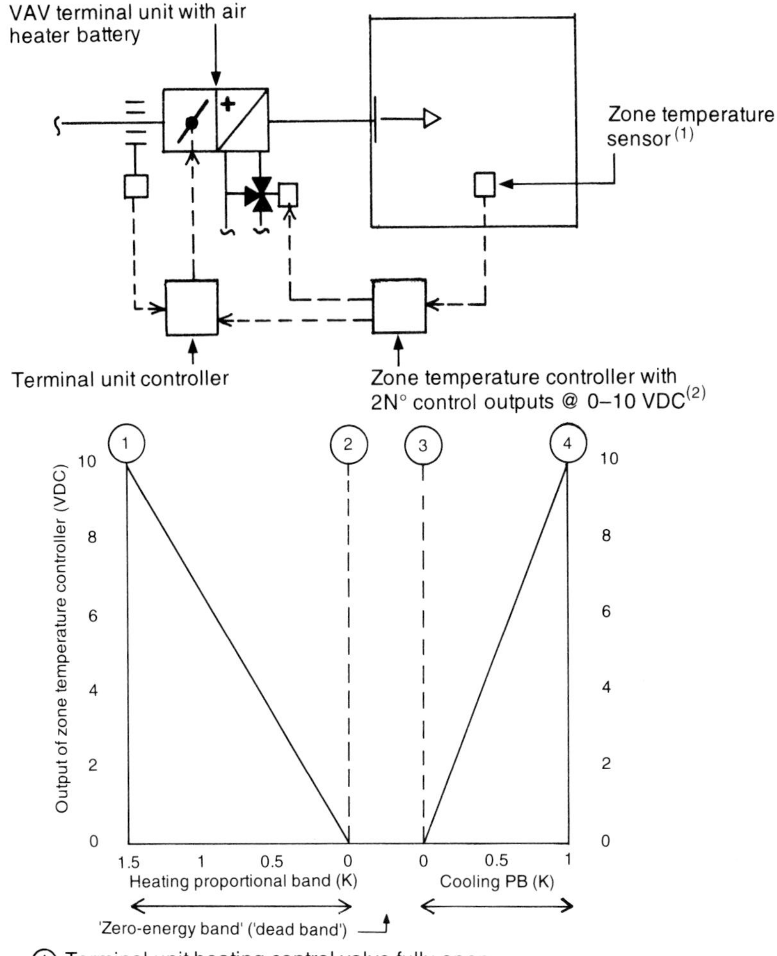

① Terminal unit heating control valve fully open.

①–② TV heating control valve progressively moves to fully closed as controller heating output decreased from 10 to 0 VDC(2).

①–③ Terminal unit cooling control damper modulates only to maintain specified minimum limit ($\dot{V}_{MIN}$) of volume flow rate to zone as set-point.

③–④ Terminal unit cooling control damper modulates to maintain set-point flow rate to zone which is reset between specified minimum ($\dot{V}_{MIN}$) and maximum ($\dot{V}_{MAX}$, design value) limits by increase in cooling output of zone temperature controller form 0 to 10 VDC(2).

④ Terminal unit cooling control damper modulates only to maintain specified maximum limit of volume flow rate to zone at maximum cooling output from zone temperature controller.

Notes:

(1) Zone temperature sensor and controller may be combined within a single unit located within the space.
(2) Operating range of output may differ (2–10 VDC, 1.5–9.5 VDC being examples).

Fig. 7.11 Operation and control schematic for pressure-independent single-duct throttling terminal unit: cooling and heating operation.

establishing a two-way communications channel for central monitoring and reset of terminal unit control and feedback parameters.

Direct digital control (DDC) has been mentioned several times in the foregoing discussion. The term DDC is synonymous with microprocessor-based control. A schematic of a basic DDC controller is shown in Fig. 7.12. Conventional analogue function blocks are replaced by instructions to the microprocessor which comprise one or more mathematical or logic statements in a suitable control language. When DDC control was introduced to HVAC systems the initial tendency was for it to be used only for principal items of central plant. The VAV terminal units retained their own analogue electronic or pneumatic control systems, which operated independently of the DDC system, but in parallel with it. Temperature sensors may have been positioned throughout the building, but these were rather to facilitate the monitoring and reporting of space conditions to the DDC system, as a basis for adjustment of the central plant items. While the 'central plant' approach remains popular, the DDC approach has extended to 'full DDC' systems, which integrate the control of both terminal units and central plant. Such a system may need to analyse several orders of magnitude more sensors than 'central-plant-only' systems, and a higher level of performance is required. In particular the level of communications offered between zone and supervisory

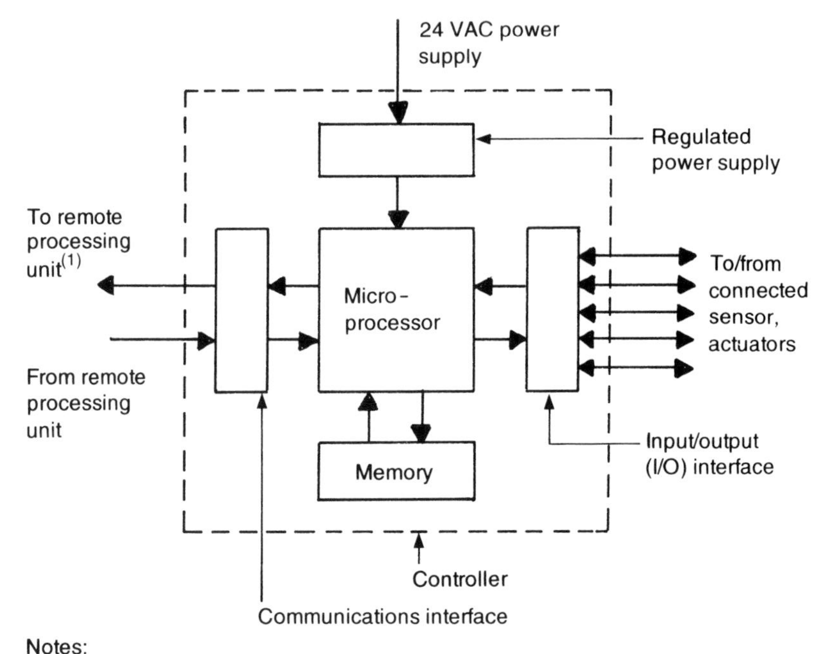

Fig. 7.12 Schematic of a basic DDC controller.

controllers puts a large burden on the system's communications network and data processing capacity.

More effort is also required on the part of the the HVAC designer if such systems are to be clearly specified, properly installed, and successfully commissioned. Unless the result goes beyond simply repackaging conventional control strategies in a new guise, the extra effort is unlikely to be worthwhile. However, the evidence so far suggests that integrated control has the potential to yield substantial improvements in both comfort and energy efficiency when applied to modern buildings.

An early approach to integrated control in a VAV environment is the terminal regulated air volume or TRAV concept[1,2,3]. This is summarised in Fig. 7.13. In this approach there are no set low and high limits of zone volume flow rate, not even in the form of software control algorithms. Below the space heating set-point the set-point of volume flow rate at the terminal unit is adjusted to maintain the minimum ventilation requirements of the zone, heating as necessary to maintain zone temperature within its design range. This requires information on the outside air fraction at the central AHU, which must of course be maintained as system volume flow rate increases to meet the ventilation-based demand of terminal units.

During system operation on full outside air the set-point of volume flow rate at the terminal unit is adjusted according to outside air temperature, upwards in warm weather and downwards in cool conditions. This provides an element of pre-cooling or pre-heating, in anticipation respectively of the arrival of warmer weather or cooler weather, which requires the use of an outside air temperature prediction algorithm.

As zone temperature rises above the cooling set-point, the set-point of volume flow rate at the terminal unit rises steeply over a narrow proportional band to the zone design value. If zone temperature continues to increase while this design flow rate is maintained, a high limit temperature is reached at which the terminal unit volume flow rate set-point is 'boosted' to ensure that the terminal unit damper has modulated to fully open. This ensures that the maximum flow rate of conditioned air is delivered to the zone, to try to limit any further rise in temperature.

In addition to their demands for a high-speed communications network and extensive data processing capability, such an approach to system control also raises questions about the nature of the programming languages used, and the extent to which system designers and operators should be involved in the actual design of the software control strategies themselves.

In any event, the ability of a DDC system to provide accurate control of rapidly changing or highly interactive systems is directly dependent on its *scan rate*. This is the rate at which the system sequentially reads input data from sensors, calculates the results, and then finally outputs the appropriate control signals. Scan rates may be from 1 to 0.05 Hz (once per second to once

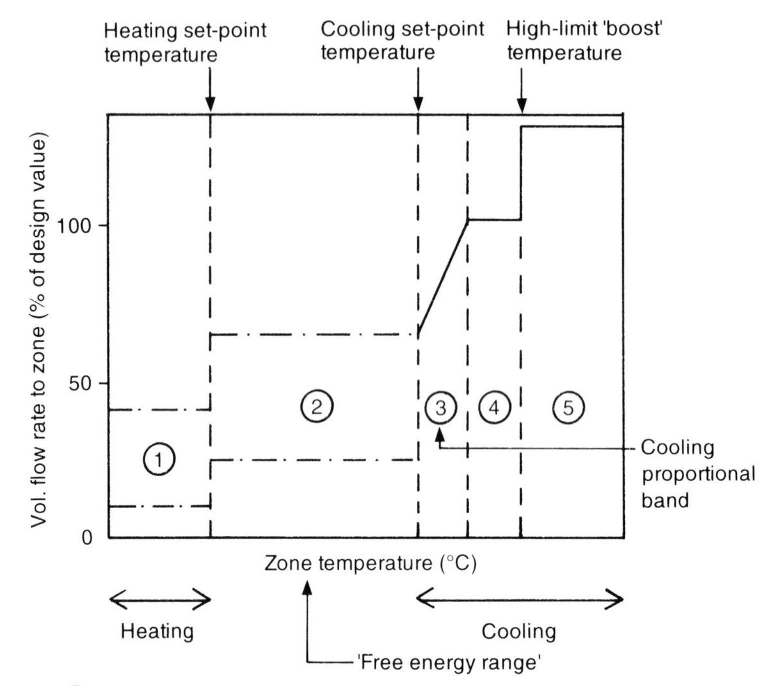

(1) TU volume flow rate set-pont is reset to maintain zone ventilation requirements.

(2) Central plant supplies full outside air. TU volume flow rate set-point is reset according to outside air temperature.

(3) TU volume flow rate set-point is reset upwards to the design flow rate for the zone as zone air temperature rises over a defined (narrow) proportional band.

(4) TU volume flow rate set-point remains at design value for zone; zone temperature 'floats' upwards if zone cooling load continues to rise above design value.

(5) TU volume flow rate set-point stepped up ('boosted') on zone temperature reaching a specified high-limit value. Aim is to ensure damper is fully open, and zone can receive maximum volume flow rate that system can provide, to limit further increase in zone temperature.

Fig. 7.13 Integrated direct digital control of VAV boxes: the TRAV approach.

in 20 seconds). Effective scan rates must take into account the rate at which supervisory elements of a system can interface with field devices, allowing for sensor time constants, actuator speeds, transducer response and distance–velocity lags in air and water flow systems. Unless DDC controls can respond faster than the system they are to control is changing, it is unlikely that they will be able to keep up and achieve adequate control. Indeed theoretical considerations typically associate increased delay times with a tendency towards system instability. Static pressures and zone volume flow rates in

VAV systems change relatively quickly, while system volume flow rate and fan speed are relatively slow to change.

In principle DDC control can allow the occupant of a space to open a window, provided that the $\dot{V}_{MIN}$ of the terminal unit serving the zone is reset to zero on activation of a window switch. The use of *wireless* zone sensors is a development of potentially great benefit where DDC terminal unit control is employed. A wireless zone sensor is an indoor spread-spectrum transmitter designed to send zone temperature and other status information to a local receiver, which may be located up to 300 m away. Hence receivers may conveniently be co-located with VAV terminal units. Spread-spectrum technology makes use of a narrow-band FM signal. This is mixed with a unique digital code which causes the FM signal to be repeated at many different frequencies at the same time. At the receiver, the transmitted signal is reconstructed to the original FM 'message' by combining it with a reference signal having the same digital code. Any interference received is combined with the reference signal, and is hence also spread out. This improves the reliability of radio frequency (RF) communications, reduces interference effects from other RF signals and existing electrical 'noise' in the building. On the other hand, signal modulation techniques do not cause interference because a spread-spectrum transmitted signal has random noise characteristics.

Wireless zone sensors are assigned to the terminal unit(s) in their area. Without any need for wiring, they allow maximum flexibility in positioning of zone sensors during commissioning, or following changes in partitioning arrangements (zoning) or occupancy. Repositioning can be carried out simply and without affecting the fabric and finishes. This is particularly relevant in speculative commercial developments, where re-zoning may be necessary to suit the requirements of potential tenants, and wherever minimal disturbance to fabric and finishes is required. A back-up sensor may be allocated to specified (critical) terminal units. A single zone temperature sensor may be assigned to more than one terminal unit or, alternatively, a single DDC terminal unit controller may receive signals from more than one wireless zone sensor.

7.5 Dual-duct VAV box

7.5.1 Design and construction

VAV boxes are available with dual inlet spigots and damper assemblies for connection to the hot and cold supply ducts in dual-duct VAV systems. The general layout of typical dual-duct terminal units is illustrated in Fig. 7.14. As might be expected, where the dual inlet spigots are both located on the

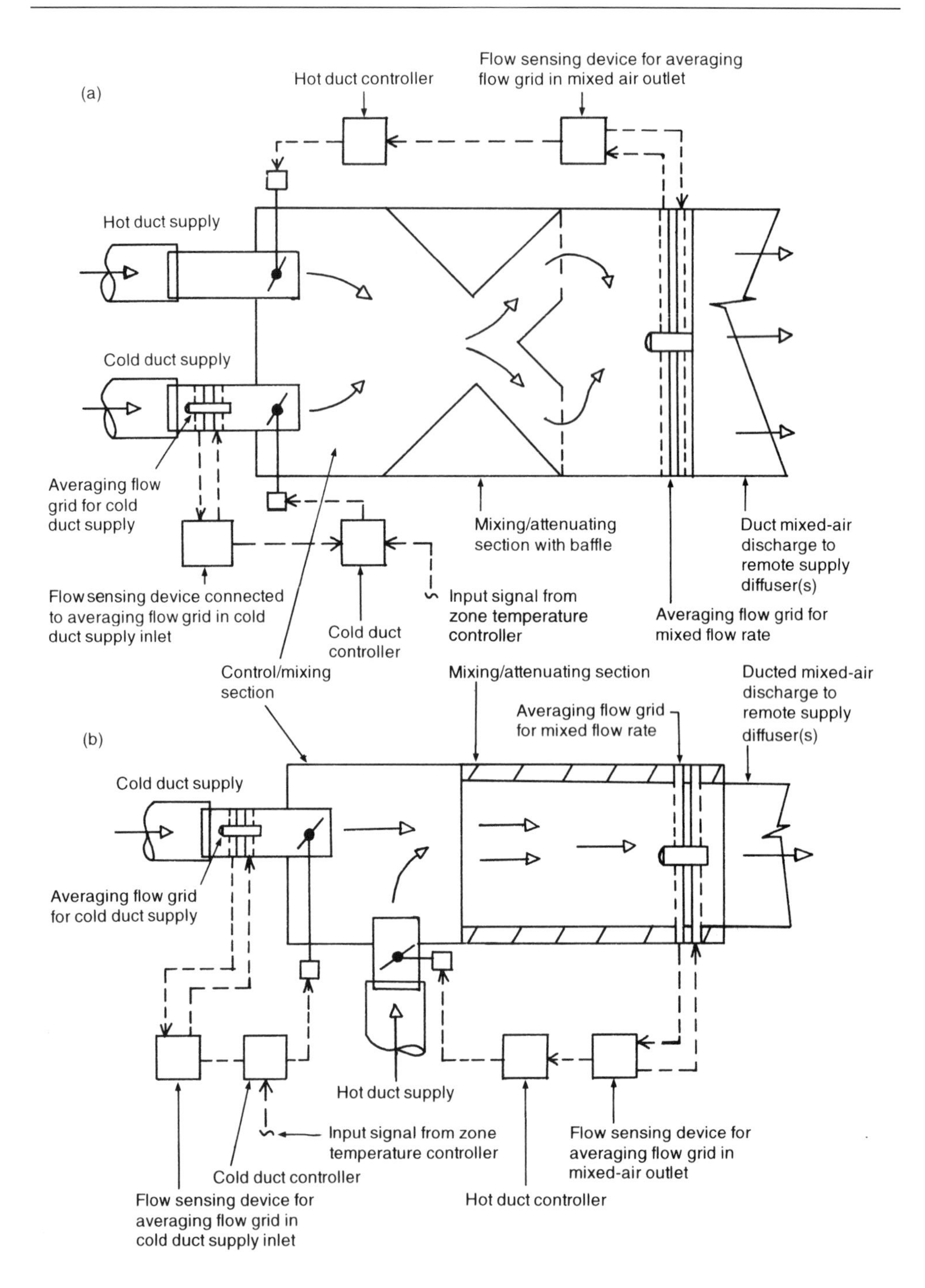

Fig. 7.14 Dual-duct VAV box with variable-constant-volume ('pressure-independent') action and remote supply diffuser(s).

front of the terminal unit, the casing dimensions are typically greater than for equivalent single-duct units. In particular their width may be of the order of twice that of an equivalent single-duct terminal unit. Where the dual inlet spigots are arranged 'front-and-side', the increase in width of the terminal unit is minimised.

Depending on the mode of control employed, both cold and hot duct inlets may be equipped with individual flow sensing devices. These are of the same basic types as employed in single-duct terminal units, and operate on the same principles. Alternatively flow may be sensed only in the cold duct inlet, in which case a further flow sensing device is located in the mixed air outlet from the terminal unit. In this case (Fig. 7.14) the attenuating section also performs the necessary role of cold/hot flow mixer.

7.5.2 Operation and control

Typically a minimum static pressure differential must be maintained across a dual-duct terminal unit to ensure that pressure differences between the hot and cold ducts cannot result in reverse flow between them. Two control options are possible, although the fundamental characteristics of one of these is such that it is unlikely to be encountered in UK design practice (although the same is not necessarily true of practice elsewhere in the world). The first is shown in Fig. 7.15, and requires flow sensing in the cold duct inlet spigot and in the mixed air discharge. The sensor in the mixed air discharge controls the position of the hot duct damper. Under design cooling conditions the air supplied to the zone is 100% cold air. This is the design volume flow rate supplied to the zone for cooling, and the $\dot{V}_{MAX}$ or high-limit setting of both the cold duct and the terminal unit as a whole. As the demand for cooling decreases, the zone temperature drops down through the controller's proportional band. By the time that it reduces to the set-point temperature of the zone, the cold duct volume flow rate has been reset down to the $\dot{V}_{MIN}$ or low-limit setting of the mixed supply to the zone. However, in contrast to the case for typical single-duct types, as zone temperature drops below set-point, the cold duct volume flow rate continues to be reset downwards until it reaches zero (with the cold air damper at shutoff). It is also the maximum hot duct volume flow rate. Since the flow sensor which controls the position of the hot duct damper actually senses the rate of mixed air discharge from the terminal unit, $\dot{V}_{MIN}$ is maintained by progressively opening the hot duct damper as the cold duct damper moves to full shut off.

Under design heating conditions the hot duct damper is at its maximum open position, and the flow rate to the zone is 100% from the hot duct. This is the design volume flow rate supplied to the zone for heating, and the $\dot{V}_{MIN}$ or low-limit setting for the terminal unit. Since the cold duct is fully shut off, the hot duct must supply an outside air fraction equal to that of the cold duct,

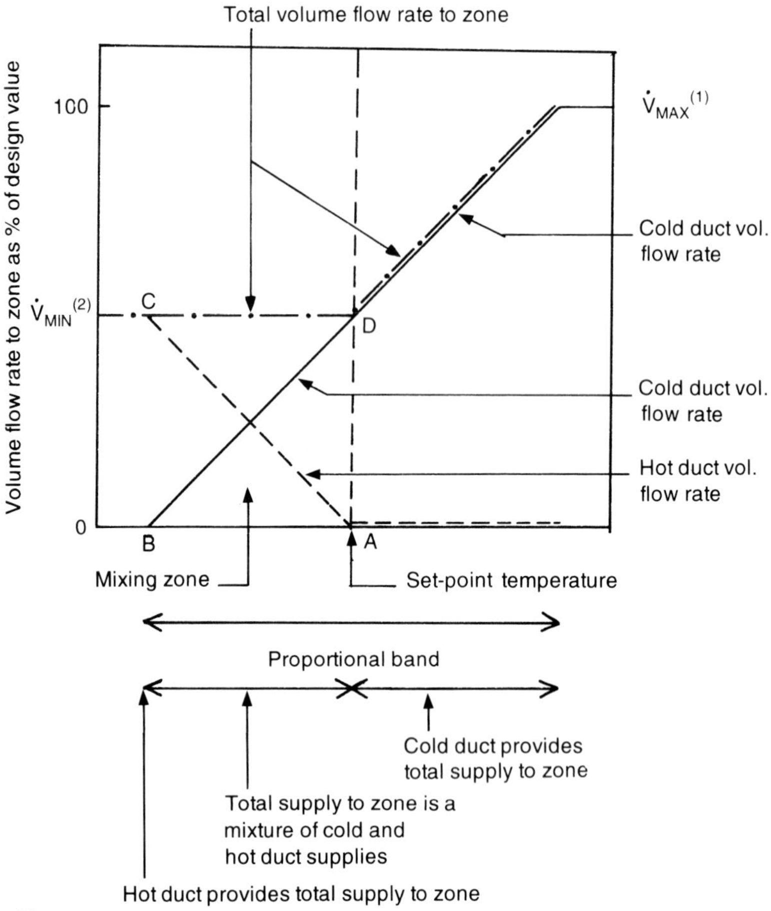

Notes:

(1) Of cold duct and terminal unit; equals design flow rate to zone for cooling. Cold duct $\dot{V}_{MIN}$ setting = 0.

(2) Of hot duct and terminal unit; equals design flow rate to zone for heating. Constant-volume operation of hot duct controller at $\dot{V}_{MIN}$ set-point in absence of any set-point input signal (zone temperature controller not linked to hot duct controller).

Fig. 7.15 Dual-duct VAV box – cold duct and mixed air control.

if zone ventilation is to be maintained at a minimum level. The required outside air fraction should be determined at $\dot{V}_{MIN}$. Since air velocity is inevitably lower in the mixed-air discharge from the terminal unit than in the inlet spigots, accuracy of flow sensing may again be problematic. The hot duct damper remains at full shutoff for all cold duct flows above $\dot{V}_{MIN}$, so that mixing of hot and cold air streams only occurs in the relatively small zone marked ABCD in Fig. 7.15, and total flow rate to the zone is not increased by this.

In order to avoid wasteful mixing under cooling conditions, which would also increase the supply volume flow rate to the zone, it is important that the hot duct damper should be of a low-leakage design that can achieve as near to a positive shutoff as is practical from an economic manufacturing perspective. The same is true of the cold duct damper, since any leakage through this at full shutoff will result in a shortfall in design heating performance of the terminal unit. Mixing of the cold and hot air streams initially cancels the remaining cooling potential of the supply air stream to the zone. Furthermore, from Fig. 7.15 it can be seen that, if $\dot{V}_{MIN}$ is less than 50% of $\dot{V}_{MAX}$, the part of the proportional band which generates heating output will be less than that which generates full cooling. As in the case of single-duct terminal units, the performance of the room air distribution design must be considered when operating at $\dot{V}_{MIN}$. In a variant of this control strategy the hot duct shares the $\dot{V}_{MIN}$ of the cold duct, but has a $\dot{V}_{MAX}$ setting intermediate between this and the design cold duct volume flow rate. The part of the proportional band which generates a heating output is increased, and room air distribution performance may be improved during heating operation.

The alternative mode of control for 'pressure-independent' dual-duct terminal units is illustrated in Fig. 7.16. This is based on flow sensing in both cold and hot duct inlet spigots of the dual-duct terminal unit. Both cold and hot ducts have maximum flow limits which correspond to the zone supply volume flow rate under design cooling and heating conditions respectively. While these $\dot{V}_{MAX}$ values may be identical, a heating $\dot{V}_{MAX}$ of around 40–50% of the cooling value would typically be anticipated. If the design heating loads required substantially more than this, alternative approaches to space heating may prove more appropriate. The minimum flow limit for both cold and hot ducts is zero, i.e. both dampers close to full shutoff. While the absence of any mixing between cold and hot duct supplies is the obvious attraction of this approach to dual-duct VAV control, it is also the source of the fundamental problem with this control approach, since this implies that a zero-flow condition will exist at the zone temperature set-point value or in any 'dead band' on either side of it. No ventilation air can reach the zone served through the dual-duct system under these conditions, or indeed over the proportional band on either side of this until an adequate supply volume flow rate is established to the zone (the magnitude of which depends on the outside air fraction of the cold and hot duct flows). Unless this situation can be tolerated, this approach to dual-duct terminal unit control can only realistically be employed where ventilation requirements are satisfied by alternative means. If, as is most likely, this involves an independent ducted system (which will certainly require central plant heating capability and, depending upon the application, may or may not incorporate central plant cooling and dehumidification capability), the capital cost, space require-

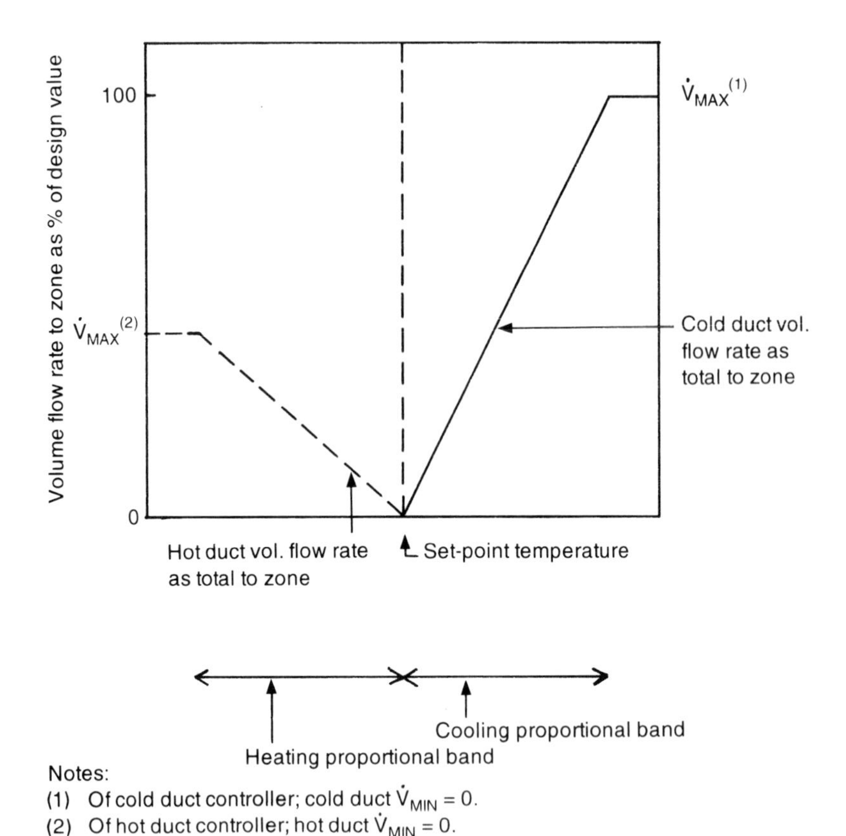

Fig. 7.16 Dual-duct VAV box – cold duct and hot duct control.

ments and complexity of the HVAC design may be so greatly increased, over and above the already costly and space-hungry dual-duct approach, as to render it impractical in any realistic application (at any rate those that are likely to be encountered in the UK!). Furthermore, in the absence of any supply to the zone, there is no control over humidity within the space.

Since in the dual-inlet-sensor approach the hot duct flow sensor only has feedback on what is going on in its own inlet spigot, the only alternative to this state of affairs would be to reintroduce some form of $\dot{V}_{MIN}$ limit on either or both supply ducts.

If the former approach (i.e. cold duct $\dot{V}_{MIN}$) follows a pattern of implementation typical for single-duct VAV boxes, it would again result in a mixing of cold and hot duct supplies throughout the heating operation, and would require an increased hot duct temperature to offset its effect under design heating conditions. Such a solution offers no benefit over cold duct and mixed air control.

The second option (i.e. cold and hot duct $\dot{V}_{MIN}$ limits) may be practical if a changeover sequence can be effected at a specific point on the 'zero-energy band' (mid-point or bottom end would appear to be the most obvious choices). At this point the cold duct damper must be disabled from its $\dot{V}_{MIN}$ operation and motored *to* full shutoff, while the hot duct damper is enabled *from* full shutoff with a $\dot{V}_{MIN}$ set-point. Again, sufficient outside air must be available from both hot and cold ducts if the system is to achieve its aim of maintaining minimum zone ventilation needs. DDC controllers are, of course, eminently suited to the implementation of such relatively complex terminal unit control strategies.

7.6 Fan-asssisted terminals: series-fan types

7.6.1 Design and construction

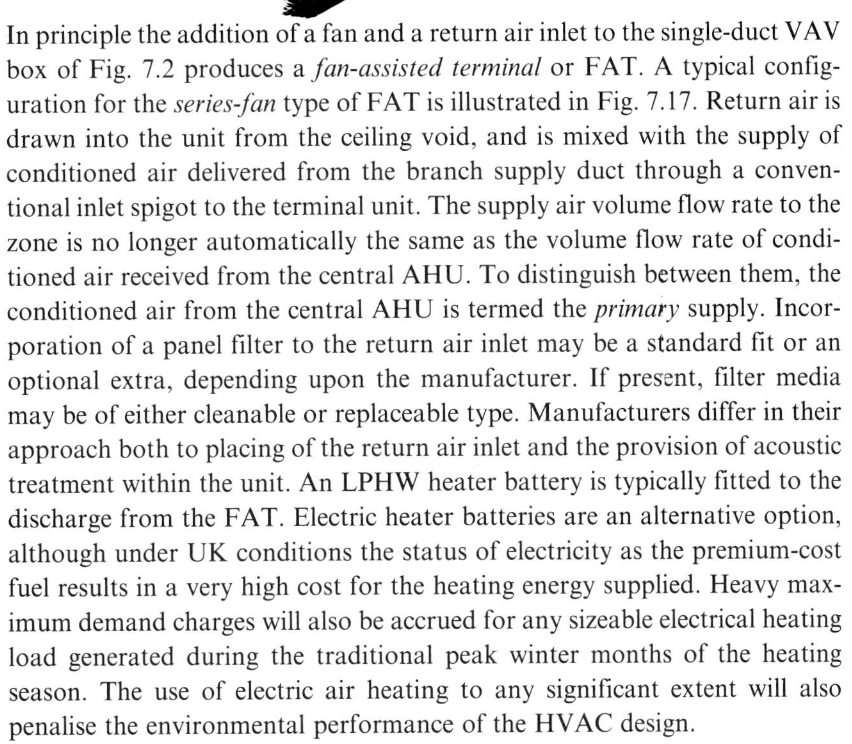

In principle the addition of a fan and a return air inlet to the single-duct VAV box of Fig. 7.2 produces a *fan-assisted terminal* or FAT. A typical configuration for the *series-fan* type of FAT is illustrated in Fig. 7.17. Return air is drawn into the unit from the ceiling void, and is mixed with the supply of conditioned air delivered from the branch supply duct through a conventional inlet spigot to the terminal unit. The supply air volume flow rate to the zone is no longer automatically the same as the volume flow rate of conditioned air received from the central AHU. To distinguish between them, the conditioned air from the central AHU is termed the *primary* supply. Incorporation of a panel filter to the return air inlet may be a standard fit or an optional extra, depending upon the manufacturer. If present, filter media may be of either cleanable or replaceable type. Manufacturers differ in their approach both to placing of the return air inlet and the provision of acoustic treatment within the unit. An LPHW heater battery is typically fitted to the discharge from the FAT. Electric heater batteries are an alternative option, although under UK conditions the status of electricity as the premium-cost fuel results in a very high cost for the heating energy supplied. Heavy maximum demand charges will also be accrued for any sizeable electrical heating load generated during the traditional peak winter months of the heating season. The use of electric air heating to any significant extent will also penalise the environmental performance of the HVAC design.

In the series-fan configuration the fan operates continuously, offsetting pressure losses downstream of the terminal unit, and credit for this may be taken when calculating the fan total pressure for the system (primary air) supply fan. The supply air volume flow rate to the zone is substantially constant, within a typical tolerance band of $\pm 7.5\%$ in volume flow rate to the zone over the full range of the primary air supply. All the air is handled by the fan. If the primary supply is at low temperature, its volume flow rate may

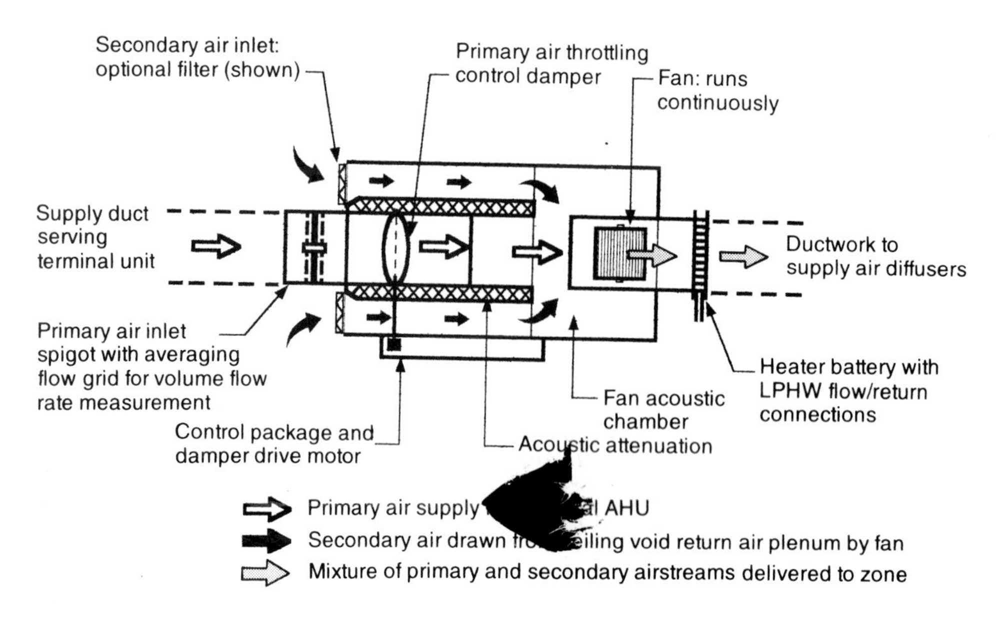

Fig. 7.17 Typical configuration for series-fan FATs.

be significantly less than that to the zone, even under design conditions. Fans are of the forward-curved centrifugal type, and are directly driven by single-phase induction motors of the permanent split capacitor type. These are typically provided with solid-state stepless speed control, based on varying the voltage suppled to the fan motor, although adjustment over a limited range of, say, 50–100% of full speed may be adequate. Fan speed may therefore be manually adjusted to match the fan duty to the pressure losses in the distribution ductwork downstream of the terminal unit. To prevent flow into the ceiling void in 'series' units the fan must be adjusted to match the maximum cooling airflow setting – to ensure that the primary air flow does not exceed the fan's capacity. Backdraft dampers may not be necessary if terminal unit fans are started before the primary supply fan in the central AHU.

Fan-assisted terminal units are also of the variable-constant-volume or 'pressure-independent' type, this characteristic being achieved in the same way as described for the basic throttling-only VAV box described in Section 7.4.

Series-fan FATs are suitable for use with a low-temperature primary air supply and conventional supply air diffusers. In this case thermally insulated shrouds (with a vapour barrier) may be provided to the inlet spigots of the terminal units, to avoid condensation which would otherwise occur on the exposed metal surface of the spigots.

FATs are available to handle zone design supply air volume flow rates between approximately 100 and 1000 l s^{-1}. Again associating a supply air temperature differential of 8 K with these volume flow rates gives a range of zone design sensible cooling load between approximately 1 and 10 kW. Casing dimensions are typically greater than for an equivalent simple throttling terminal unit, width needing to reflect the additional return air inlet, while depth and length must reflect the addition of the terminal unit fan.

7.6.2 Operation and control

The operation of a typical series-fan terminal unit is shown in Fig. 7.18. In principle, control of the primary air supply is no different from that for the purely throttling terminal unit. Again the terminal unit controller provides preset maximum and minimum volume flow rate limits for the primary air supply. For complete flexibility in providing both cooling and heating simultaneously in different zones, the zone temperature controller typically provides two continuous 0–10 VDC modulating outputs. One of these provides the control signal to the terminal unit controller for positioning of the primary air supply volume flow rate control damper. The second provides the control signal for positioning a three-port modulating control valve (a mixing valve in a diverting pipe circuit is typical) controlling the output of the LPHW heater battery at the discharge from the terminal unit. After the primary air volume flow rate reaches the minimum permitted for ventilation, the zone air temperature is allowed to float down over a 'dead band' or 'zero energy band' before the three-port valve on the heater battery starts to open.

In the event that an electric heater battery is used, this must be automatically disconnected in the event of a loss of airflow. This may be achieved either by means of a low-pressure switch with a pressure tube connected to a tapping on the terminal unit fan, or alternatively by using a bimetallic cutout near the element itself. The risk of primary air spilling back into the ceiling void at system start-up may be prevented by a direct interlock that starts the terminal unit fans before the main supply fan in the central AHU. Alternatively, an indirect interlock may be achieved using either a differential pressure switch mounted in the primary supply airstream or an anti-backward rotation device on the fan. Without such a device, backward rotation of the fan will occur if the primary airstream moves through unpowered forward-curved fan blades, which can build up an inertia that may be difficult for the fan motor to overcome. Since at design stage terminal units will never be selected such that their normal fixed running speed is the maximum setting of the (manually set) speed controller, a 'hard start' facility may be provided to apply full power to the motor for several seconds on start-up, generating an initial high torque to overcome any backspin. This may be achieved by shorting across the motor's voltage controller to apply maximum voltage for a timed delay, following which motor voltage returns to its

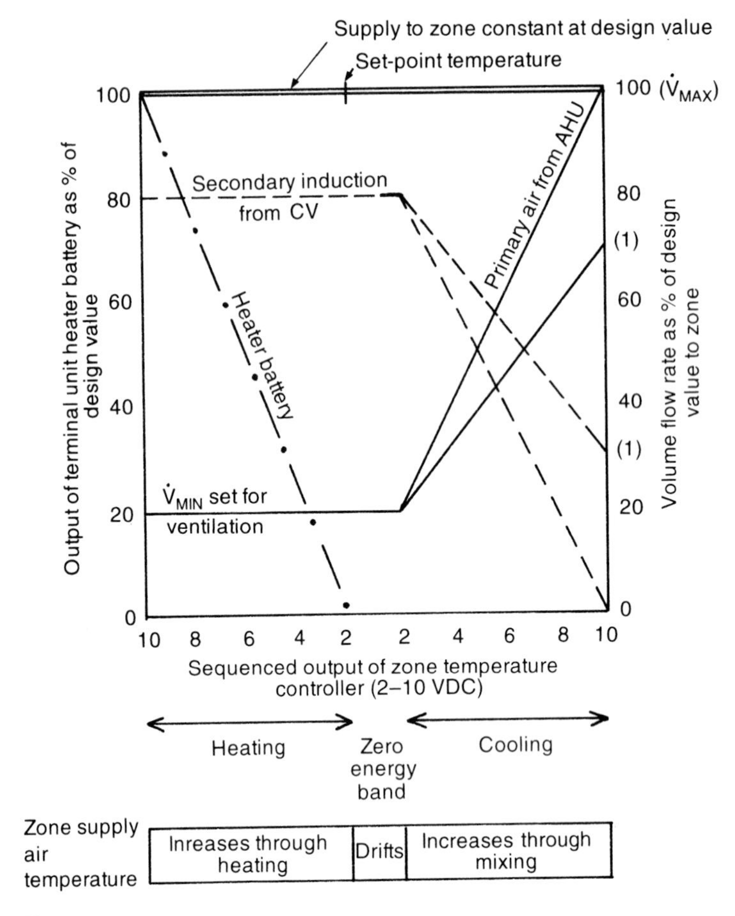

Fig. 7.18 Operating diagram for series-fan FATs.

preset value for normal running. The duration of full motor power is too short to significantly increase the discharge velocity, and thus noise, from the supply diffuser(s) served by the terminal unit.

The addition of a differential pressure sensor in one of the induction (return air inlet) ports of the terminal unit shown in Fig. 7.17 can provide an output signal for flow monitoring, since the total induced (return) air flow may be determined if a suitable calibration exists.

Although the densities of the primary supply, the recirculation drawn from the ceiling return air plenum and the discharge from the FAT naturally differ, for almost all practical purposes this may be ignored. During purely mixing operation (i.e. without any assistance from the terminal unit's LPHW or electric heater battery) the discharge air temperature from a series-fan FAT may therefore be expressed as:

$$t_{disch} = [(\dot{V}_P\, t_P + \dot{V}_C\, t_C) \div \dot{V}_{disch}] + \Delta t_{fan} \tag{7.3}$$

where

$\dot{V}_{disch}$, t_{disch} = discharge volume flow rate, air temperature from FAT for mixing operation only ($l\,s^{-1}$, °C)

$\dot{V}_P$, t_P = volume flow rate, temperature of the primary air supply ($l\,s^{-1}$, °C)

$\dot{V}_C$, t_C = volume flow rate, temperature of the air drawn from the ceiling plenum ($l\,s^{-1}$, °C)

Δt_{fan} = temperature rise across the terminal unit fan, which is located in/cooled by the discharge airstream; determined conventionally (°C)

While fan speed in 'series' FATs is conventionally held constant throughout cooling and heating operation, a variable-speed control option is in principle possible[4]. The justification for this is that 'series' FATs selected to achieve a specified noise rating under design cooling conditions may be significantly larger than required for the cooling load. With variable-speed control (Fig. 7.19) the FAT is still selected on acoustic criteria, but under an average cooling load. As cooling load increases above this 'design average', fan speed is similarly increased. Noise level at the supply diffusers will inevitably also increase, but may not be objectionable to occupants if comfort conditions are maintained under the typical combination of hot and sunny weather which generates the highest cooling loads, as long as the occurrence of these operating conditions is relatively infrequent. The perceived attraction of the approach is in the reduced capital cost of smaller 'series' FATs and any associated 'knock-on' reductions in size and cost of the electrical distribution system serving them. Against this must be offset the additional cost of controls and the increased complexity which is introduced into the HVAC design.

The proposed method of achieving the required fan duty modulation is by phase-cut techniques, which can be applied to the split capacitor motors common to FATs. Phase cutting delays the turn-on of electrical power after each 'zero point' of the sinusoidal voltage waveform. The greater the delay, the lower will be the average voltage available to the motor. However, with a sinusoidal voltage waveform simple control of delay time does not yield linear speed control, and a digital controller is required that can adjust the delay period to achieve this, allowing the form of control to be a simple *open loop*. Fan power is reduced below cooling design flow rate, although fan and motor efficiency are also likely to be lower and adequate motor cooling must be assured. As with other non-linear electrical loads, the effect of *harmonic*

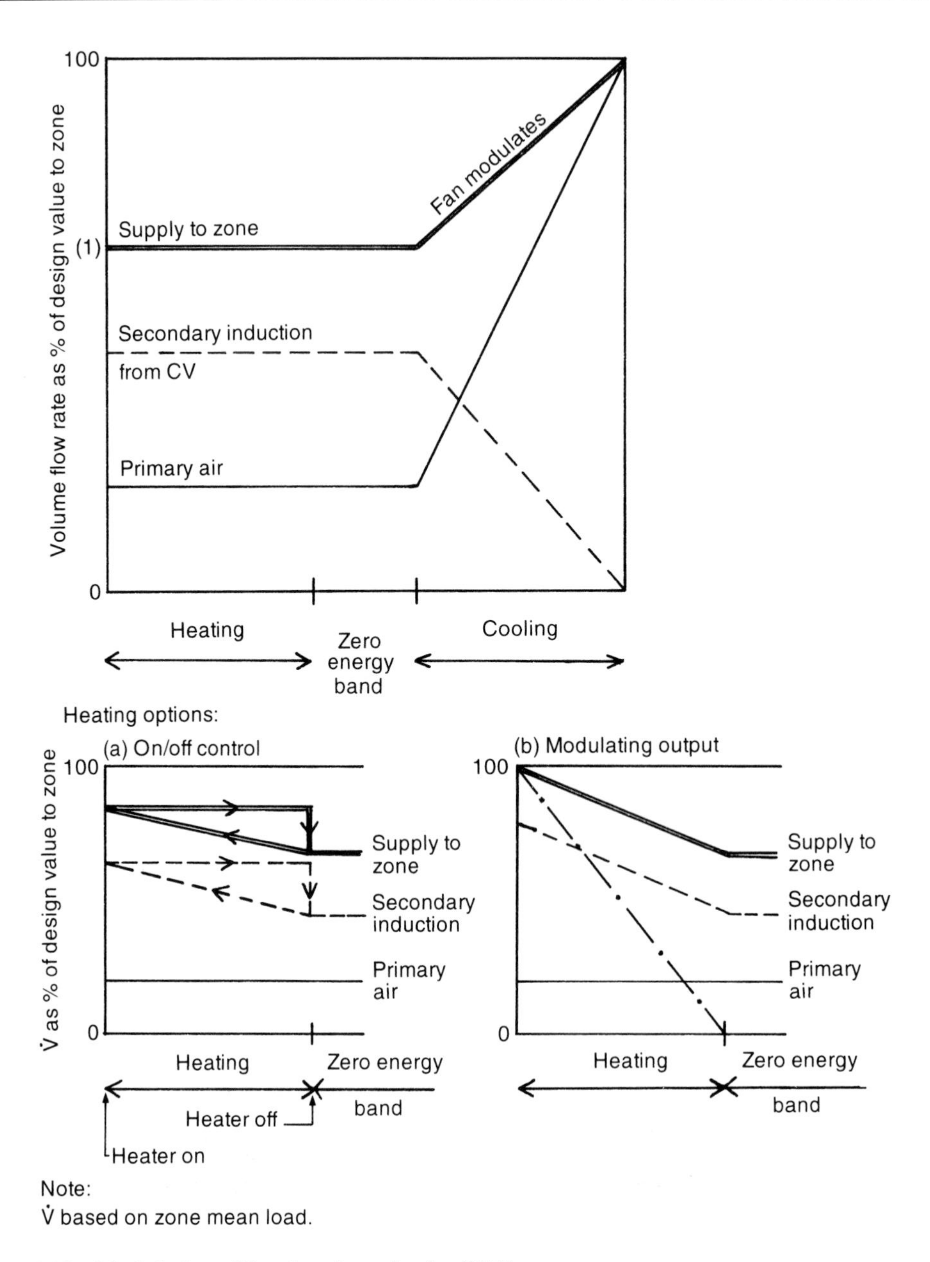

Fig. 7.19 Modulation of fan duty in series-fan FATs.

generation on the electrical supply must be considered. The implications on performance of the room air distribution design of varying the discharge volume flow rate from the FAT must also be considered. A similar control philosophy is possible under heating conditions, whereby the duty of the terminal unit fan and the heater battery output may be modulated together to achieve an overall optimum performance (acoustic, fan energy, room air distribution and heating).

7.7 Fan-assisted terminals: parallel-fan types

7.7.1 Design and construction

Fan-assisted terminals as such appear to have originated in this general configuration in Houston, Texas during the mid-1970s, as a solution to heating and zoning requirements[5]. A typical configuration for a *parallel-fan* FAT is shown in Fig. 7.20. This type of FAT differs from series-fan types in the following respects. The most important difference concerns operation of the terminal unit fan itself. This only operates intermittently – at low zone sensible cooling load, and when heating is required. As a result, fans in 'parallel' FATs typically only handle between approximately one-half and two-thirds of the design supply air volume flow rate to the zone, and are thus smaller and quieter than those in equivalent 'series' FATs, although of the same general type and specification. This makes the 'parallel' FAT a more efficient solution from the point of view of terminal fan energy consumption.

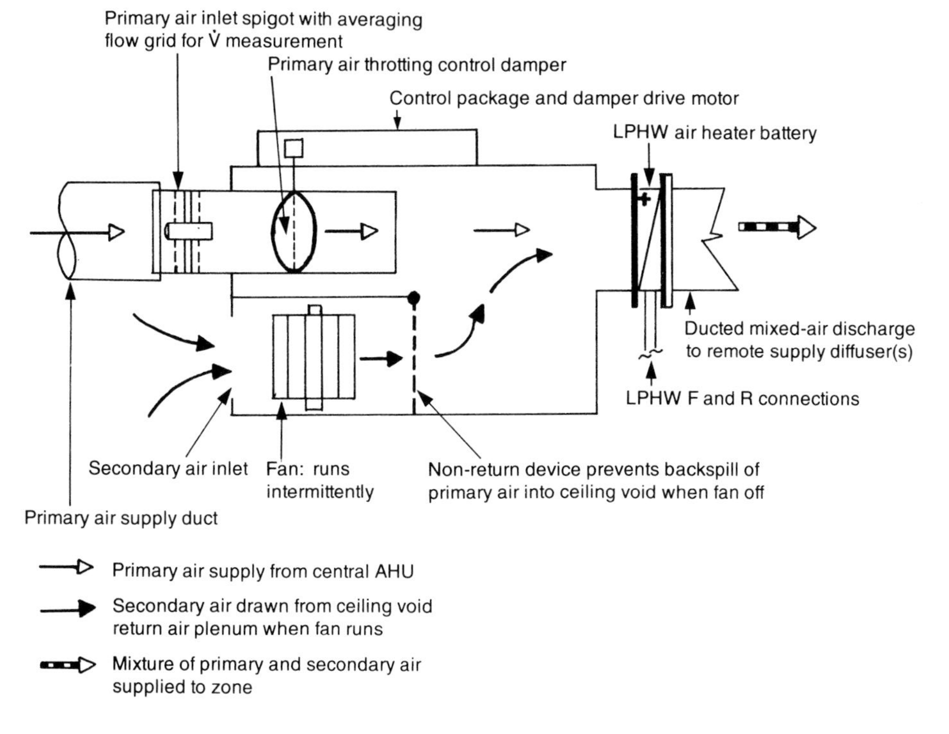

Fig. 7.20 Typical configuration for parallel-fan FATs.

Since the terminal unit fan will not be operating under zone design conditions, the fan total pressure of the system (primary) supply fan must include pressure losses downstream of the terminal unit. However, this is not a disadvantage where system pressure can be more efficiently developed by the primary supply fan, which should be the case for much of the system's annual operating period. A backdraft damper is definitely required in 'parallel' FATs to prevent primary conditioned air short-circuiting into the ceiling void when the fan is off. The characteristics of 'series' and 'parallel' FATs are compared in Table 7.2.

7.7.2 Operation and control

The operation of parallel-fan terminal units is shown in Fig. 7.21. Comparing this with Fig. 7.18 clearly shows the different mode of operation of the parallel-fan type. Fan operation is initiated by the terminal unit controller. Since the cooling capacity of the discharge airstream is reduced on initiation of fan operation, fan cycling can occur if the zone cooling load is near the fan switching point. This may generate noise complaints from the occupants of the space served, since changes of noise are inherently much more noticeable (and therefore less likely to be tolerated) than even a higher steady noise level. In this case a switching differential may be used to reduce the rate of cycling.

Again ignoring differences in density between the supply, return and discharge airstreams, the discharge air temperature from a parallel-fan FAT during purely mixing operation (i.e. without any assistance from the terminal unit's LPHW or electric heater battery) may therefore be expressed as:

$$t_{disch} = [\dot{V}_P\, t_P + \dot{V}_C\,(t_C + \Delta t_{fan})] \div \dot{V}_{disch} \tag{7.4}$$

where

$\dot{V}_{disch}$, t_{disch} = discharge volume flow rate, air temperature from FAT for mixing operation only ($l\,s^{-1}$, °C)

$\dot{V}_P$, t_P = volume flow rate, temperature of the primary air supply ($l\,s^{-1}$, °C)

$\dot{V}_C$, t_C = volume flow rate, temperature of the air drawn from the ceiling plenum ($l\,s^{-1}$, °C)

Δt_{fan} = temperature rise across the terminal unit fan, which is located in/cooled by the return airstream; determined conventionally (°C)

When running, the terminal unit fans in 'parallel' FATs conventionally operate at constant speed, although again an alternative operation and control philosophy based on modulation of fan speed is in principle possi-

Table 7.2 Comparison of the characteristics of 'series' and 'parallel' fan-assisted terminals.

Characteristic	Series fan configuration	Parallel fan configuration
Fan operation	Continuous	Intermittent
Volume flow rate to zone	Constant	Variable during cooling operation Constant during heating operation, at 50–67% of cooling design flow
Supply temperature to zone	Variable	Constant during cooling operation Variable during heating operation
Static pressure at primary air supply inlet	Lower ($\leq$ 50 Pa). Hence lower fan total pressure for system (primary air) supply fan	Higher ($\geq$ 100 Pa). Hence higher fan total pressure for system (primary air) supply fan
Risk of primary supply air 'spilling' into ceiling void	Yes, if fan not adjusted to maximum cooling air flow rate. However, backdraft dampers may not be required if terminal fans started before primary supply fan	Yes. Backdraft damper required.
Suitable for low-temperature primary supply air (VAV with 'cold air distribution')	Yes	Not typically
Fan size	To maximum cooling air flow rate	To heating air flow rate, at 50–67% of cooling design flow
Acoustic implications	Fan noise continuous	No fan during positive cooling operation (moderate zone cooling load factor and above). Fan noise continuous during positive heating operation (moderate zone heating load factor and above). Fan cycles in 'intermediate' operating region between low zone cooling and heating load factors; cycling typically more noticeable than continuous operation.
Physical size	Greater than equivalent single-duct throttling terminal unit	Wider than 'series' configuration

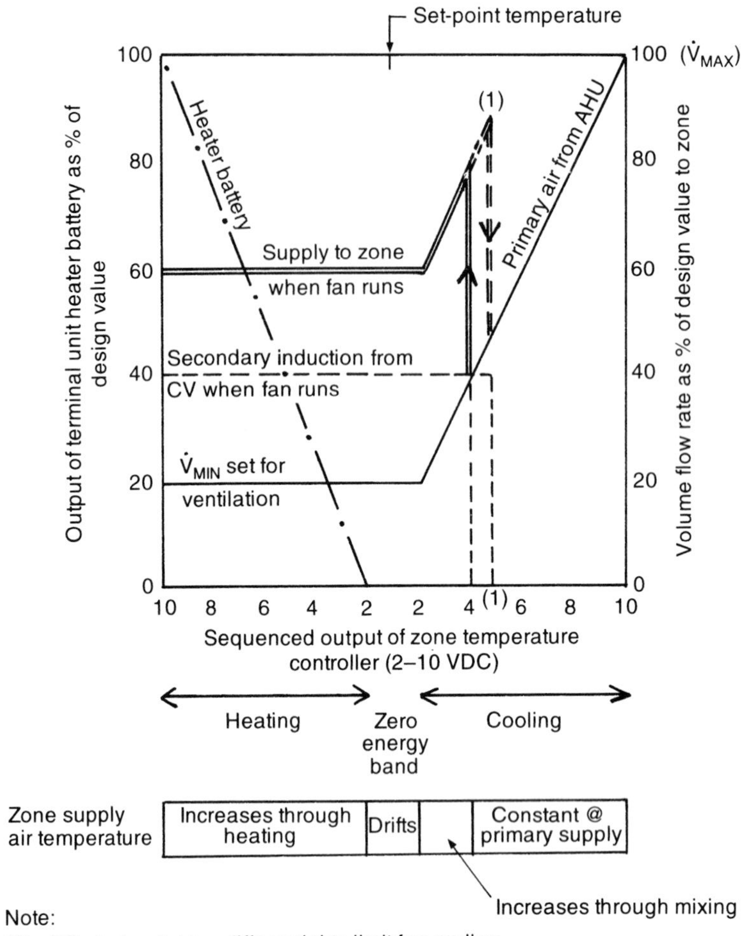

Fig. 7.21 Operating diagram for parallel-fan FATs.

ble[4]. This is shown in Fig. 7.22. The benefits claimed for this approach include a reduction in fan cycling through start-up at low duty, although a small switching differential may still be required. Where simple on/off electric heating is employed, modulation of fan duty can provide a first stage of heating. Again where modulating LPHW heating is employed, both fan duty and heater battery output may be modulated together to achieve an overall optimum performance (acoustic, fan energy, room air distribution and heating). Speed modulation of the fans in 'parallel' FATs is also proposed as a more economic alternative to 'series' FATs where constant-volume zone operation is desired. Again DDC-controlled phase-cut is the proposed means of achieving fan speed modulation.

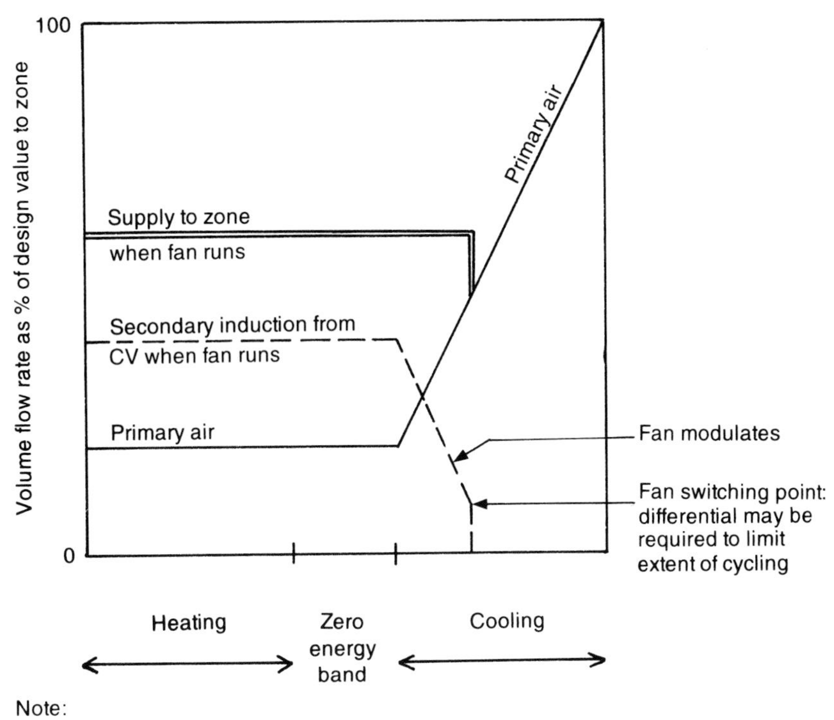

Fig. 7.22 Modulation of fan duty in parallel-fan FATs.

7.8 Induction VAV terminals

VAV boxes are also available in which the induction of return air from the ceiling void into the unit, when part-load conditions exist in the zone, is achieved by discharging the primary supply of conditioned air through a purpose-designed *nozzle* within the unit. The general layout of an induction VAV terminal is illustrated in Fig. 7.23. The principle of operation requires a relatively high static pressure to be maintained at the primary air inlet to the terminal unit. This increases the total pressure requirement and annual energy consumption of the supply fan, and may require a larger, more expensive fan. Apart from the implications for supply fan energy consumption, attenuation of noise developed at the primary air inlet nozzle must also be considered in the light of the particular application. Achievement of an acceptable noise level in the space served will typically require static pressure at the induction nozzle(s) to be kept to a maximum of approximately 500 Pa. This type of terminal unit is thus inherently less suited to applications where a low design noise rating is required.

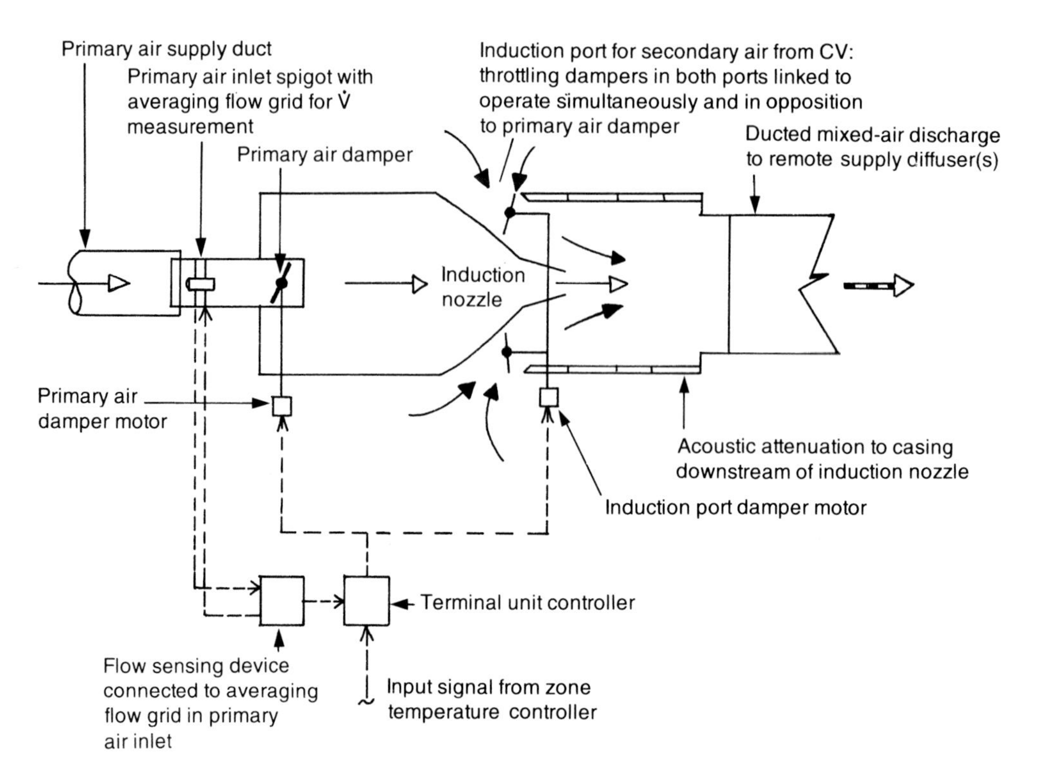

Fig. 7.23 Typical configuration of induction VAV terminal units.

Considerations of pressure loss suggest that induction units are unlikely to be suited to the addition of an LPHW air heater battery for full heating duty, although heat transferred to the ceiling plenum via return air light fittings may be recovered, and supplementing this with limited electric air heating at the terminal unit may be practical. Air induced from the ceiling return air plenum is typically not filtered, which inherently has an implication for IAQ within the spaces served.

Overall, induction terminal units probably bear better comparison with FATs, if cold air distribution can be employed. The increased fan energy consumption associated with the greater fan pressure development required by the induction nozzles may then be offset against the consumption of the terminal unit fans, particularly since these are conventionally of the series-fan type where FATs are combined with cold air distribution. The operation of a typical induction VAV terminal unit is shown in Fig. 7.24.

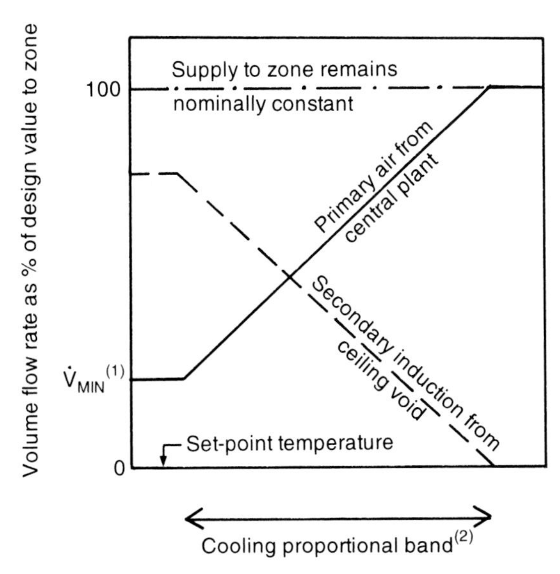

Notes:
(1) Set for ventilation.
(2) Induction VAV terminal units are unlikely to be suitable for other than limited electric heating at the terminal unit.

Fig. 7.24 Operating diagram for induction VAV terminal units.

7.9 System-powered terminal unit with variable-constant-volume action

7.9.1 Design and construction

The terminal unit shown in Fig. 7.25 illustrates a rather different approach to terminal unit design. While still of the variable-constant-volume or 'pressure-independent' type, it is *system-powered*, and also discharges the supply airstream directly into the space served through an integral supply air diffuser. This type of terminal unit is almost invariably used in ceiling void installations, although in some circumstances sidewall applications may be feasible.

The diffuser is invariably of the linear slot type, turning the supply airstream through 90°, and depending on the Coanda effect to produce a horizontal discharge across the ceiling. The supply air jets may discharge across the ceiling on one or both sides of the diffuser axis (one- or two-way 'blow'), and it is claimed that mixing of the supply airstream with room air may be achieved within 300 mm of the slot outlet. This would imply that generous supply air temperature differentials could be used to advantage without discomfort to the occupants of a space, and it is suggested by one manufacturer that up to 14 K may be acceptable. Terminal units are intended to complement a modular system design approach, and diffuser lengths are

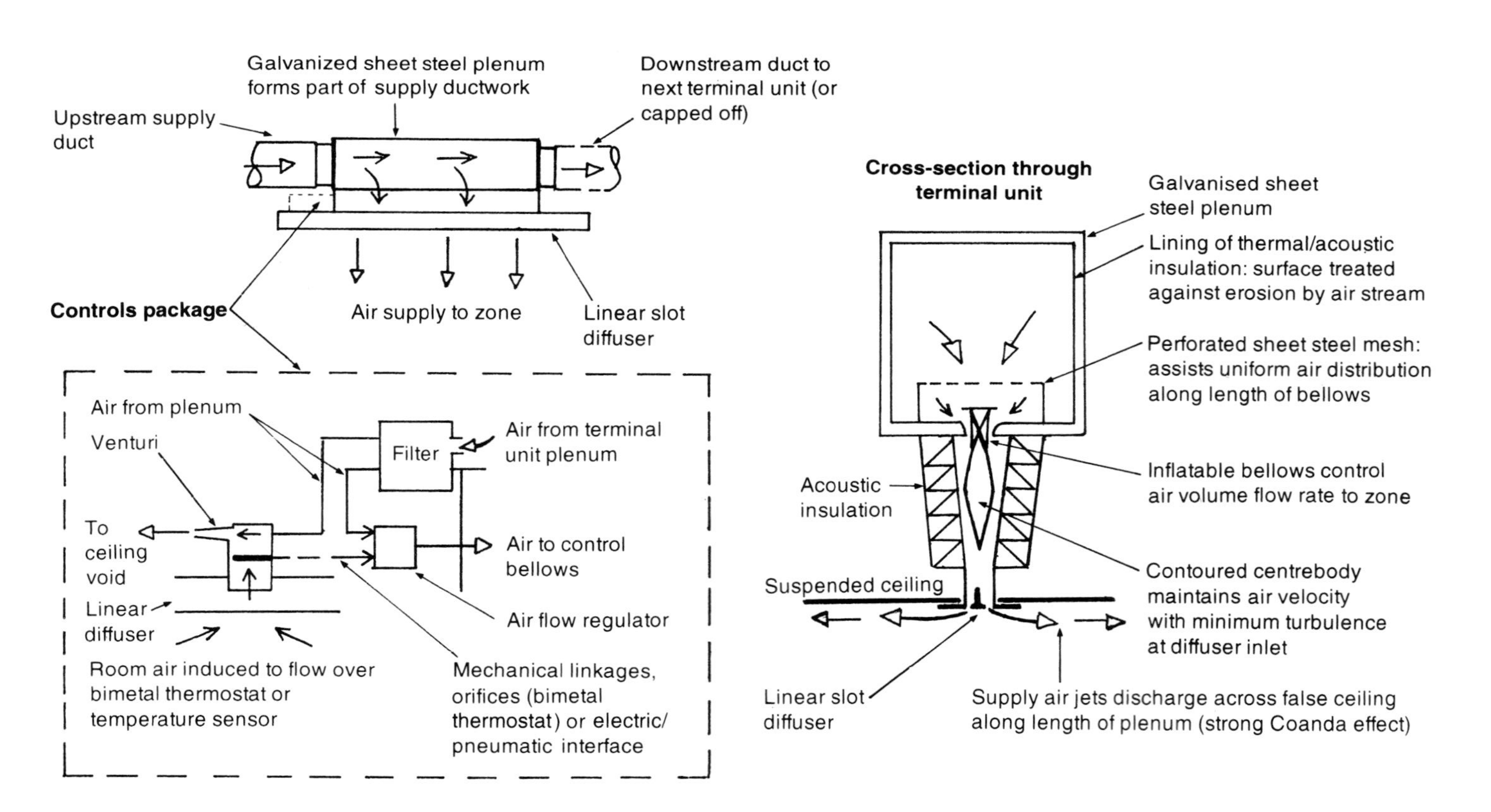

Fig. 7.25 Configuration of system-powered VAV terminal unit with variable-constant-volume ('pressure-independent') action and integral supply diffuser.

typically standardised at either 1200 or 1500 mm, although 600-mm lengths may be available for the lowest volume flow rates.

The casing of the terminal unit comprises a galvanised sheet steel plenum of square cross-section. Each capped end has a circular inlet or outlet duct spigot that can either be used to make a supply duct connection or be blanked off. The plenum is lined with acoustic insulation (fibreglass, neoprene-coated against corrosion is normal), and mounting lugs are provided for threaded rod or wire hangers.

The description of the terminal unit as 'system-powered' follows from its total reliance on the air pressure *within the plenum* itself to actuate the damper controlling the volume flow rate of air supplied to the zone served. Neither electric motor nor pneumatic actuator are required by the damper. Air can flow out of the plenum to the supply air diffuser through a lengthwise slot in its base. The flow is driven by the difference in air pressure between the plenum and the space served. Uniform air distribution along the length of the slot is typically encouraged by a perforated sheet steel baffle in the bottom of the plenum. The volume flow rate of air via this slot to the supply air diffuser is controlled by two neoprene bellows which fit lengthwise along the centreline of the slot. These bellows are inflated by static pressure tapped from the plenum itself. The bellows act as a volume flow rate control damper. As they inflate, the area of the slot available for air to flow through is reduced on either side of the bellows, which effectively form two variable-area orifices of high aspect ratio. For a constant pressure available in the unit plenum, volume flow rate is reduced, and vice versa.

A control assembly is mounted at one end of the plenum. This typically comprises an arrangement of chambers and orifices, by the action of which the pressure available to the bellows is varied. This either inflates them or allows them to deflate, as required to supply the volume flow rate of conditioned air demanded by the zone. As might be expected, the throttling action of the bellows is non-linear. The basic control assembly is typically *direct-acting*, although modern versions may typically be fitted with an optional interface to accept either standard pneumatic or electrical control signals.

The flow to the zone is divided into two airstreams around the bellows, and this is maintained almost to the supply diffuser by a centrebody running the length of the flow passage. The smooth aerofoil-like contours of this centrebody create profiled passages for the airstreams on either side of the bellows. Air velocity is maintained at a high level, and the turbulence created by the throttling action of the bellows is damped out (or at least kept to a minimum). This is an important consideration due to the strictly limited distance between the bellows and the supply diffuser. As usual, the aim is to establish a strong Coanda effect at the outlet from the diffuser slot. The outside walls of the flow passages are typically of double-skinned sheet metal construction, with the intervening gap being filled with acoustic insulation to attenuate the noise generated by the throttling action of the bellows. The plenum pressure required for bellows operation also maintains the static pressure above the diffuser as volume flow rate to the zone reduces under part load – and hence maintains the discharge air velocity and room air distribution performance. Some units allow positioning of the supply diffuser directly along the line of a partition, serving the spaces on either side from a single terminal unit. In this case each bellows is individually controlled to the requirements of one of the spaces served.

While all VAV terminal units are *components* of the ducted air distribution system, the terminal unit of Fig. 7.25 forms *part of the ductwork itself*, since a number of terminal units may be linked in series by flexible or spirally wound ducts of circular cross-section. The supply airstream to any terminal unit in a series has passed through all the upstream units in the series. In some circumstances a facility for side inlet of air to the plenum may be feasible.

The pressure independence of the terminal unit in Fig. 7.25 does not rely on sensing of the volume flow rate to the zone. Rather it is *inherent* in the system-powered principle. Changes in plenum pressure that would change the volume flow rate to the zone, at any given degree of inflation of the bellows, also change the pressure available to inflate them. The changes act in a corrective sense – that is, a drop in plenum pressure also deflates the bellows, reducing the resistance they offer to volume flow to the zone, and enabling the level set by the terminal unit controller to be maintained at the reduced plenum pressure (and vice versa).

Individual system-powered terminal units are typically available to handle zone design supply volume flow rates between approximately 25 and 200 $l\,s^{-1}$. Associating these volume flow rates with a supply air temperature differential of 8 K gives a range of zone design sensible cooling load between approximately 0.25 and 2 kW. Plenum size must also take into account the 'through' volume flow rate to downstream members of a series of terminal units. Commercially available terminal units offer plenum sizes in the range 180–280 mm square, to handle total volume flow rates at inlet to the leading plenum in a series ranging from approximately 200 to 500 $l\,s^{-1}$ respectively.

The provision of space heating with such terminal units is more problematic, although in some cases an LPHW or electric heater battery can be incorporated at the inlet to the plenum of the terminal unit. Whether this is feasible or not as a means of providing space heating in practice depends on achieving a suitable layout of terminal units. In any event it is unlikely that the results will match those that can be achieved by alternative means, including terminal units utilizing electric or pneumatic actuation. However, even with system powered terminal units that only provide cooling during normal operation, it is still possible to operate a *morning warm-up cycle*, prior to occupancy of the building, using the main heating coil in the central AHU.

At the start of the warm-up cycle each terminal unit connected to the system is activated to constant-volume operation by a temperature-sensitive switch. This bypasses the normal cooling control thermostat when warm air is received from the central AHU. The warm-up cycle lasts until the common return air temperature to the central AHU reaches a specified value, when the supply temperature reverts to that associated with normal cooling operation. The thermal switch in each terminal unit switches its operation back to the normal cooling thermostat when it receives air at this temperature from the central AHU.

Terminal units can be of the 'normally open' or 'normally closed' type. In the former system pressure is used to close the control bellows, in the latter to open them. If a 'normally open' terminal unit receives inadequate duct static pressure, the bellows will remain fully open, and controlled operation of the terminal unit will not be established. This is particularly relevant during system start-up, especially where a pre-occupancy warm-up cycle is employed. In the event that 'normally closed' terminal units are used, care must be exercised to ensure that excessive duct pressure is not built up on start-up, which might cause actual damage to the ducts. The risk of such an occurrence would be minimised either by starting the system supply fan at minimum speed (or at the minimum setting of whatever form of duty modulation is employed) or by installing a high-limit duct static pressure override at the fan. In any event this has the added benefit of protecting the supply ductwork from the effects of accidental closure of a fire damper during normal system operation.

In some examples access to the terminal unit's control assembly can be gained via a drop-down diffuser section, for situations where local access via removable ceiling panels is unavailable.

Units with independent dual plenums are available for dual-duct systems – typically one plenum providing VAV cooling, the other CAV heating. In various versions the system-powered principle has also been applied to the conventional 'VAV box'.

7.9.2 Operation and control

In its basic form the system-powered terminal unit shown in Fig. 7.25 requires no external electrical or pneumatic control signal. The terminal unit controller is typically mounted on an inactive extension of the supply diffuser (i.e. not used for the supply airstream to the zone) at one end of the unit casing. A small volume flow rate of supply air is bled from the unit plenum, filtered, and exhausted into the ceiling void through a small plastic venturi (inset Fig. 7.25). The depression in the throat of this draws a flow of air from the room through the diffuser below and over a bimetallic element. Differential expansion or contraction of the metals creates a movement of the element with a change in temperature of the air drawn from the room. Through a mechanical linkage this moves a sliding plate across an orifice. The rate at which air is bled off through this orifice controls the pressure available to inflate the bellows controlling the volume flow rate of supply air to the zone.

The control mechanism typically offers an adjustable maximum limit on supply air volume flow rate, which may be preset, but no minimum limit. The bimetal thermostat may alternatively be replaced by a temperature sensor, the output from which goes to a separate zone air temperature controller. In its turn the output from this controller then acts through an interface unit to move the sliding plate and vary the orifice area as before. Typically either electronic or pneumatic zone controllers can be used, if appropriate interfaces to the terminal unit controller are available.

An adequate ductwork design is essential to successful system-powered VAV operation, since failure to achieve sufficient duct static pressure at terminal units distant from the central AHU will not simply result in a degree of undercooling, but will rather result in a total loss (or indeed absence) of control – since the control bellows will not inflate. A minimum duct static pressure in the range 100–150 Pa is required at each terminal unit in the system to ensure that full control of volume flow rate is available.

7.10 Pressure-dependent VAV terminal unit

7.10.1 Design and construction

The terminal unit shown in Fig. 7.26 is a *pressure-dependent* design. There is no averaging flow grid, velocity sensor, or orifice-type baffle in the inlet spigot to the unit, and hence no provision for the sensing of volume flow rate at the terminal unit. The terminal unit has a simple variable-volume action which positions the control damper according to the dictates of the local zone temperature controller. Since the volume flow rate supplied by the terminal unit will vary with fluctuations in static pressure at inlet to the terminal unit, separate provision must be made to limit such fluctuations to an acceptable level, either by general system design or by the provision of specific pressure control elsewhere in the system.

Pressure-dependent designs in the UK are typically integrated with the supply diffuser. In the example shown this is of the linear slot type, for a ceiling void installation of the terminal unit. This dictates the basic layout of the terminal unit as a linear plenum, again typically fabricated from galvanised sheet steel, with side-mounted inlet spigot for conditioned air from the central AHU. In some instances the terminal unit may need only to sit on the linear diffuser when this is installed in the suspended ceiling, rather than for the diffuser to be an integral feature of the terminal unit itself.

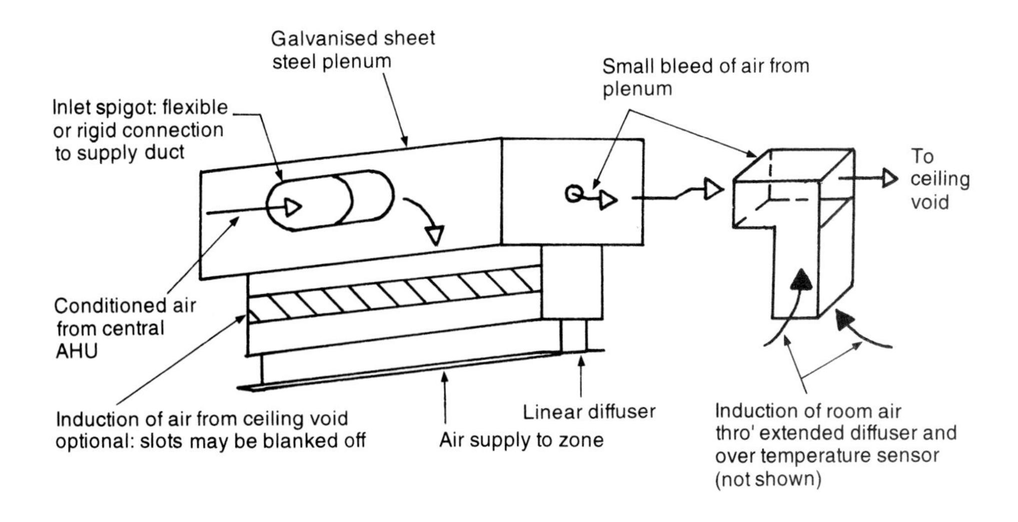

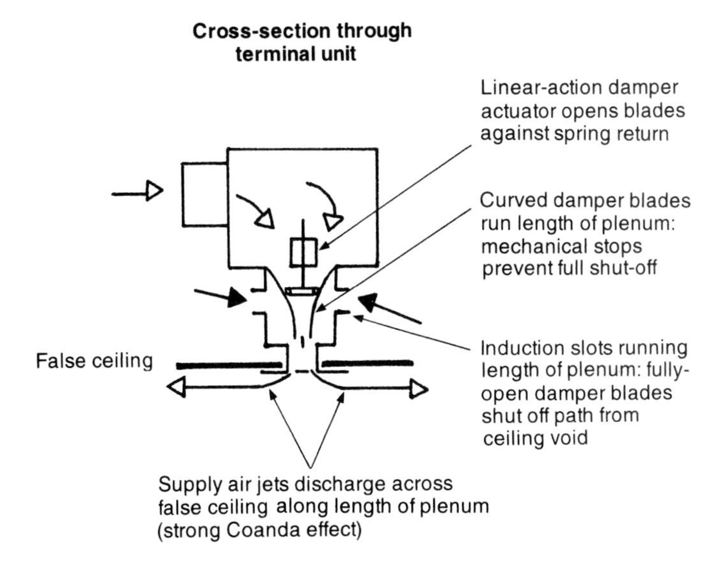

Fig. 7.26 Configuration of a pressure-dependent VAV terminal unit.

In the example shown, the volume flow rate control device comprises a pair of opposed curved blades running the length of the plenum. These are typically prevented from closing to full shutoff by a mechanical stop between the blades. The damper blades are actuated by an electric motor, but using a linear-drive (screw-jack) action, rather than the rotary action typical of single-blade and multileaf damper types. This linear driving action pushes the damper blades open against a spring action that closes them when the driving action is reversed. As elsewhere, pneumatic actuation typically remains available as an alternative option. The smoothly curved blades are important in creating a throttling action without generating turbulence at such close proximity to the supply diffuser. As the blades close they form a smoothly profiled flow passage of progressively decreasing throat area. Static pressure in the terminal unit plenum above the damper blades increases with reducing volume flow rate through the unit. The velocity of the supply airstream leaving the flow passage between the blades is therefore maintained as its volume flow rate is reduced by their closing action. This helps to maintain room air distribution as zone volume flow rate reduces under part-load conditions.

The airstream discharging from the flow passage formed by the damper blades also has a capacity to induce return air from the ceiling void. To this end induction slots run the length of the control section on both sides, adjacent to and behind each damper blade. Mixing with the conditioned supply air handled by the terminal unit, this induced air lessens the variation in supply air volume flow rate to the zone under part-load conditions. Since the velocity of the primary airstream leaving the flow passage is maintained as the damper blades close, which closing action also opens the passage for induced air, the induction effect is greatest when the blades are in their minimum open position. This is of course when it is of greatest potential benefit to room air distribution. In the fully open position the damper blades close off the induction slots. If the facility is not required, the slots can simply be blanked off.

An LPHW heating element may be fitted within the terminal unit. Heating operation is typically associated with a *change in the pattern* of air discharge (Fig. 4.4). This is designed to maintain the discharge velocity of the supply airstream at a more appropriate level for space heating, which is invariably carried out at the minimum supply air volume flow rate to the zone. A unit using a 'two-way blow' discharge pattern under cooling operation (both towards and away from the building perimeter) would typically change to a 'one-way blow' towards the perimeter under heating operation. Alternatively, a unit mounted adjacent to the building perimeter would change from blowing towards the interior of the space during cooling operation to discharging vertically down the perimeter glazing during heating operation. The changeover is typically achieved mechanically within the unit. This may require an additional heating/cooling changeover damper, and must in any event be reflected in the internal configuration of the terminal unit and the design of the supply air diffuser.

Overall width and height of a cooling-only unit may each typically be of the order of 300 mm, with various lengths of plenum available up to a typical maximum of 1200 mm. Maximum supply air volume flow rates (primary air) between approximately 40 and 180 l s^{-1} may typically be expected, according to length of plenum and static pressure at inlet to the unit. Associating these volume flow rates with a supply air temperature differential of 8 K gives a range of zone design sensible cooling load between approximately 0.4 and 1.8 kW.

Pressure-dependent designs are also available which are based on the use of high-induction 'swirl' diffusers (Fig. 7.27). These both turn the supply airstream through 90° and split it into a large number of supply air jets, which spread out across the ceiling in a combined radial and tangential pattern away from the diffuser. Dimensions of these units are typically of the order of 600 mm square by 350 mm high. A unit of this size might be capable of a maximum supply air volume flow rate between approximately 100 and $250\,l\,s^{-1}$, according to static pressure at the inlet spigot. Again associating these volume flow rates with a supply air temperature differential of 8 K gives a range of zone design sensible cooling load between approximately 1 and 2.5 kW.

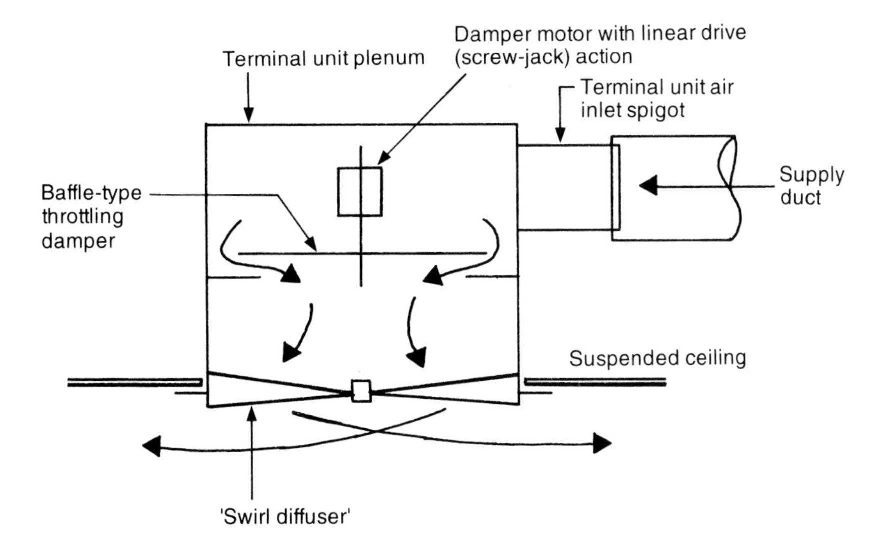

Fig. 7.27 'Swirl-type' pressure-dependent VAV terminal unit.

7.10.2 Operation and control

In the pressure-dependent terminal unit shown in Fig. 7.26 the control damper in the terminal unit is simply positioned according to the control signal from the zone air temperature controller. There is no feedback control of volume flow rate by the terminal unit controller, which is thus unable to provide constant-volume control at any given demand from the zone controller.

The volume flow rate to the zone is dependent not only on the demand from the zone controller, but also on the *pressure available* at the inlet spigot of the terminal unit. In a VAV system this will inevitably fluctuate, to a greater or lesser extent, as the distribution of volume flow rates throughout the system varies with the changing pattern of zone sensible cooling loads. However, an increase or decrease in the volume flow rate to any zone that is produced by a change in the pressure available at the terminal unit inlet spigot will ultimately result in a change in the temperature sensed by the zone controller. This will be reflected in a change in the control signal to reposition the terminal unit control damper in a corrective sense. Under these circumstances fluctuations in zone air temperature may be expected even at constant zone cooling load. There are no preset maximum or minimum volume flow rate limits, although the terminal unit is prevented from going to full shutoff by the mechanical stop between the damper blades which limits their maximum closure.

A high degree of pressure-independence, together with effective maximum and minimum volume flow rate limits, are typically achieved in practice by *regulating the pressure available* at the terminal unit inlet spigot. The supply air distribution layout is designed so that groups of about six terminal units are each supplied by an individual branch duct, and a pressure regulating damper is installed in this duct. A controller positions the blades of this damper to maintain a nominally constant duct static pressure at a sensor position downstream of the damper. While static pressure sensor location is treated in Chapter 8, the control set-point must be sufficient to provide adequate static pressure at each of the terminal units in the group under design conditions. For this to be a minimum careful attention should be paid to the design of the ductwork downstream of the pressure regulating damper. Ideally the loss of total pressure should be as nearly identical as possible between the static pressure sensor and the inlet spigot to each terminal unit in the group. Pressure-dependent control is shown schematically in Fig. 7.28, and a general block diagram is given in Fig. 7.29. The combination of a group of pressure-dependent terminal units served by a common pressure regulating damper effectively shares the 'pressure-independent' feature between the units in the group, rather than providing it to each unit individually.

An alternative type of electronic control which is available for pressure-dependent terminal units is described as *rate-aided proportional control* (RAPC). In this the signal generated by the controller is based not only on the measured deviation of zone air temperature from its set-point value, but also on its rate and direction of change. This type of control therefore inherently requires the capabilities of a DDC controller. The output signal from this varies the time and direction of travel of the terminal unit damper. The direction of damper travel is determined by the amount and direction of the change in zone temperature in each sampling period. With RAPC a direction of travel that is in the correct sense for the measured deviation from set-point value (e.g. damper to open for zone temperature above set-point value) may be 'overruled' and reversed by a sufficiently high rate of change of

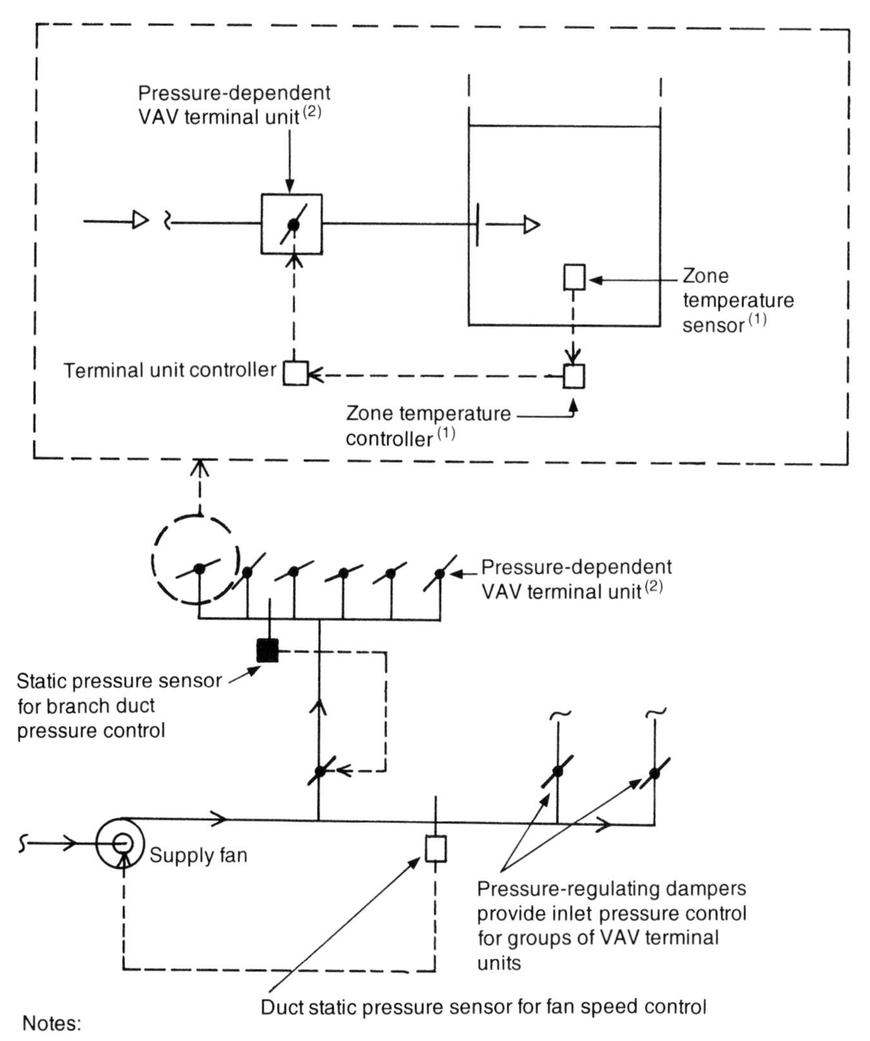

Fig. 7.28 Control of pressure-dependent VAV terminal units.

zone temperature in the opposite sense (decreasing in the example given). The controller determines a new damper travel time and direction at intervals which are several times the maximum travel time permitted during any single operation. A controller update interval of 30 seconds, with a damper motor energised for up to 7.5 seconds in either direction, are typical of commercially available terminal units.

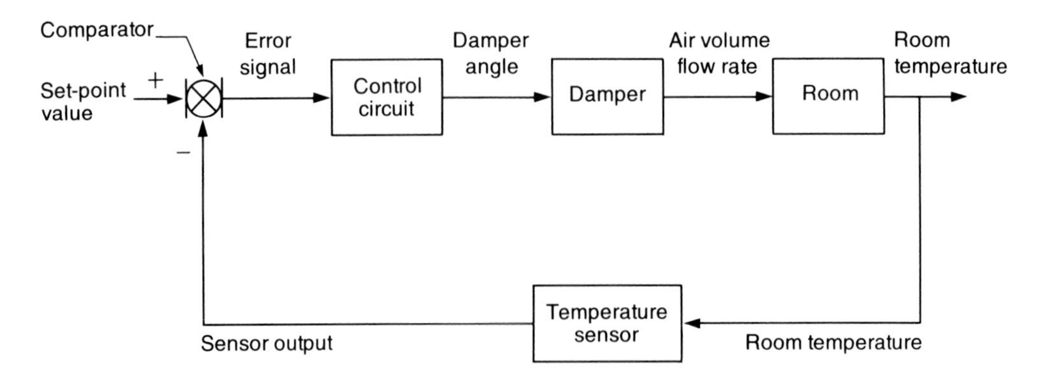

Fig. 7.29 Block diagram for pressure-dependent control action.

7.11 Thermally-powered VAV diffusers

7.11.1 Design and consruction

The terminal unit shown in Fig. 7.30 is again a pressure-dependent type. Described as a *thermally powered diffuser*, this is effectively a supply outlet equipped with a volume flow rate control device and self-acting controller. A circular disc or baffle is fitted closely below the circular neck of the supply outlet. Moving this control disc, which is typically hidden from view by a protective and decorative faceplate, towards or away from the neck respectively reduces or increases the volume flow rate delivered to the zone.

Electric heating is available for some units, using an element mounted upstream of the neck of the basic unit. Thermally powered units may have neck diameters between approximately 150 and 350 mm, typically corresponding to ceiling openings between 300 and 600 mm square. Maximum supply volume flow rates may typically be expected to lie within a range of approximately 40 to 400 l s^{-1}, according to neck diameter and static pressure at inlet to the terminal unit. Associating these volume flow rates with a supply air temperature differential of 8 K gives a range of zone design sensible cooling load between approximately 0.4 and 4 kW.

7.11.2 Operation and control

Under the Coanda effect the supply airstream from the thermally-powered diffuser spreads out across the ceiling, inducing room air to mix with it. Specifically a flow of room air is also induced upwards towards the underside of the terminal unit. This flows over a wax-filled temperature-sensing element, which may either be located on the underside of the terminal unit or in a low-pressure area behind the edge of the faceplate. The wax in this element is formulated to melt at room temperature, respectively expanding or

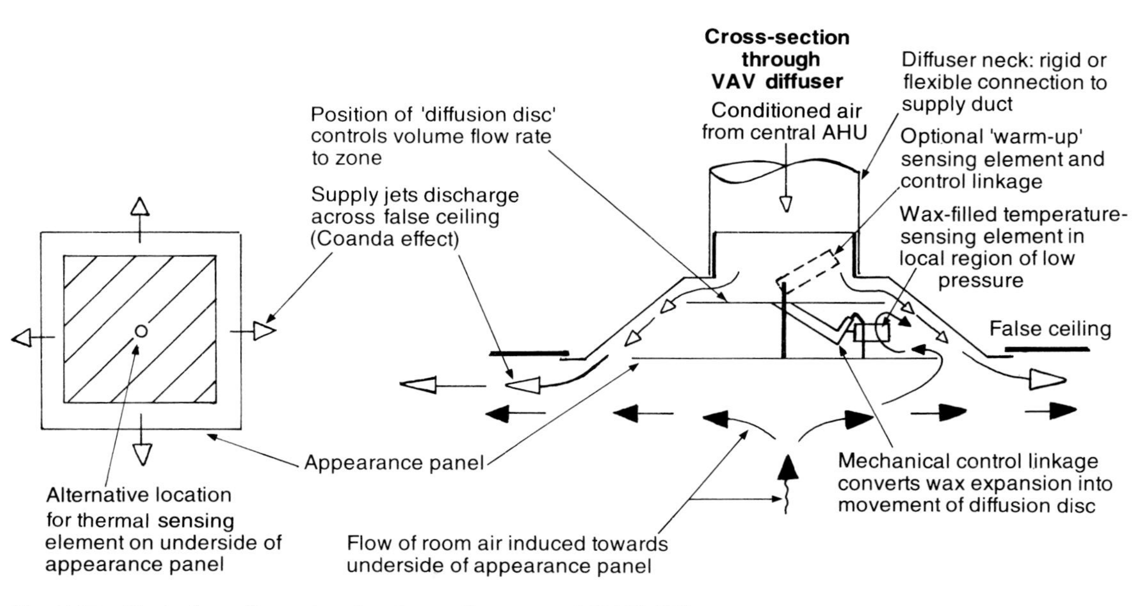

Fig. 7.30 Typical configuration for thermally-powered VAV diffusers.

contracting with an increase or decrease in the temperature of the room air induced over the element. This expansion or contraction is utilised to move a diaphragm and piston in the sensing element, and this movement is converted and amplified by a mechanical linkage to position the volume flow rate control baffle respectively away from or towards the neck of the unit.

7.12 Extract VAV terminal units

All of the VAV terminal units described thus far have have been, either explicitly or implicitly, associated with the *supply side* of VAV systems. This is not of necessity so. However, of all the terminal unit types considered in the preceding sections, only 'VAV boxes' are in practice suitable for extract use, in the modified form shown in Fig. 7.31. The modifications from a dedicated supply terminal unit reflect the need both to keep the acoustic attenuating section room-side and to keep the averaging flow grid upstream of the control damper.

Owing to the cost premium involved, terminal units are only likely to be found on the extract side of VAV systems in *industrial* air conditioning applications where close control of space pressure is required. To achieve this either the set-point volume flow rate of the terminal unit controller is reset directly according to the output of a space static pressure sensor, or alter-

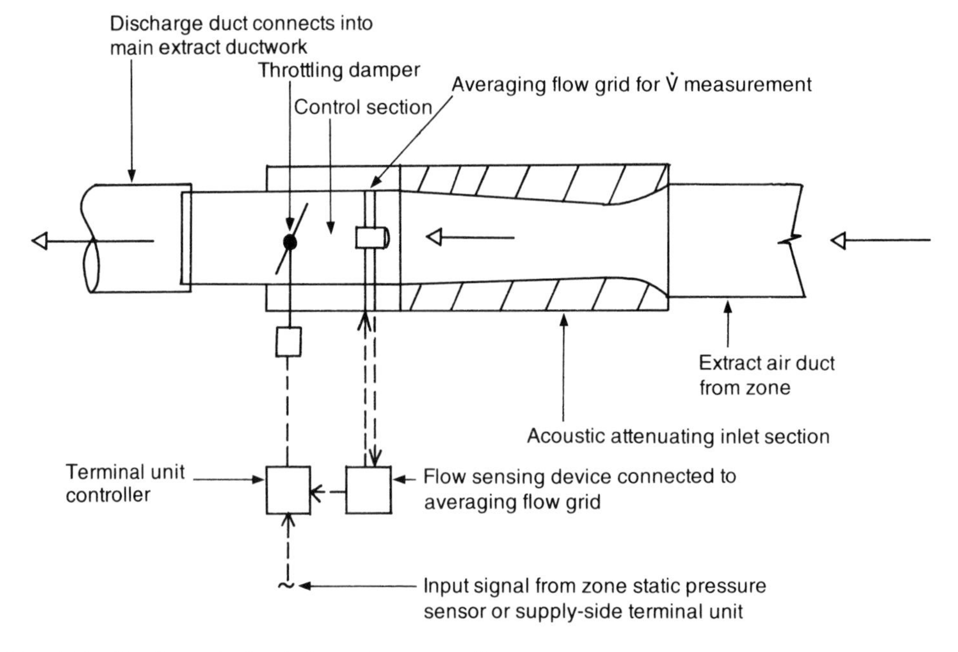

Fig. 7.31 VAV terminal unit for extract operation.

natively the extract terminal unit is made to track the supply unit serving the common zone. Both techniques are shown schematically in Fig. 7.32. In the former case the output of the space static pressure sensor simply takes the place of the zone temperature controller on the supply side. In the latter the supply terminal unit controller outputs a control signal to the extract unit controller which is based on the 'actual value' of volume flow rate sensed at the supply unit. Depending upon the type of controller fitted to the extract terminal unit, this may be manipulated to give tracking either with a fixed differential between supply and extract or with a fixed ratio of extract to supply.

7.13 Terminal unit performance

7.13.1 Aerodynamic performance

The aerodynamic performance of pressure-dependent terminal units is typically expressed in terms of the volume flow rate which results when a specified static pressure exists at the inlet to the terminal unit, with the damper fully open. For variable-constant-volume ('pressure-independent') types it is typically expressed in terms of the minimum static pressure differential between measuring points upstream and downstream of the unit for a given volume flow rate. This is the minimum requirement for the terminal unit to be able to pass that volume flow rate of air with its damper fully open.

The pressure drop across a VAV terminal unit under steady-state conditions may be modelled conventionally in terms of a velocity pressure loss factor. Naturally there are an infinite number of these loss factors, each of which is associated with a unique position of the terminal unit's control damper. The difficulty in attempting to derive a model of steady-state terminal unit performance is in following the strongly non-linear increase in this loss factor with increasing damper angle (or whatever parameter is used to measure the degree of damper opening or closure).

One approach suggests that a basic steady-state model for the control damper in a terminal unit of the 'VAV box' type may have the general form[6]:

$$\ln k_\theta = a + b\,\theta^n \tag{7.5}$$

where

k_θ = velocity pressure loss factor at a damper angle θ from fully open, based on the velocity pressure in the terminal unit inlet spigot.

a, b, n = experimentally determined coefficients.

For one example of this type of terminal unit, with an elliptical single-blade damper having a range of angular movement of 61.5° from a fully open

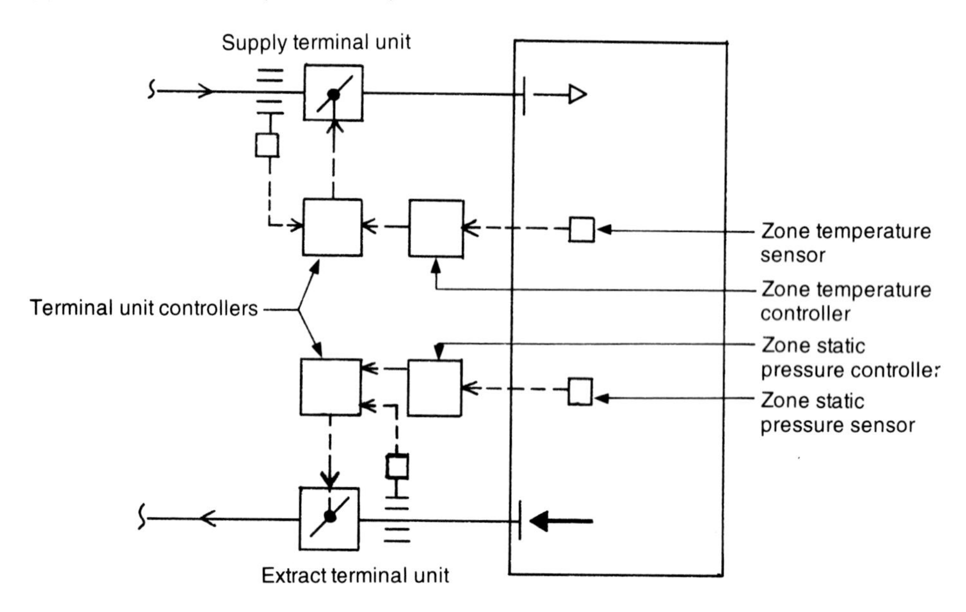

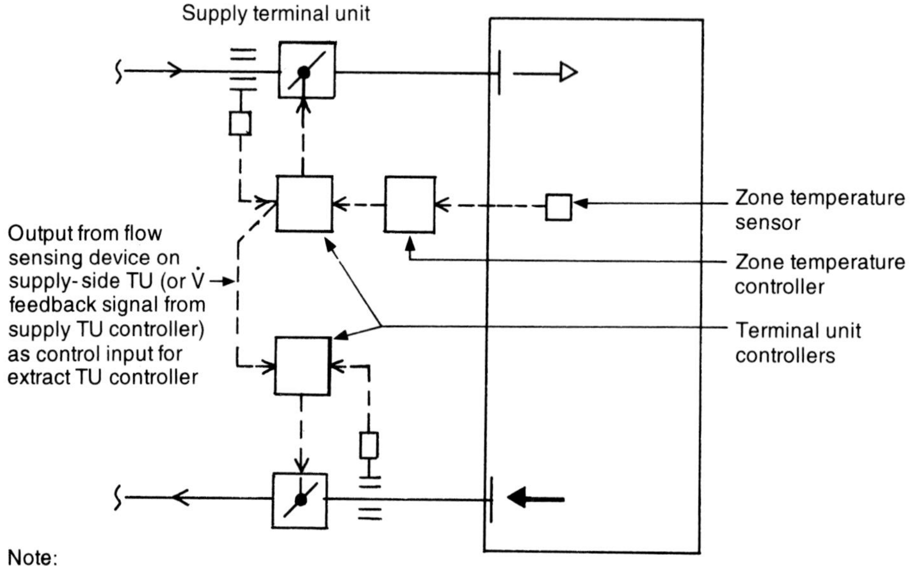

Note:

(1) Tracking control based on supply-extract $\Delta \dot{V}$ or $\dot{V}$ ratio is possible according to controller type and model.

Fig. 7.32 Control of VAV extract terminal units.

position lying along the horizontal axis of the terminal unit, adding the quoted coefficients gives:

$$\ln (k_\theta \times 10^{-6}) = -13.5 + 0.00437\ \theta^{1.77} \quad (7.6)$$

Hence applying the velocity pressure loss factor derived from above to the velocity pressure in the terminal unit inlet spigot, and substituting for the latter in terms of volume flow rate and cross-sectional area:

$$\Delta P_t\,(\mathrm{TU}) = 0.9727\ d^4\ \dot{V}^2\ e^{(-13.5\ +\ 0.00437\ \theta\hat{}1.77)} \quad (7.7)$$

where

ΔP_t (TU) = loss of total pressure across terminal unit (Pa)
d = diameter of terminal unit inlet spigot (m)
$\dot{V}$ = volume flow rate in the inlet spigot ($m^3 s^{-1}$).

However, because of the strongly non-linear characteristic of such control dampers, it is unlikely that a basic model of this form will hold much beyond approximately half of the damper's range of angular movement. Completely different coefficients may apply to different configurations of flow control damper. The measured steady-state performance of such a terminal unit is shown in Fig. 7.33[7].

7.13.2 Control range and accuracy of flow sensing

Variable-constant-volume terminal units also have a specified volume flow rate control range. The upper limit of this is the manufacturer's nominal maximum rating for the unit, which will be termed $\dot{V}_{NOM}$, and the lower limit is typically expressed as a percentage of this. This is the lowest volume flow rate the terminal unit can control down to, and represents its performance-limited maximum turndown. It is logical to expect the value to be a function of two considerations. The first is the accuracy with which volume flow rate or air velocity can be sensed at the terminal unit at low volume flow rates. The second is the leakage past the damper in the fully closed position when high differential pressure exists across the terminal unit. Since the latter may typically be limited to the order of a few litres per second using conventional sheet metal construction techniques and the judicious use of flexible edge seals, the accuracy of sensing at the terminal unit is typically the most significant factor.

The accuracy of flow sensing at the terminal unit is typically quoted as a percentage of the set-point value, and should be expected to decrease substantially at low volume flow rates. Best performance is typically of the order of $\pm 5\%$ of the set-point volume flow rate, when this is above approximatley 60% of the unit's maximum rating (i.e. from $\dot{V}_{NOM}$ to 0.6 $\dot{V}_{NOM}$). Below this the potential sensing error is typically assumed to increase linearly to perhaps

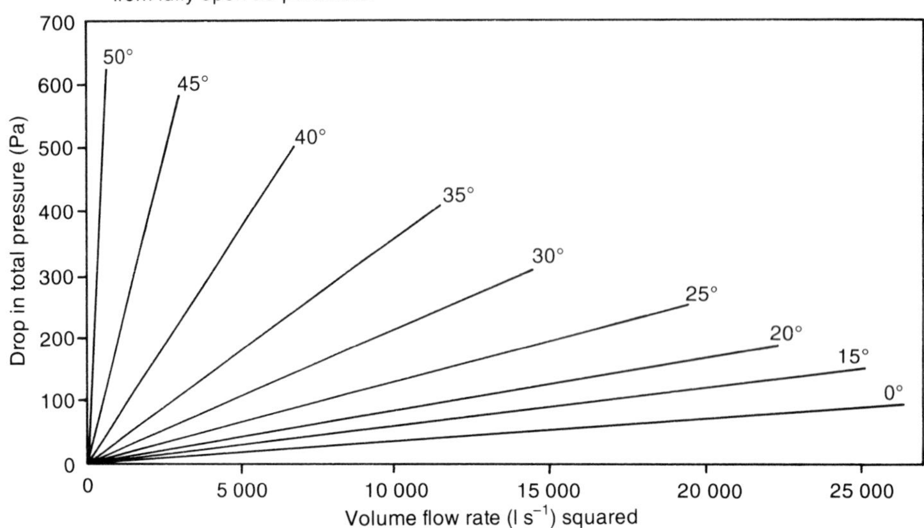

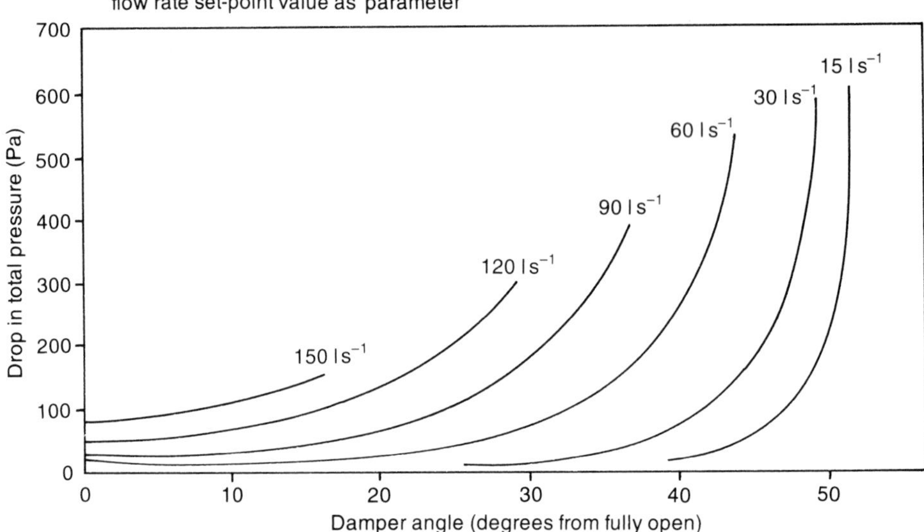

Fig. 7.33 Steady-state performance of a VAV box with variable-constant-volume ('pressure-independent') action.

$\pm 10\%$ of set-point value by the time this has dropped to around 40% of the unit's maximum rating (i.e. 0.4 $\dot{V}_{NOM}$), and to more than $\pm 15\%$ of set-point value when this has dropped to around 20% of the unit's maximum rating (i.e. 0.2 $\dot{V}_{NOM}$). If set-point volume flow rate decreases further, accuracy of sensing may be expected to deteriorate rapidly. It must be borne in mind that

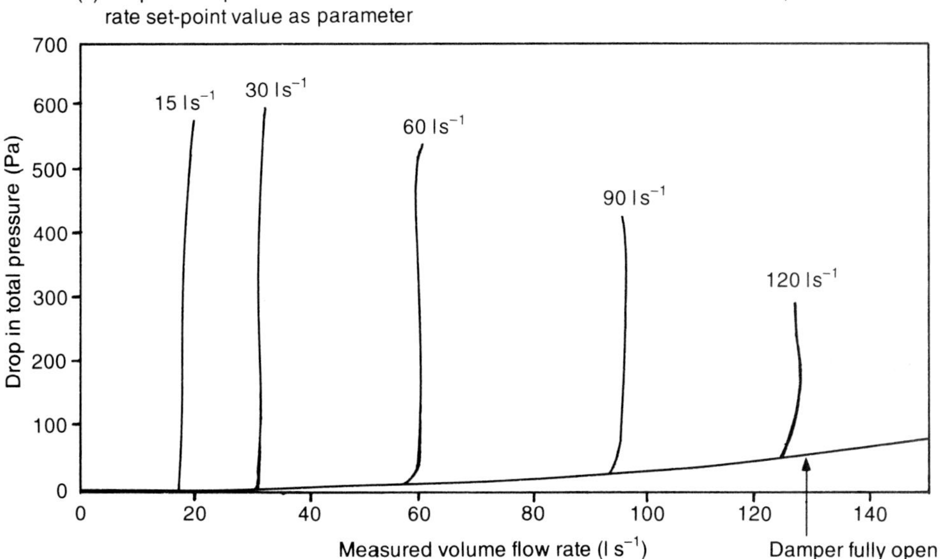

Fig. 7.33 (Continued).

a much greater proportion of operating time annually will be spent at zone volume flow rates below 60% of the unit's maximum rating (and hence below best sensing accuracy) than above it. The lower limit of the terminal unit's volume flow rate control range may be dependent on the type of control employed – whether analogue or digital, electronic or pneumatic. A reduced value is typically quoted for electronic controls. In any event this performance-limited lower end of the volume flow control range should not to be confused with $\dot{V}_{MIN}$, the minimum volume flow rate set-point for the the terminal unit's variable-constant-volume controller. The latter is specified by the system designer for each terminal unit, to suit the needs of the particular application – typically in terms of maintaining either a minimum ventilation rate or adequate room air distribution performance (or both).

Achieving the specified volume flow rate control range will typically require the static pressure differential across the terminal unit to be within a range defined by the minimum requirement (damper fully open) and a specified maximum value (typically of the order of 1500 Pa). An excessive pressure differential across the terminal unit may lead to leakage of conditioned air from the casing and its connections, and to the possibility of physical damage to the unit and its controls. Depending on its pressure–volume flow rate characteristic, the damper may operate near the fully closed position, resulting in a potential for control instability, generation of disturbed flow patterns in both the inlet spigot and downstream of the terminal unit, and in high noise generation by the damper. Where the supply diffuser

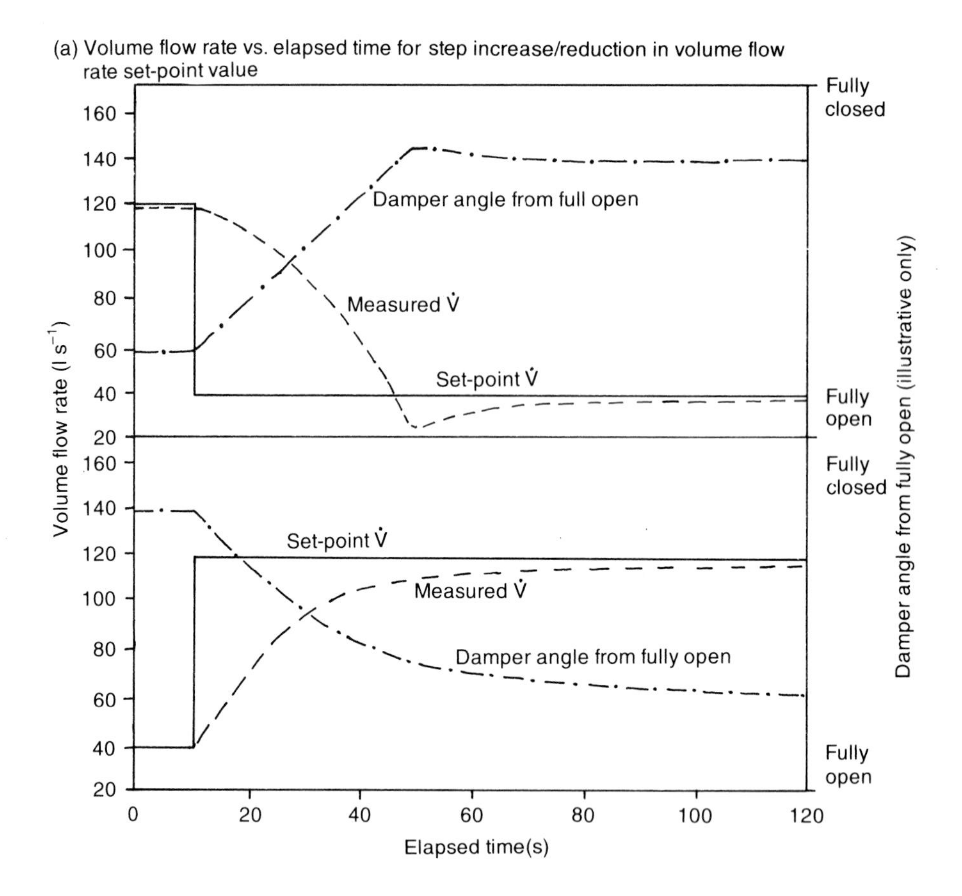

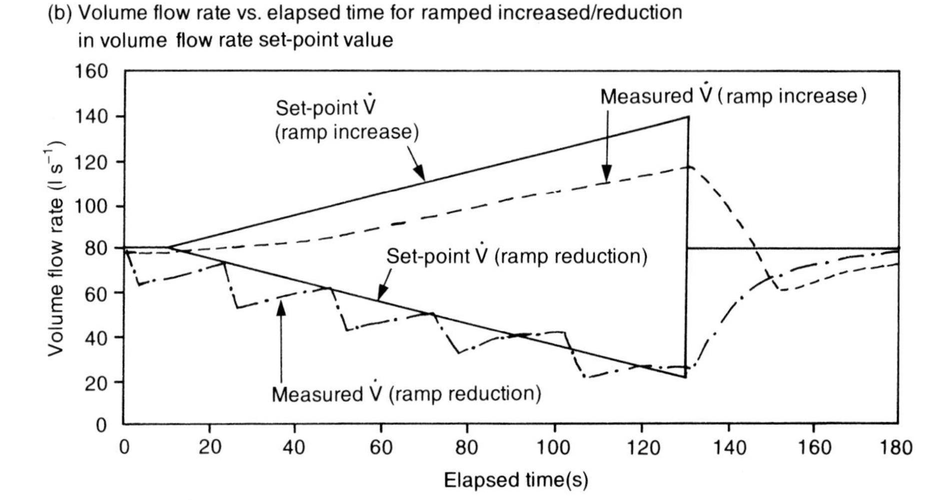

Fig. 7.34 Dynamic performance of a VAV box with variable-constant-volume ('pressure-independent') action.

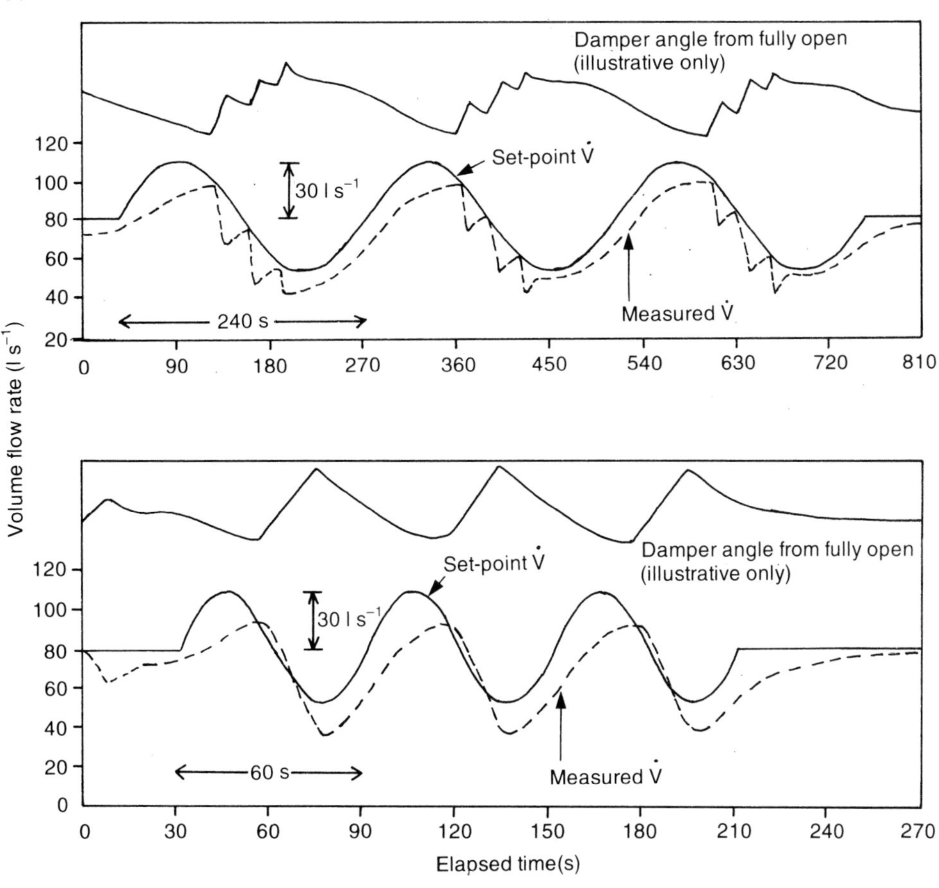

Fig. 7.34 (Continued).

serving the zone is an integral component of the terminal unit, ensuring that an adequate, yet not excessive, throw is achieved forms part of the terminal unit selection procedure. The performance of a VAV box, under test conditions, in terms of maintaining its set-point volume flow rate with increasing pressure drop across the unit, is shown in Fig. 7.33[7]. Such a terminal unit also has a dynamic performance (Fig. 7.34).

With fan-assisted terminals the required volume flow rate and downstream pressure losses are additional performance characteristics. Fan performance is typically quoted in terms of the static pressure loss downstream of the terminal unit. Conventionally equating the downstream loss of total pressure to these values allows for a loss of total pressure equivalent to one velocity pressure in the connection between the terminal unit discharge spigot and the downstream ductwork.

Methods for the aerodynamic testing of VAV terminal units are described in British Standard 4979:1986.[8] Single-duct throttling and induction types are covered, as are dual-duct and constant-volume units, but fan-assisted terminals are not included.

7.13.3 Acoustic performance

The acoustic performance of non-fan-assisted terminal units is typically expressed in terms of noise radiated from the casing of the unit and air-generated noise propagated along the ductwork upstream and downstream of the unit. This is typically given in terms of octave-band sound power levels at specified volume flow rates and static pressure differentials across the terminal unit. Sound pressure levels and NR or NC (noise rating or noise criteria) levels may also be quoted. However, care should always be taken in interpreting these types of data, since they are typically based on an assumption that 8 dB of room attenuation will be achieved across all octave bands in the actual installation.

With fan-assisted terminal units of the series-fan type, sound power levels are typically given for the combined noise radiated from the casing and through the induction ports for secondary air drawn from the ceiling return air plenum. The two extremes of operation are 100% primary air (induction ports closed) and 100% secondary air (fan noise only). Noise propagated downstream of the terminal unit is likely to be determined by the fan, irrespective of the volume flow rate through (or pressure drop across) the primary supply air damper. Where NR or NC levels for case-radiated and induction port noise are quoted, the values may include a combined allowance of up to 14 dB for ceiling transmission loss and room attenuation, again across all octave bands. As with most HVAC equipment, there is a considerable variation in the format and detail in which both aerodynamic and acoustic performance data are presented by the manufacturers and distributors of VAV terminal units. Performance considerations for VAV terminal units are summarised in Table 7.3.

7.13.4 Sizing

Diameter of the inlet spigot is a typical measure of terminal unit size. The smaller the terminal unit in any range, the higher will be the minimum pressure differential required across the unit for any given volume flow rate. The flow sensing unit of a variable-constant-volume terminal unit of the type shown in Fig. 7.2 is calibrated between zero flow and the $\dot{V}_{NOM}$ rating specified by the manufacturer. Volume flow rate cannot therefore be increased beyond $\dot{V}_{NOM}$ simply by increasing the pressure at the terminal unit inlet. In any product range the number of terminal unit sizes available is inevitably limited by commercial considerations. Hence both the required

Table 7.3 Performance considerations for VAV terminal units.

Performance characteristic	Expressed as
Aerodynamic	*Pressure-dependent terminal unit* Volume flow rates at specified static pressures at inlet to terminal, with damper fully open *Variable-constant-volume or 'pressure-independent' terminal unit* Minimum static pressure differentials across terminal unit for specified volume flow rates, with damper fully open *Fan-assisted terminals or FATs* Volume flow rates at specified static pressures at downstream outlet from the terminal unit
Pressure differential across terminal unit under steady-state conditions	May be expressed as a velocity pressure loss factor, which is a non-linear function of damper position
Operating pressure limit	Maximum static pressure differential across terminal unit (typically 1500 Pa)
Control range	Upper limit ($\dot{V}_{NOM}$) Lower limit as percentage of upper
Accuracy of flow sensing (variable-constant-volume or 'pressure-independent' terminal units only)	Percentage of set-point volume flow rate. Varies with flow rate (expressed as proportion of upper limit of control range for the particular terminal unit, $\dot{V}_{NOM}$), e.g. $\pm 5\%$ for $\dot{V} \geq 0.6\ \dot{V}_{NOM}$, decreasing linearly to $\pm 10\%$ over range $0.4 \leq \dot{V} \leq 0.6\ \dot{V}_{NOM}$, and to $\pm 20\%$ at $\dot{V} = 0.2\ \dot{V}_{NOM}$
Room air distribution (where terminal unit discharges directly to space via integral supply diffuser)	Throw
Acoustic	*Throttling-only terminal units* Case-radiated noise; air regenerated noise in ductwork on low pressure (downstream) and high pressure (upstream) side of terminal unit. Defined in terms of octave band sound power levels at specified volume flow rates and static pressure differentials across terminal unit. *Fan-assisted terminals or FATs* Fan-only: discharge noise and combined case-radiated and induction port noise. Defined in terms of octave band sound power levels at specified volume flow rates and static pressures at downstream outlet from the terminal unit. Fan and 100% primary supply air: combined case-radiated and induction port noise. Defined in terms of octave band sound power levels at specified volume flow rates and static pressures at downstream outlet from the terminal unit, for specified static pressures at primary supply air inlet to terminal unit.

zone design volume flow rate ($\dot{V}_{MAX}$) and the designer's specified minimum zone volume flow rate ($\dot{V}_{MIN}$) must fall within a realistic control range for the terminal unit. In this respect it should be remembered that $\dot{V}_{MIN}$ is typically expressed as a percentage of $\dot{V}_{MAX}$, rather than of $\dot{V}_{NOM}$. If $\dot{V}_{MAX}$ is 80% of $\dot{V}_{NOM}$ and $\dot{V}_{MIN}$ is 30% of $\dot{V}_{MAX}$, then $\dot{V}_{MIN}$ is equivalent to 30% of 80% of $\dot{V}_{NOM}$, i.e. 24% of $\dot{V}_{NOM}$. If two sizes of terminal unit in a range both satisfy these criteria, a comparison suggests that the smaller unit will:

(1) be the cheaper;
(2) exhibit improved accuracy in volume flow rate measurement;
(3) require a higher static pressure at the inlet spigot for any volume flow rate;
(4) probably generate more noise.

The higher inlet static pressure requirement implies that less throttling action will be needed from the damper for volume flow rate control, improving the quality of control effected by the terminal unit.

References

1. Hartman, T. (1989) TRAV: a new HVAC concept. *Heating, Piping and Air Conditioning*, **61**, July, 69–73.
2. Hartman, T. (1993) Terminal regulated air volume systems. *ASHRAE Transactions*, **99**, Part 1, 791–800.
3. Hartman, T. (1993) *Direct Digital Controls for HVAC Systems*, McGraw-Hill, New York, 110–120.
4. Haessig, D. (1994) Variable air volume controls for VAV fan terminals. *AIRAH Journal*, **48**, November, 29–32.
5. Graves, J.R. (1986) Evolution of intermittent fan terminal. *ASHRAE Transactions*, **92**, Part 1B, 511–518.
6. Khoo, I., Levermore, G.J. and Letherman, K.M. (1996) Findings of a UMIST VAV research project. In *Variable Flow Control*, General Information Report 41, BRECSU, Garston/Watford.
7. Shepherd, K.J. (1995) Variable-air-volume systems for optimised design and control, M.Sc. thesis, UMIST.
8. British Standards Institution (1986) *Aerodynamic Testing of Constant and Variable Dual or Single Duct Boxes, Single Duct Units and Induction Boxes for Air Distribution Systems*, British Standards Institution, London, BS 4979: 1986.

Chapter 8
Supply Fans and Air Distribution in VAV Systems

8.1 Fans in VAV systems

VAV systems typically utilise two fans – a supply fan and a return fan. Depending on the needs of the application and the overall design approach, these may be either single units or duplicate fan sets (run and standby). Almost invariably the fans are of three types – forward-curved centrifugal, backward-curved centrifugal or axial, although mixed-flow types are possible. The principal areas of interest in the present work are the energy consumption and control of fans in VAV systems. The general design, construction and operating characteristics of each of these fan types have been fully discussed and well documented by other authors, and it is therefore not proposed to dwell on these basics here. Any reader new to the subject, or seeking to refresh himself or herself on the basics of fan engineering is referred to the standard industry reference on the subject in the UK, *Woods Practical Guide to Fan Engineering*[1], to any standard text on air conditioning[2,3], or to the latest editions of either the CIBSE Guide (Volume B) or ASHRAE Handbook (Equipment).

8.2 The significance of fan energy consumption

Globally air handling and pumping as a whole may account for as much as 30–40% (3–4 × 10^9 MWh) of all electricity consumed for all uses[4]. At around four to five times the cost of natural gas or oil per kWh of delivered energy, the nature of electricity as the premium-cost fuel is reflected in the high cost of fan and pump power. In the UK alone, AC electric motors driving fans and pumps for building services are estimated to consume approximately £850 million worth of electricity annually[5]. The environmental impact of this is thus particularly significant, since each kWh of electricity consumed by an electric motor involves the release of approximately 3.4 times as much carbon dioxide into the atmosphere as each kWh of natural gas, and approximately

2.5 times as much as each kWh of oil. Typical figures are respectively 0.7, 0.29 and 0.2 kg kWh^{-1} for electricity, oil and natural gas.

Recent research carried out on behalf of the Department of the Environment showed that, over a limited sample of 55 non-domestic buildings, mechanical ventilation and air conditioning systems on average accounted for approximately 13% of total deivered energy, 14% of primary energy and approximately 14% of total CO_2 emissions. While overall air conditioning in buildings may still account for a relatively small proportion of total energy consumption in the UK, fans and pumps typically account for approximately 40% of the combined annual consumption for HVAC and lighting, or up to 20% of total delivered energy[6]. This is approximately twice the consumption by refrigeration plant, and in cost terms may represent between one-quarter and one-third of total delivered energy cost in buildings typical of the national stock of air conditioned office space as a whole. In a VAV system the main supply and return fans will typically account for between one-half (at least) and two-thirds of the total energy consumption of all fans and pumps associated with the system. The annual energy consumption of the system supply fan typically dominates this, accounting for up to four times that of the return fan.

Since they produce much greater savings in fossil fuel consumption at the point of generation, reductions in electricity consumption on site are therefore of much greater benefit, both economically and environmentally, than equivalent reductions in fossil fuel consumption on site.

Undoubtedly the best way to reduce the energy consumption for air conditioning in buildings is to design or refurbish them to avoid the need for air conditioning, or at least to reduce its extent. However, the next most promising route to reducing the energy consumption of air conditioned buildings is clearly to minimise the consumption by system fans.

8.3 The need for control of VAV supply fans

Taken on its own, the cumulative effect of throttling action by individual VAV terminal units is sufficient to control system volume flow rate. Without any further action, however, the supply fan would simply follow (or 'ride') its characteristic. This typically results in an increase in fan total pressure with a reduction in volume flow rate. The increase in pressure may be of the order of 20% for a reduction in volume flow rate of 10% of the design value, although the effect may be minimised if a fan with a relatively flat pressure–volume flow charateristic can be employed. However, it is of principal concern in modern HVAC design that this inevitably results in wasteful fan energy consumption, since the system actually needs less fan total pressure with reducing volume flow rate. The excess must simply be throttled off somewhere in the system, either at the terminal units or at pressure-reducing dampers.

In addition to energy waste, there are other potential disadvantages which can result if fan total pressure is unconstrained. These include the risk of physical damage to the system, increased noise levels and unstable fan operation. Hence conventional design practice has adopted a duct static pressure control loop in VAV systems.

8.4 Methods of fan duty control

Maintaining duct static pressure by modulating a simple throttling damper positioned in series with the fan in the main supply duct was an early approach, and capable of avoiding the risk of physical damage to the system associated with unconstrained fan operation. However, the throttling action is itself a potent source of noise generation, and any potential instability associated with riding the fan curve remains. More importantly, as far as present-day VAV systems are concerned, throttling of unrequired excess pressure still represents wasteful fan energy consumption. Where duct static pressure control is employed, the relationship between system pressure loss and volume flow departs from the simple 'square law'. However, for simplicity some idea of the potential extent of wasteful throttling with damper control of fan duty is given by considering that, in such a 'square-law' system, a reduction from full design duty to 70% of this might require throttling of the order of 40% of design fan total pressure. Since fan power is directly proportional to fan total pressure, a significant waste of fan energy consumption is clearly implied. Thus conventional practice includes the VAV supply fan itself in the duct static pressure control loop. This either modulates fan *speed*, in the case of centrifugal fans, or *blade pitch angle*, in the case of variable-pitch-in-motion axial fans, to maintain a set-point value of static pressure at some specified point in the ductwork supplying the VAV terminal units. This approach is shown schematically in Fig. 8.1. Speed control of axial fans is also possible. However, it is seldom favoured by HVAC designers since equivalent performance may be achieved at lower capital cost by both VPIM axial fans and the centrifugal fan/AC inverter drive combination.

Damper control retains some potential in VAV applications employing fan bypass techniques. These are typically confined to low-duty fans serving a relatively small number of zones by comparison with mainstream VAV applications. The approach originated to allow VAV techniques for individual zone control to be extended to the kind of light commercial applications that could not afford the capital cost premium or complexity of speed control or variable-pitch-in-motion fans. The principle is illustrated in Fig. 8.2,

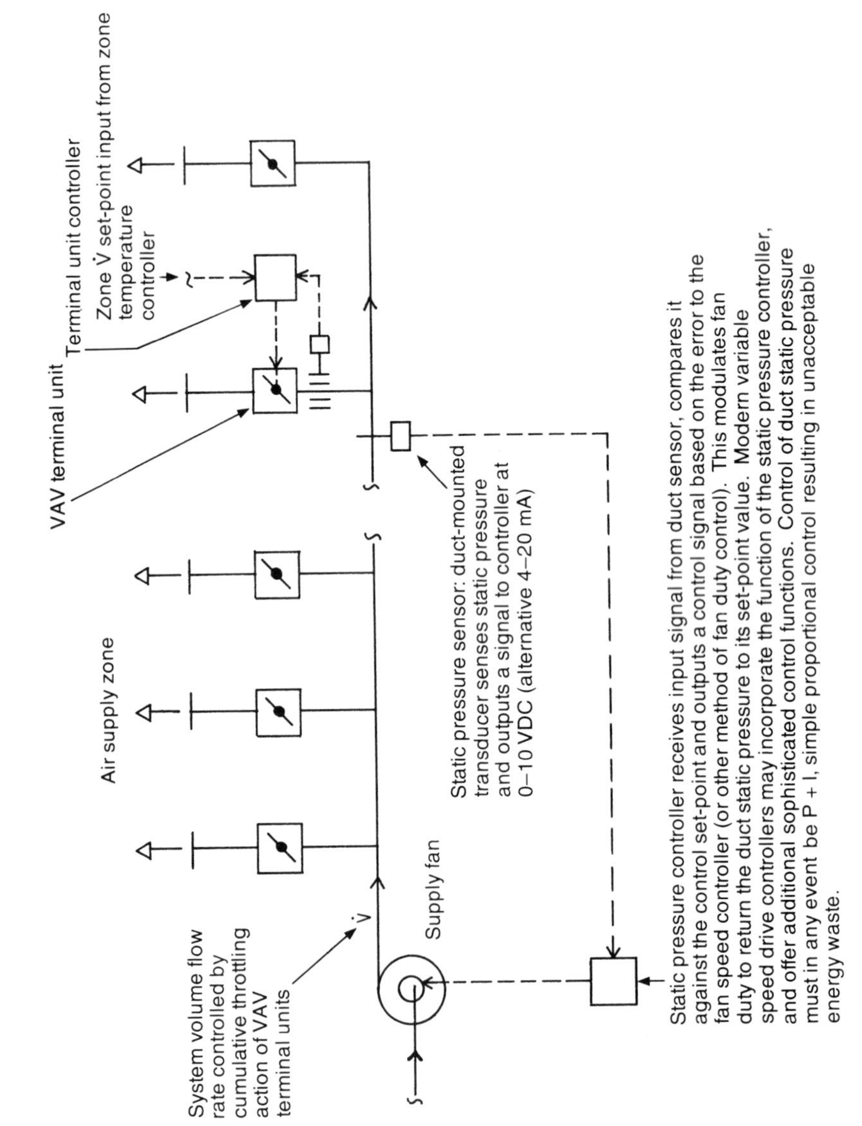

Fig. 8.1 Duct static pressure control.

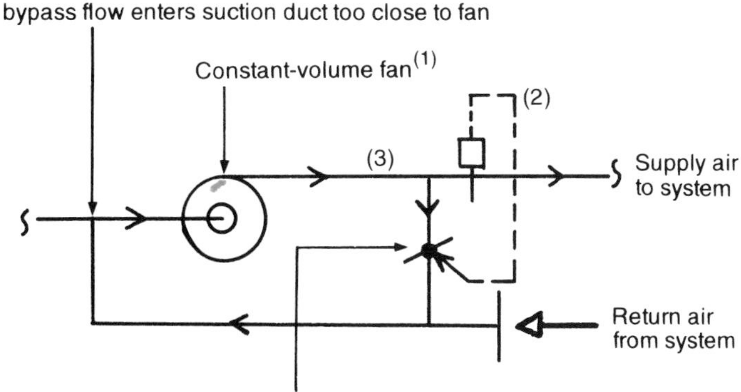

Bypass duct with throttling damper allows air to short-circuit back to fan inlet, bypassing the downstream system (2)

Notes:

(1) If fan will 'ride' its pressure~volume flow rate characteristic to any significant extent, a forward-curved centrifugal type may be unsuitable on account of its rising power characteristic.

(2) Modern implementations of the bypass damper technique typically modulate the damper to maintain constant static pressure at a sensor location close downstream of the bypass.

(3) Any non-uniformity or disturbance in the main flow approaching the bypass duct may introduce substantial pressure losses at the branch and adversely affect flow conditions on to the bypass damper.

Fig. 8.2 Bypass VAV systems.

where the damper modulates to bypass air between the main supply and return air ducts. Since there is no system return fan, this is in effect a bypass from fan discharge to fan suction. Depending on the control technique, the fan may or may not be allowed to ride its curve. If so, fan duty will increase with reducing system load, perhaps reaching a level of the order of 130% of design for a combined flow to the zones of 70% of design value (the remainder bypassing direct to the fan suction). In this event some saving in fan power is likely, but since fan power is directly proportional to volume flow rate, the waste inherent in the technique is clearly evident. If the control technique maintains the fan operating point, operation is constant-volume with respect to the fan, and the system offers no saving in fan energy consumption through VAV operation at the zones.

The development of solid-state inverter technology has lead to the combination of a centrifugal fan equipped with an AC inverter *variable-speed drive* (VSD) increasingly becoming the first-choice for VAV systems, although theoretical considerations suggest that the *variable-pitch-in-motion* (VPIM) fan most closely approaches the ideal power input–volume flow characteristics, whether the system operates at constant resistance, constant pressure or (as in the case of VAV supply systems) a combination of the two.

A variety of methods of controlling fan speed are available (Table 8.1).

Table 8.1 Methods of fan speed control and their principal characteristics

Characteristics	AC inverter	Switched reluctance drive	Eddy current coupling
Method of variable-speed control	Solid-state electronic control of frequency/voltage output to fan motor	Solid-state electronic control of pulsed DC output to fan motor	Varying the slip in an electro-magnetic drive coupling between motor and fan.
Type of motor	Standard 'squirrel cage' AC induction or wound-rotor types	Special brushless motor	Standard 'squirrel cage' AC induction motor
Approximate power range	Up to 300 kW	Up to 75 kW	Up to 90 kW
Typical speed range (% full speed)	10–100% (PWM; minimum higher for CSI)	*c*.1–100%	10–95% (decreases as drive rating increases, fan speed always $<$ motor, as always slip)
Speed control accuracy and response	Good (to $\pm 1\%$ of set speed, typically sufficient for HVAC); very good (to $\pm 0.1\%$) with tachogenerator feedback. Good response with vector control	Best accuracy (to *c*. $\pm 0\%$ of set speed) Very good response	Very good accuracy. Adequate response for typical HVAC needs
Multiple-motor control capability	Yes (PWM, not CSI)	Facility not available	Facility not available
Typical efficiency at full speed[1]	86% (PWM); 81% (VSI, CSI)	91%	84%
Typical efficiency at 50% speed[1]	82% (PWM); 70% (VSI, CSI)	86%	40%
Power factor	PF ≈ 1 at all loads if suitable DC link filtering	PF ≈ 1 at all loads	PF varies with load
Motor sizing	No derating with modern IGBT transistors; 5–10% with older bipolar type	No derating	No derating
Starting	'Soft-start' on drive at *c*.110% FLC; require separate starter to bypass failed drive	'Soft-start' on drive at *c*.25% FLC	Standard DOL or star-delta; start with motor unloaded limits starting current duration
Isolation	Conventional for modern types; older may require power supply contactors	Conventional	Conventional
Motor power cabling	Three-core SWA cable required; four-core may be recommended (fourth core as earth)	Six-core SWA cable required	No special requirements; as normal 'squirrel cage' induction motor
Generation of harmonics	High	High	None

Electromagnetic compatiblity (EMC)	Unacceptable levels of RFI possible Segregate cables; SWA as screen; keep motor cable short	Needs to be considered	Good
Noise and vibration	Controller facility to avoid natural resonant speeds typical Higher noise from motor, cooling fans; modern types use higher carrier frequencies, tuning of variable switching frequencies to improve	Typically no facility to avoid resonant speeds Characteristic pulsing of drive gives relatively higher noise level, though improved on modern types	No facility to avoid resonant speeds
Reliability and maintenance	Modern types considered reliable Failed drive can be bypassed if additional motor starter fitted Minimal maintenance	Motor inherently simple, reliable; failed controller cannot be bypassed Not affected by loss of 1 ph of supply. Minimal maintenance	Mature technology, proven, reliable Bypass of coupling not possible; full speed possible (modern types) on failed controller if 240 V direct to coupling Minimal maintenance
Commissioning	Specialist procedures required	Specialist procedures required	Specialist procedures required
Control	Speed compared against reference set from 0–10 VDC or 4–20 mA control signal, or via DDC software Sophisticated functions typical (max/min speeds and accel./decel. rates, resonant speeds, etc.) Serial link comms facility for BMS/DDC typical (if suitable protocol)	Output from tachogenerator on output shaft compared against reference set from 0–10 VDC control signal Sophisticated control functions generally available as for inverter Serial link comms facility for BMS/DDC typical	Output from tachogenerator on output shaft compared against reference set from 0–10 VDC control signal Control functions typically less sophisticated than for inverters, SRDs (accel./decel. rates, torque limit possible) Serial link facility untypical
Space requirements	Additional plant room space for inverters No need for drive to be in line with motor, but local position optimum for EMC	Motor/controller size compatible with conventional 'squirrel cage' induction motor and inverter	In-line configuration of drive increases plant room space required 'Piggy-back' possible at cost of additional losses (drive belts)
Typical capital cost	Higher than ECC in new installations Typically lowest for retrofits	Highest in new and retrofit applications	Lowest in new installations Typically highest in retrofits
Applications[2]	New; specially suited to retrofits	New; cost restricts retrofit potential	New ('niche' if no harmonics and good EMC important); retrofit restricted by space needs

Notes
(1) Source: Energy Technology Support Unit and FEC Consultants Ltd. Energy Efficiency Office Good Practice Guide 14 (1991) *Retrofitting AC Variable Speed Drives*. Garston/Watford: Building Research Energy Conservation Support Unit, BRECSU. Efficiencies are for complete VSD and motor system.
(2) Comparisons are restricted to typical commercial-type HVAC applications. The implications of applying VSDs in hazardous environments vary, and should always be the subject of confirmation by the manufacturers or suppliers of motors and drives, and supported by specific risk assessments.

Prior to the widespread adoption of AC inverter VSDs, the use of variable-speed *drive couplings* was popular for VAV fans. These were typically based on using an *eddy-current coupling* to provide *variable slip* between the driving motor and the fan. The technique remains a valuable alternative to inverter drives in applications where the additional *harmonic distortion* associated with the latter is unacceptable. Of the remaining techniques, only *swiched reluctance drives* are applicable to typical VAV applications, and there is as yet little published data on their use, either in HVAC applications in general, or with VAV fans in particular.

Another formerly common technique of control for centrifugal VAV fans was by modulating the position of guide vanes radially-disposed around the inlet to a backward-curved impeller, relying on changing the impeller aerodynamics rather than varying its speed. Modulating the *inlet guide vanes* (IGVs, also known as variable inlet vanes or VIVs) imparts a controllable degree of swirl to the airstream entering the impeller. This alters the pressure–volume characteristic of the fan, rather than simply throttling the air flow. A relatively low-cost technique in comparison with methods of speed control formerly available, or with the alternative of a VPIM axial fan, its former popularity means that the technique may continue to represent a significant proportion of existing VAV installations for some time. The principle is illustrated in Fig. 8.3. Decreasing angle α_1 adds swirl in the same direction as impeller rotation, and reduces the pressure rise across the impeller. Since some of the torque that would otherwise have to be applied by the impeller to

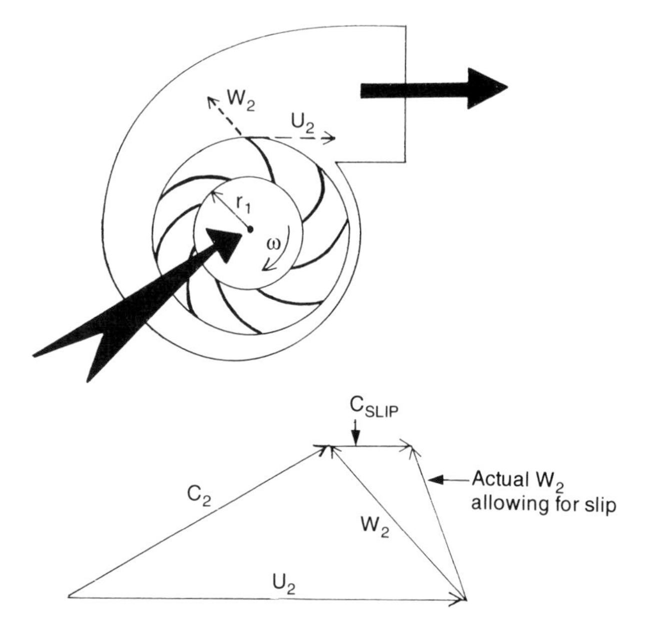

Fig. 8.3 Inlet guide vane control of fan duty.

achieve a given pressure rise is applied by the inlet guide vanes, which apply it without consuming power, impeller power is decreased for a given pressure rise. However, the flow conditions at inlet to a centrifugal fan with partly-closed IGVs are certainly more complex than Fig. 8.3 can suggest, and poorly designed vanes may exert unpredicted effects on fan performance.

Fan modulation using inlet guide vanes is typically capable of turndown to low system load factors, although unstable flow conditions may develop at system volume flow rates below approximately 30% of design value. In any event, fan efficiency drops more rapidly than for speed control or VPIM axial fans, so that power input diverges markedly from levels associated with these techniques as system load factor reduces. Using the standard vehicle for comparison of a pure 'square-law' system, power input with inlet guide vanes at 50% of design volume flow rate may be of the order of 2.5 times that with speed control. However, the technique would formerly have been comparable in performance to the combination of a constant-speed motor with eddy-current coupling at system loads greater than approximately 60% of design.

Inlet guide vanes may add some pressure loss to the system when fully open under design conditions, although the effect on the normal fan characteristic should be minimal if the guide vane assembly is purpose-made and manufacturing standards are adequate. As with dampers generally, performance will be adversely affected by poor installation and play in the operating linkages, whether resulting from poor manufacturing quality or wear. Regular inspection and maintenance is necessary for reliable and accurate operation. High torque requirements may necessitate the use of high-pressure actuators in pneumatic control systems. DIDW fans require two sets of IGVs, leading to a complex linkage.

Alternative mechanical methods of varying the duty of a centrifugal fan include *variable-ratio pulley systems*, which are mechanically complex, and *disc throttles*. The latter offers the means of altering the effective width of the impeller (Fig. 8.4), and allows controlled turndown to very low system flow rates. Disadvantages of the technique include the pressure loss associated with the throttle and the double linkage required for DIDW fans (although respectively lower and simpler than for IGVs). Again regular maintenance is required to ensure reliable operation. There is little evidence to suggest that the disc throttle technique has been used in VAV systems to any significant extent.

Standard comparison between methods of controlling fan duty are typically given for three ideal cases – constant system resistance (i.e. exhibiting a pure 'square-law' relationship between system pressure loss and volume flow rate), constant system total pressure and constant system volume flow rate. As noted previously, a VAV system employing duct static pressure control does not follow the classic 'square-law' relationship between system total

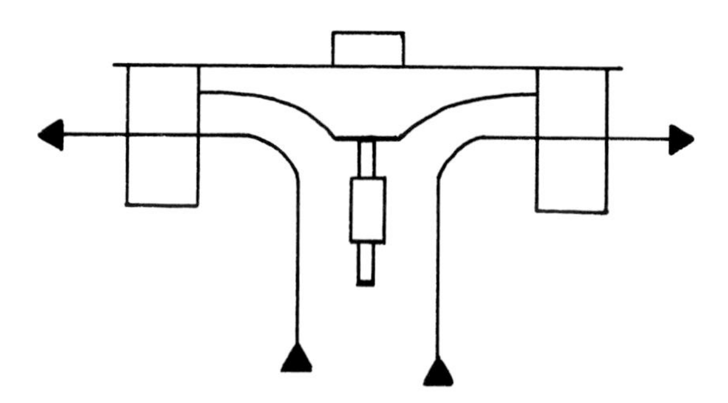

Fig. 8.4 Disc throttle control of fan duty.

pressure loss and volume flow rate. This point must be clearly understood, since otherwise any assessment of the energy-saving potential of the various techniques available for fan duty control in a VAV system will be significantly in error. For reference the 'square-law' case is shown in Fig. 8.5. A number of comparisons and studies of methods of modulating centrifugal and/or axial fans are available in the literature[7–12]. All add their weight to the supremacy in energy-saving potential of either the combination of a centrifugal fan with inverter drive or a variable-pitch-in-motion axial fan for VAV systems. Potential annual energy savings of the order of 50% were shown for these types of duty modulation in comparison with inlet guide vane control, making the refurbishment of existing IGV installations with AC inverter drives an attractive proposition.

Since efficiency may vary widely between fans, selection of a high-efficiency model is always desirable. Furthermore, since VAV fans typically spend so much of their annual operating time at relatively high turndown, it is important to consider efficiency in this operating range when selecting a fan. Belt drives allow the permanent connection of standby motors, together with location of the motor ouside of the airstream. This gives the advantage of both reduced heat gain to the conditioned airstream and lower noise transmission, but at the expense of increased energy losses in the belt drive itself. Belt losses are typically quoted at 5–10% of input power, and can be significantly more than this for small drives, or when belts are incorrectly fitted and inadequately maintained. Cogged V-belts offer improved drive efficiency, while further improvements are claimed for flat or synchronous belts.

The process of fan selection typically results in an oversized fan. Up to 20% more duty at the design system pressure loss could easily result from the use of a modest safety factor and the 'next size up' from a manufacturer's catalogue, and would be unlikely to cause any eyebrows to be raised. Cases of greater oversize than this may certainly be encountered. VSDs allow final adjustment of fan duty to the design requirements without the need for the

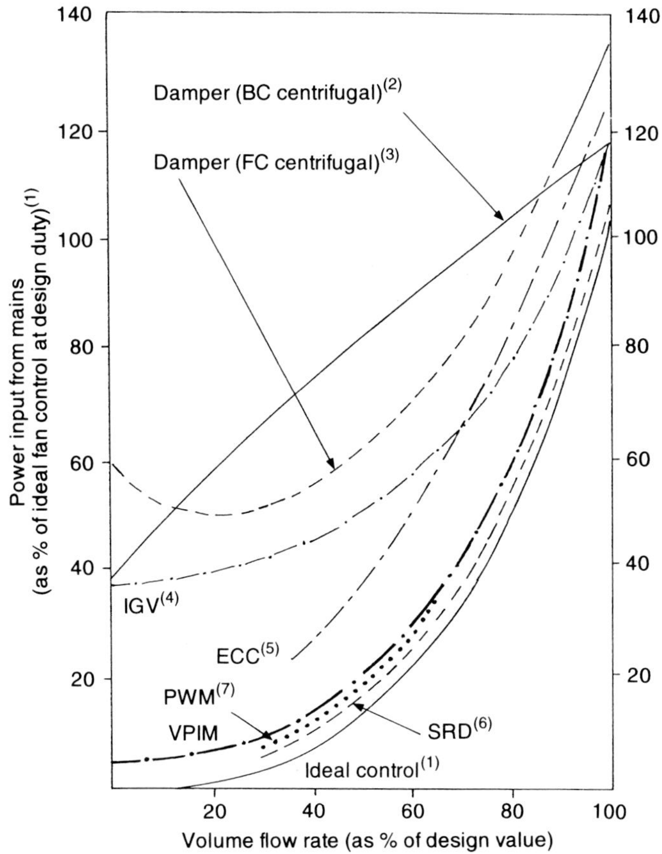

Notes:

(1) 100% power input @ 100% (design) volume flow rate.
(2) Backward-curved centrifugal fan.
(3) Forward-curved centrifugal fan.
(4) Inlet guide varies.
(5) Eddy current coupling.
(6) Switched reluctance drive.
(7) Pulse width modulation inverter.

Fig. 8.5 Comparison between methods of fan duty control for a 'square law' system (constant resistance). Source: Building Research and Energy Conservation Support Unit (BRESCU) (1996) *Variable Flow Control*. General Information Report 41. Garston, Watford.

traditional belt and pulley change. Hence the technique is equally applicable to direct-drive fans. While an oversized fan will be compensated by the duct static pressure control loop, this will typically be at the expense of poorer control, potentially reduced turndown (from actual duty under design conditions) and increased risk of fan instability at high turndown. Since power factor suffers where induction motors are oversized, gross oversizing of any fans should naturally be avoided. Shared-duty (parallel) operation of dual fans is less likely to be encountered (or cost-effective) in air systems than in water systems, where run and standby pump sets are common.

8.5 Variable-pitch-in-motion control of axial fans

Where minimum system pressure is to be maintained, as is typical in a VAV system, a centrifugal fan must inherently operate at a higher point on its fan curve. This may be a reduced-speed curve, but it is still a higher point on the curve, and typically associated with reduced efficiency. An alternative approach is offered by an axial fan employing the VPIM technique. This has the inherent ability to turn down to very low system load factors, and may typically be expected to maintain high fan efficiency over a wide range of load, perhaps decreasing by only 10% (absolute) from the maximum efficiency while turning down to 50% of design volume flow rate. However, care is required in fan selection to avoid *blade stall* over the full range of fan operation. Repeated instances can cause shearing of fan blades. Unless the risk can be avoided altogether, stall detection equipment is an option which can be linked to audible and visual alarms. In any event on detection of the onset of a stalled condition, control of a VPIM fan should be such as to pull it back and fully out of the stalled condition.

Figure 8.6 illustrates the principle of VPIM operation. The value of angle α, between the outlet edge of the blade and the tangent to the direction of rotation, is important in determining the performance of any type of fan, whether axial or centrifugal. Increasing α increases the forward component of air velocity, and hence volume flow rate, without substantially changing the 'swirl' component on which pressure development largely depends. The higher-frequency noise generated by axial fans is inherently more easily attenuated in duct systems.

Operation of the VPIM mechanism may be pneumatic, electro-pneumatic or electric. Pneumatic operation does not present any particular problems where a supply of instrument-standard clean and dry air is available. This typically requires a duplex air compressor set with receiver, filtration and refrigerated air drier. The cost premium of this additional equipment will be minimised in multiple-fan installations where the same compressor set can serve a number of fans. Reliability will suffer if any dirt and/or moisture is present in the pneumatic supply. Air leaving the cooling coil at close to dew-point temperature may adversely affect exposed elements of the operating linkage. The use of *positioners* is recommended to avoid hysteresis in the pitch control mechanism resulting from friction in the control linkage. The positioner will typically require a higher supply pressure than a conventional pneumatic control signal. Where VPIM actuation is pneumatic, the fan duty controller will almost certainly be an analogue electronic device, requiring a

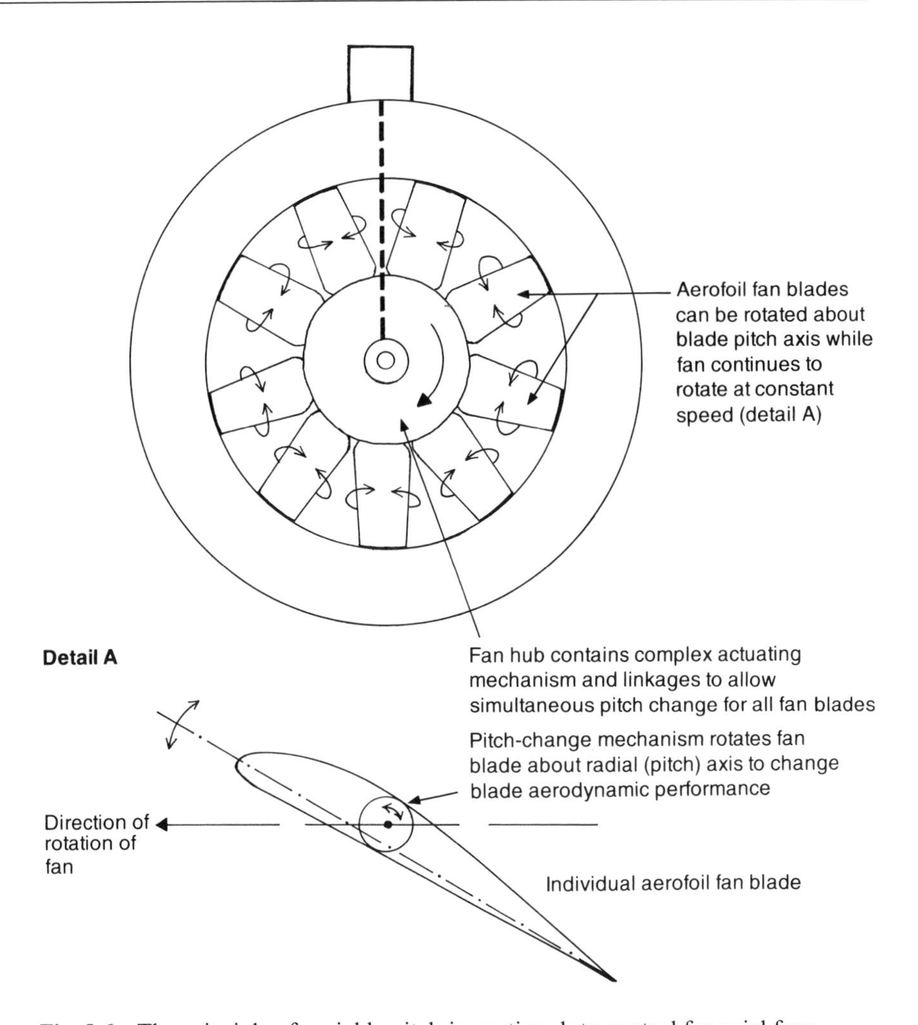

Fig. 8.6 The principle of variable-pitch-in-motion duty control for axial fans.

transducer to convert the 0–10 V controller output to the required 3–15 psi pneumatic range. Minimum start time, together with reduced starting current, can be achieved by ensuring that the fan blades are returned to their minimum pitch setting on shutdown.

In the past, torque requirements have made pure electrical actuation of VPIM fans rare, although this situation may be changing. Electro-pneumatic actuation typically utilises three-point control, allowing the blade pitch either to increase, decrease or stay the same. Direct control, via electro-pneumatic relays, of the mini-compressor supplying the pneumatic pitch-change actuation results in unstable operation. The necessary corrective measures are considered likely to degrade performance in comparison with that obtainable under pure pneumatic control[5].

With its apparently more-or-less ideal control characteristics, it is logical to question why the VPIM axial fan, which was available long before solid-state inverter variable-speed drive technology became economically feasible for centrifugal fans in HVAC applications, did not come to predominate in VAV systems. The answer is probably a function of both cost and the installational ease and convenience with which centrifugal fans can be incorporated into packaged air handling units. In any event the complex mechanics of the pitch change mechanism, together with additional pneumatics, inevitably leads to a significant maintenance requirement for VPIM fans, with a corresponding potential for extended down times.

8.6 Variable-speed drive control of centrifugal fans

Potential variable-speed drives (VSDs) for VAV fans fall into three basic categories:

(1) eddy current couplings (ECCs);
(2) inverters;
(3) switched reluctance drives (SRDs).

HVAC fans and pumps typically require relatively low starting torque from a VSD, in contrast to many industrial applications. Since drives which generate high starting torque are more expensive and use more energy, their application to HVAC systems is generally considered inappropriate. Since the VSD is an intermediate stage in the conversion of electricity from the supply into the air power generated by the fan, and inherently has an efficiency less than 100%, the total efficiency of the VSD/motor/fan will be less than that of the corresponding simple motor/fan combination. While the losses associated with any practical VSD will not outweigh the benefits of the flow regulation afforded, the extent to which drive efficiency may decrease with increasing turndown of fan duty is both a vital consideration in maximising these benefits and an essential point of comparison between alternative drives or drive technologies.

An eddy current coupling (ECC) is an *electro-magnetic drive coupling* that fits between a standard 'squirrel cage' AC induction motor and the fan that it drives (Fig. 8.7). The *constant-speed* induction motor drives a ferro-magnetic drum (or tube). This surrounds a poled rotor which is attached to the drive's output shaft. The coil on the rotor may be fed by a rotary transformer to eliminate slip rings. The torque produced by the drive is dependent on the electromagnetic effects caused by *slip* in the drive coupling, and this slip is a function of the current supplied to the coil. Speed control is achieved by varying this current (and the resulting slip) according to an error signal

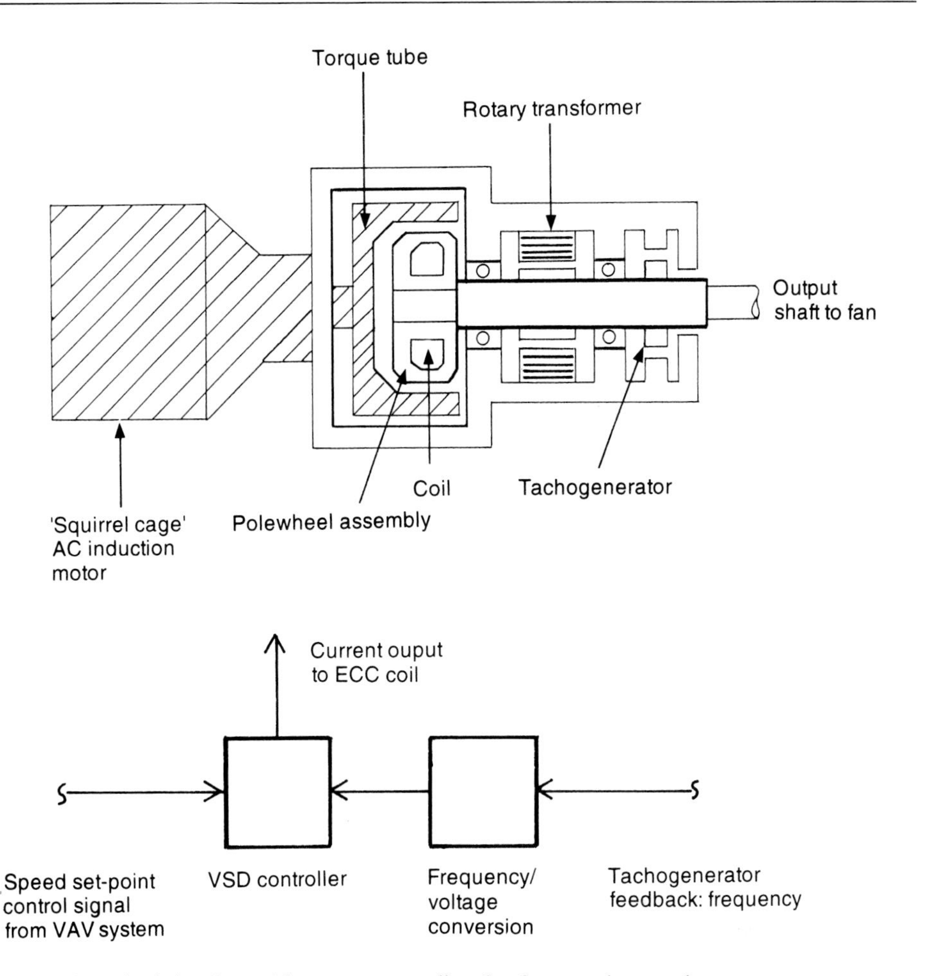

Fig. 8.7 The principle of an eddy current coupling for fan speed control.

derived by comparing a reference voltage with feedback from a tachogenerator mounted on the drive's output shaft.

ECCs represent mature technology. They are well proven and reliable. Since the technique does not rely on special motors, ECCs is may be bought separately. Because slip must *always* be present in the drive coupling, the maximum output speed of an ECC drive is always *less* than motor speed (e.g. 1300–1350 rpm for an input of 1450 rpm). There is thus an implication for fan selection, since it must achieve design duty at the maximum drive output speed. Drives are typically in-line, requiring additional space allocation in plant rooms. 'Piggy-back' configurations are available to improve on this, but require the addition of drive belts, with the additional losses that this involves. In earlier designs it was not typically possible to bypass a failed drive coupling, although modern types of ECC may have the facility to achieve full speed in this event. ECC controllers tend to be less sophisticated than their inverter drive counterparts in terms of the extent to which ancillary variables may be specified, such as acceleration and deceleration rates, maximum and minimum speeds and resonant speeds.

Athough capable of close speed control, the technique typically suffers higher losses with reducing fan volume flow rate than do inverter drives. Efficiency drops from being generally comparable to that of an inverter drive at full speed to perhaps half of this at 50% speed, dropping off rapidly as speed is further reduced. Since it is the considerable proportion of typical annual fan operating hours at low system volume flow rate that offers the key to minimising fan energy consumption, this is a considerable disadvantage of the ECC technique. However, ECCs are unaffected by harmonic distortion, which may be a significant factor in some applications.

The modern alternative to eddy current couplings is the solid-state AC inverter drive (Fig. 8.8). These control fan speed by modulating the frequency and voltage of the electrical supply to a standard 'squirrel cage' AC induction motor, which drives the fan either directly or via a fixed belt drive. The output voltage of the inverter VSD is regulated in proportion to the frequency, to maintain a relatively constant ratio between the two. This is necessary to limit motor currents and maintain the dynamic characteristics of the motor. The two principal components of an inverter drive are a *rectifier* and an *inverter*. The rectifier receives a standard 415 V/3-ph/50 Hz mains supply and converts it to either a fixed or variable DC output. The inverter takes this DC output and switches it on/off electronically, using thyristors or transistors, to produce an AC output. The pattern of this electronic switching is controlled to vary the frequency and voltage of the output, which is then

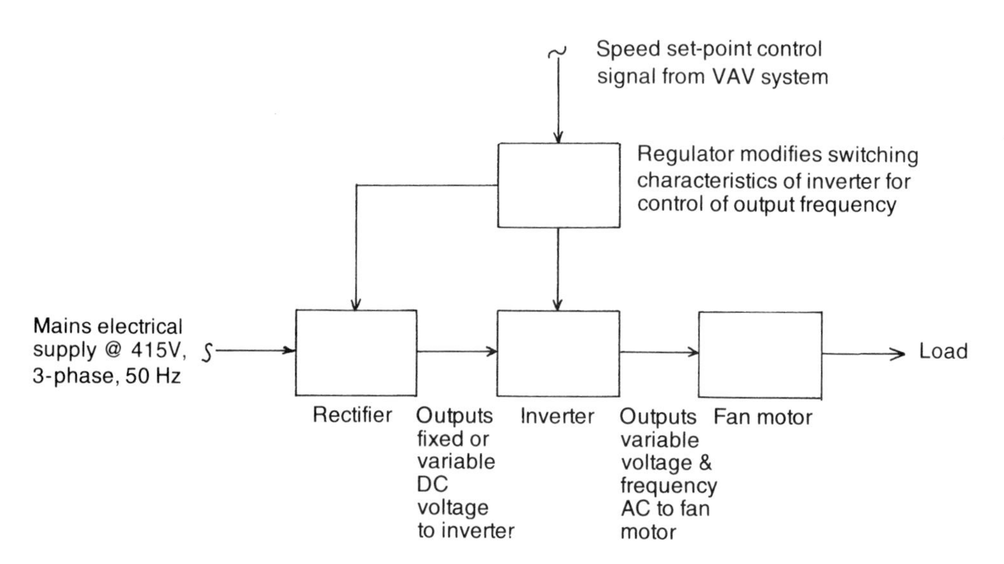

Fig. 8.8 Components of a solid-state AC inverter variable-speed drive.

fed to the motor connected to the drive. Varying the frequency and voltage of the motor supply directly varies its speed (reduced frequency resulting in reduced speed).

As their cost has fallen relative to other methods of fan duty control, solid-state AC inverters have become the first choice of VSD for most HVAC engineers. Inverters are the most efficient form of VSD, and are now generally considered to have good reliability. Their use of standard motor and drive arrangements makes inverters emminently suitable for retrofit applications.

There are three types of inverter:

(1) current source inverters (CSI);
(2) pulse amplitude modulation (PAM), also termed voltage source inverter (VSI);
(3) pulse width modulation (PWM).

A *current source* inverter varies the output voltage and frequency of a variable DC current from an adjustable source. Although having advantages of low cost and simplicity, CSIs are not generally recommended for building services applications. They are physically larger than the other types, with a lower operating efficiency. They are unsuited to both zero-load conditions and multi-motor installations, and have a power factor which varies according to load. Their use of controlled rectifiers leads to greater harmonic disturbance of the mains.

A *pulse amplitude modulation* inverter varies the output voltage and frequency of a variable DC voltage from an adjustable voltage source. The performance of PAM inverters is typically inferior to that of modern pulse width modulation types. Output voltage is varied in steps (Fig. 8.9) to give a current that approximates a sine wave. A six-pulse inverter exhibits a poor waveform. The waveform may be improved by increasing the number of pulses in each cycle, and 18 is the value recommended for PAM inverters. Use of a controlled rectifier results in greater mains disturbance through the generation of harmonics, while for PAM inverters an uncontrolled rectifier is allowed which gives near-unity power factor under all load conditions.

Pulse width modulation inverters vary the output voltage and frequency of a DC voltage from a fixed voltage source. The general configuration of a pulse width modulation inverter is shown in Fig. 8.10. In this type, which has come to be preferred over CSI and PAM inverters, the amplitude of the output voltage is not controlled, and voltage variation is achieved through rapid on/off switching. Modern insulated gate bipolar transistors (IGBTs) have allowed the frequency of this switching to be increased from a range between 2–5 kHz up to 20 kHz, allowing a current waveform that is much closer to the ideal sine wave. Voltage and current waveforms for a PWM

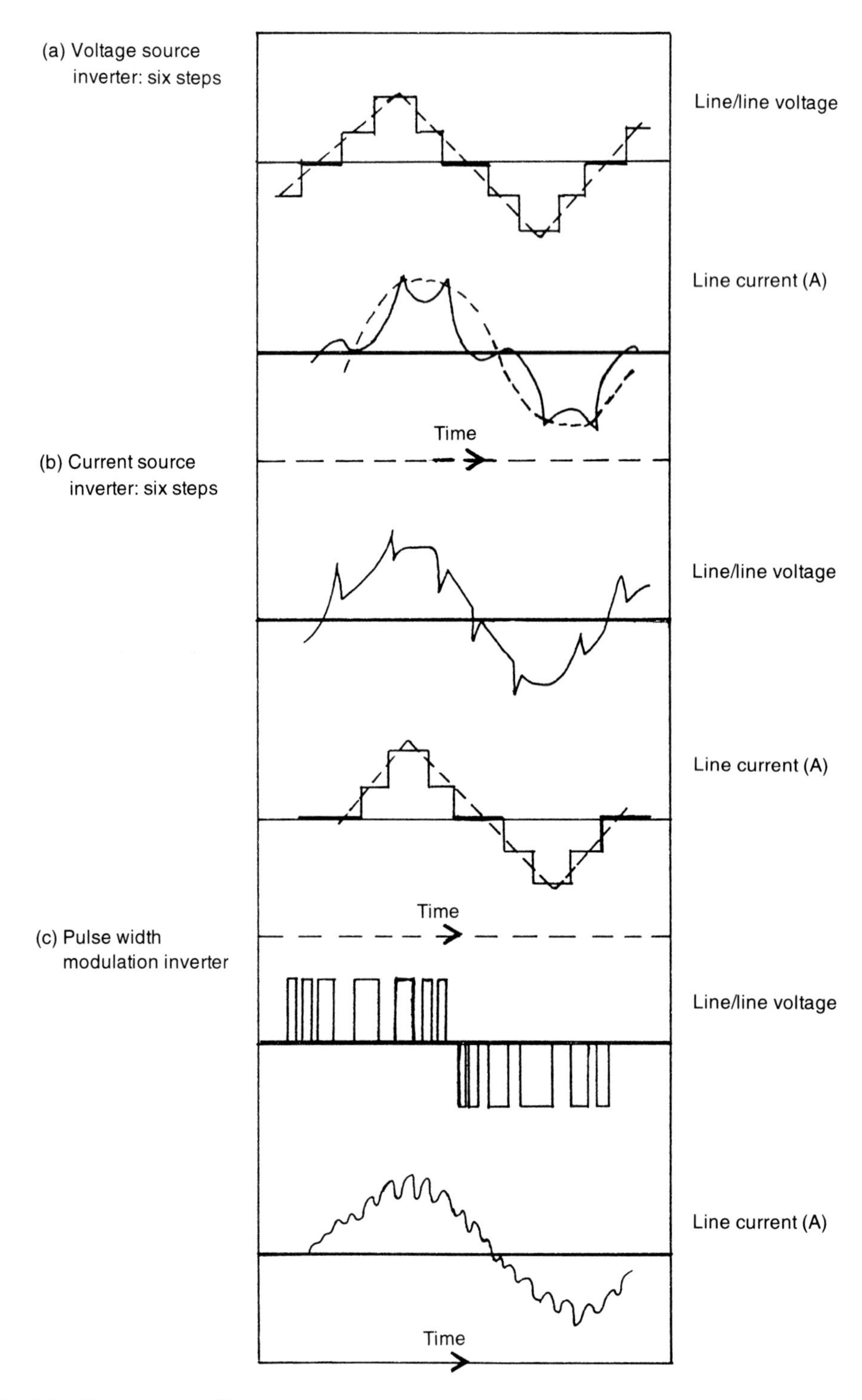

Fig. 8.9 Comparison of inverter outputs.

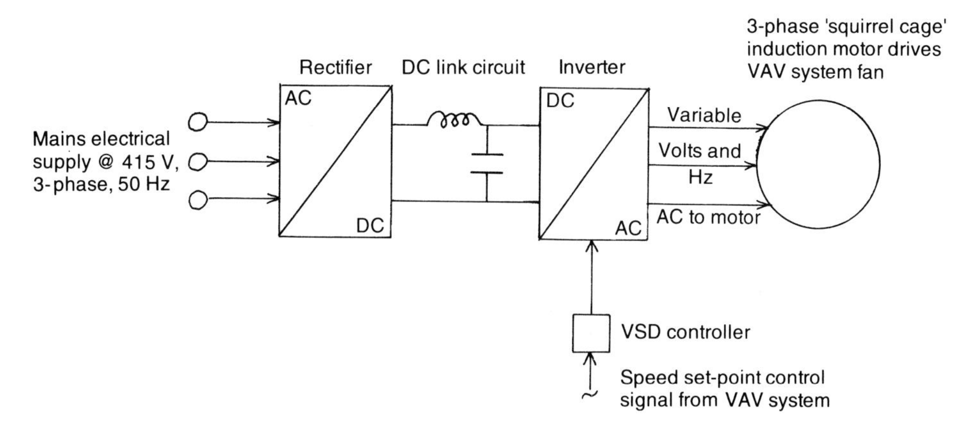

Fig. 8.10 Configuration of a PWM inverter.

inverter are shown in Fig. 8.9. Both approximate the pure sine wave that is the waveform of the mains supply. The closer to an ideal sine wave that is possible, the less likely it is that any derating of the driven motor will be necessary at full speed. Overall system efficiency is increased the closer the waveforms approximate a sine wave by reducing the losses associated with the generation of *harmonics*. While units with older bipolar transistors required derating of the associated driving motor, there is no such requirement where IGBTs are employed. Near-unity power factors can be achieved under all load conditions using suitable DC link filtering. The speed range of a typical PWM inverter may be $>10{:}1$, while speed regulation accuracy of $\pm 0.1\%$ is possible with feedback from a tachogenerator.

There have been two developments of the PWM inverter. The first is the *voltage vector control* (VCC) inverter. In this the voltage and switching frequency are both varied during the output AC cycle. Improvements in efficiency and speed control are claimed to result from this.

The second development of pulse width modulation is the *flux vector control* (FVC) inverter, also termed the 'space' or 'field vector' inverter. This technique takes into account the varying motor flux and torque as the rotor turns. Both three-phase current and feedback of rotor position are required, and the VSD must be matched to the characteristics of the motor being supplied. Vector control systems are highly responsive to control inputs, and high rates of change of output voltage are possible, giving improved motor dynamics and very good speed control – but at the expense of added cost and complexity.

Inverter VSDs typically generate noise from a number of sources, including the cooling fan, some of which are proportional to motor speed. Noise levels vary with mounting arrangments and switching frequency, and may be up to 10 dB(A) greater than for normal 50 Hz operation.

Unless it is necessary to bypass the drive in the event of failure, use of an inverter VSD eliminates the need for a separate starter, allowing a 'soft start' that typically limits starting current to approximately 110% of the motor's full load current (FLC). While star-delta starting of three-phase motors requires six wiring connections, the soft-starting inverter requires only three connections, with a consequent saving in both material and labour costs.

PWM inverter controllers typically offer the potential to set up sophisticated control configurations, including defining maximum and minimum operating speeds, limiting acceleration and deceleration rates, bypassing of resonant speeds, and the linked (on/off) control of one or more constant-speed motors. The ability to programme the controller not to stop at resonant speeds is important for variable-speed fan operation. In some cases the controller may itself provide a suitable P + I control algorithm for duct static pressure control in a VAV system, based on receiving an input from a duct static pressure sensor. An analogue (0–10 V) signal is typicaly used for speed control. Inverter VSDs may be provided with serial link communications facilities for linking with DDC controllers or BMS systems, subject to the availability of a suitable common communications protocol.

In order to avoid the risk of tripping of an inverter VSD on high temperature, or of a need to derate the drive, it is essential to provide adequate cooling of the drive's mounting enclosure. It is thus preferable to install inverter VSDs in individual panels or cubicles, each provided with its own cooling air flow. Units mounted in control panels may require a cooling fan to be fitted, and the provision of an over-temperature alarm. While some types formerly required the use of power supply contactors, isolation of modern inverter VSDs typically follows conventional practice in motor isolation. Inverter VSDs are sized in accordance with the frame size of the motor used, and should not be undersized. Motor cooling air flow is reduced at low speed, and adequate ventilation must be ensured where prolonged running at low speed is anticipated. Most inverter VSDs will tolerate an interruption of ≤ 0.5 s in the power supply before tripping.

Wiring of the motor cable is typically in three- or four-core steel wire armoured (SWA) cable, in the latter case the fourth core being used as an earth and the SWA as a radio frequency interference (RFI) screen. Instead of the HRC fuses typically used for induction motors with conventional starters, fast-acting semiconductor fuses are required to protect the VSD electronics, with HRC types at the distribution board. Current-operated residual current control devices (RCCDs) are typically unsuitable for rectified loads, since DC of sufficient magnitude can be introduced into the mains to damage the device. Voltage-operated types of RCCD are, however, generally suited to rectifier/inverter applications. From a health and safety point of view, the DC link capacitor between the rectifier and inverter sections of the VSD remains charged with a *potentially lethal* voltage for some time following mains isolation. Bleed resistors should therefore be provided by the equipment supplier, together with appropriate warning labels.

The principal potential problem areas associated with inverter VSDs are *harmonic distortion* and the adequacy of their *electromagnetic capability* (EMC). A harmonic is a current or voltage that oscillates at an integer multiple of the mains supply frequency. For example, the fifth harmonic oscillates at five times the mains frequency (50 Hz) or 250 Hz. Harmonics result from the rectification of alternating current to direct current. If the level of harmonic currents or voltages is too large in relation to the sinusoidal loads on a site, the overall current or voltage waveform will be distorted. This may result in an electrical supply that is outside the tolerance of IT or other sensitive equipment. Harmonics produced by equipment in an installation itself can travel back up the mains supply to interfere with the supplies to other consumers. Harmonic currents may be caused to flow in the neutral cabling of an installation, leading to overheating and (in extreme cases) possible burn-out of the cable.

Harmonics with even integers are not generally produced by diode bridge or fully controlled rectifiers, and on three-phase systems multiples of the third harmonic cancel each other. Harmonics above the 13th are typically negligible. Hence those of most concern are the 5th, 7th, 11th and 13th harmonics. In a perfect circuit the magnitude of a harmonic is inversely proportional to its frequency. Hence of those harmonics of most concern, the 5th is likely to be the largest, typically representing approximately 28% of line current where DC link filtering is used.

Since inverter VSDs rectify AC to DC, and then back to AC again at a different frequency, the output waveform is not a perfect sine wave, and there is thus also harmonic distortion of the outgoing supply to the driving motor. An excessive level of harmonics here can lead to overheating, insulation damage, and even (in extreme cases) the destruction of the motor itself. Formerly this was overcome by derating of the motor, although modern technology (including the use of insulated gate bipolar transistors to achieve high-speed switching) has significantly reduced the need for derating.

Carefully designed electronic filtering can be used to reduce levels of harmonic distortion. However, where there are a high proportion of filtered devices, filters may interact with undesirable consequences. Power factor (PF) correction capacitors offer a low impedance at harmonic frequencies. Absorption of harmonics and resonance at harmonic frequencies results in an increase in the voltage applied across the dielectric of the capacitor, potentially leading to overheating and premature failure. In retrofit applications there may be a need to derate existing PF capacitors to allow for the overheating caused by increased levels of harmonics.

Since the rectification of AC is an inherent element of an inverter VSD, this type of equipment is inevitably associated with harmonic distortion of the mains supply. The greater the proportion of the total connected load controlled by inverters, the greater is the potential magnitude of the problem of harmonic distortion. Since the removal of non-compliant equipment, or its modification to achieve compliance, are both potentially costly, and can be enforced by the supply authority, the implications of using a number of inverter VSDs in the same installation should be fully considered at the design stage. Some level of harmonic current will typically exist in an installation before the addition of inverter VSDs, associated with fluorescent tubes and mercury vapour or sodium discharge lamps.

In the UK maximum permitted levels of harmonic distortion of the electricity supply system are governed by recommendations of the Electricity Supply Council (specifically reference G5/3). The prospective magnitude of harmonic distortion at the point of coupling of an installation to the electricity supply system is subject to approval by the supply authority, and its establishment may require detailed calculations and, for retrofit of VSDs, measurement of existing levels of harmonics at the site in question. Permitted levels of harmonic distortion may depend on the nature of electricity use in the surrounding area. In an industrial area the supply system may already be relatively 'dirty', and stricter limits to new distortion may be applied. However, typically a maximum overall distortion of 5% is permitted for 415 V mains supplies, and 4% for 11 kV

Where the use of inverter VSDs is proposed, the electromagnetic compatibility requirements of the application must be thoroughly considered at the design stage. Since the beginning of 1996 all equipment sold in the UK has had to conform to European Community (EC) standards of EMC, and to be certified as such. Potential requirements for compliance may include use of a matching RFI filter on the supply cable, use of steel wire armoured (SWA) motor cable with high radio frequency interference (RFI) level, and limiting inverter switching frequency. Motor cables on site should preferably be as short as possible, and in any event *segregated* from other cables. The level of RFI typically increases with larger motor cables.

The final VSD of potential relevance to VAV fans, and thus far the least widely applied, is the *switched reluctance drive* (SRD), which is based on the use of a special brushless motor possessing a greater number of stator poles than rotor poles (Fig. 8.11). In effect SRDs may be considered to be a development of 'stepping' motor technology. Although utilising a special motor, the SRD is inherently simple and reliable. On/off switching of stator pairs in rapid succession, achieved electronically, produces a controlled rotation of the rotor, and a high level of speed control accuracy is claimed for the technique. The drive output to the motor is in the form of DC pulses, with motor speed and torque controlled by varying the timing and width of these current pulses. There is no derating of the motor, and this is compatible in

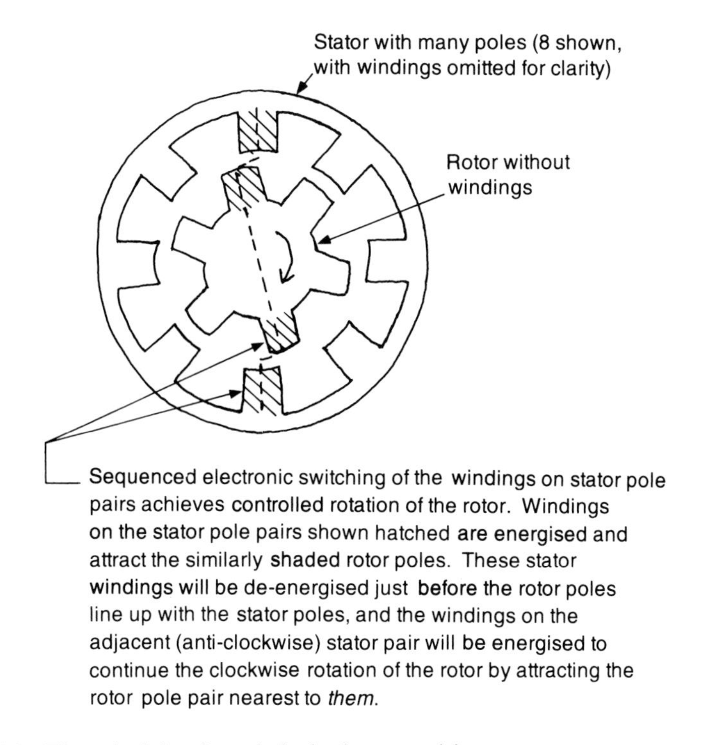

Fig. 8.11 The principle of a switched reluctance drive.

size with an equivalent 'squirrel cage' induction motor. The components of the SRD are similar to those of an inverter VSD, although the circuit and control strategy differ. The SRD controller is based on similar electronics to that of an inverter VSD, and is of similar size. It must be compatible with the motor, and offers a potential for sophisticated control configurations generally similar to those of an inverter VSD.

A motor starter is not required, the drive offering 'soft-start' capability (starting current approximately 25% of motor FLC) and near-unity power factor. Conventional motor isolation techniques are acceptable. Because of the limited in-rush current on starting, there may be no need for potentially complex phased start-up procedures in large buildings where SRDs are employed.

On the negative side, electromagnetic capability and harmonics must still be fully considered, motor noise is higher, and a six-core SWA motor cable is required. It is not feasible to bypass the controller on failure, for full-speed operation.

All VSDs require specialist commissioning, and it is particularly important that checks are made to confirm that operating temperatures are within the limits reccommended by the manufacturer. The implications of potential heat gains from plant or equipment not operational at the time of commissioning the VSD should be considered.

8.7 Duct static pressure control

All published sources agree that location of the static pressure sensor is critical to the performance of a VAV system – both in achieving stable operation and minimising energy consumption. However, uncertainty remains as to what constitutes the most appropriate position. The most common options and recommendations are summarised in Fig. 8.12.

The CIBSE Guide[13] suggests that the generally adopted position for a single static pressure sensor lies from one-half to two-thirds of the distance along the main ductwork of the index run (under design conditions) in order to enable the volume flow rate demands on this run always to be satisfied, whilst allowing maximum energy savings in operation of the supply fan. This source also cautions that the optimum sensor location is likely to alter with varying load condition. It is further suggested[14] that the sensor should be located in a position giving 'a reasonable indication of total flow requirements', and that medium-to-high mean duct velocities are required at that point to improve the sensitivity of the sensor to changes in volume flow rate.

The ASHRAE Handbook[15, 16] suggests that the sensor is typically positioned between 75% and 100% of the distance between the first and last VAV terminal units. The use of multiple sensors is recommended when the root section of the supply network (that immediately downstream of the fan discharge) serves multiple main ducts having peak total pressure requirements of similar magnitude. In any event the greatest requirement for duct static pressure must always be satisfied, which will equate to the highest potential requirement at any of the sensor locations. If the outputs of the multiple sensors are all compared with this as a *common* set-point value, then the *lowest* duct static pressure is selected for control, as it represents the smallest *excess* over and above the common set-point pressure, and thus the maximum that can be removed (by reducing fan speed, decreasing blade pitch, or whatever other method of fan duty regulation is adopted), while still maintaining at least the common set-point value at all sensor locations. It is of course inherent in this approach that all sensor locations will still have excess pressure over and above their potential worst-case requirements if the lowest duct static pressure is not at the sensor location which defines the common set-point value. It is thus preferable for the outputs of the multiple duct static pressure sensors to be individually compared against their own *sensor-specific* set-point values (again the greatest potential duct static pressure requirements at each respective sensor location). Again the lowest excess duct static pressure is the most that can safely be removed at all sensor

(a) Basic configuration – duct static pressure control at S.

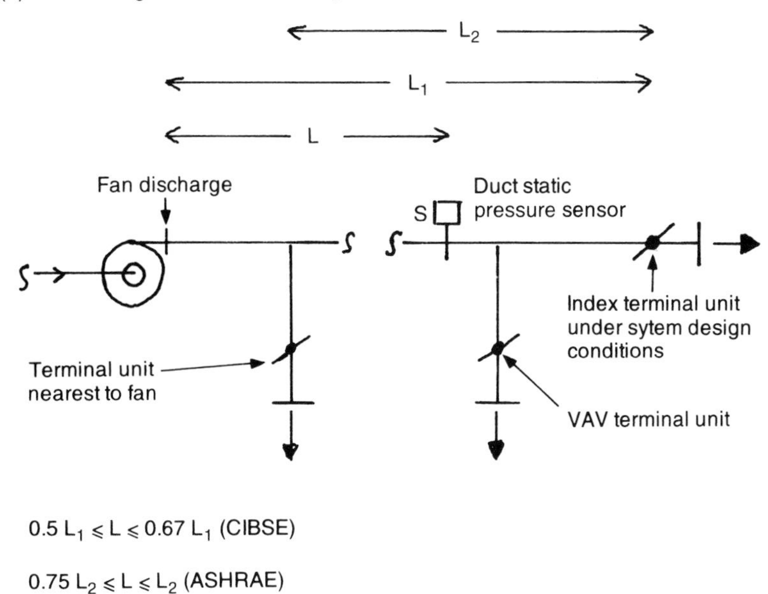

$0.5\,L_1 \leqslant L \leqslant 0.67\,L_1$ (CIBSE)

$0.75\,L_2 \leqslant L \leqslant L_2$ (ASHRAE)

(b) Reset configuration – sensor PS2 is used to reset the set-point for duct static pressure control at PS1.

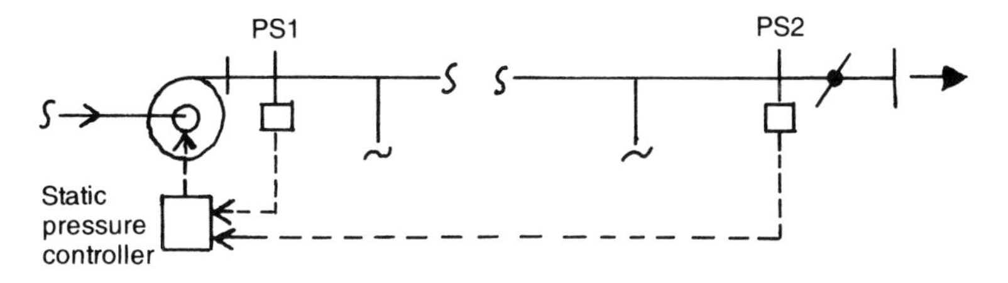

(c) Multiple duct static pressure sensors – selector compares inputs from sensors with set-point(s); controller output always to suit least-favoured sensor.

Static pressure sensor

Static pressure controller

Static pressure sensor

Selector

Fig. 8.12 Positioning the duct static pressure sensor.

locations by reducing fan duty. However, the lowest excess static pressure will not necessarily correspond with the lowest *absolute* value, and selection for control should be on the basis of the lowest *error signal.*

The location of the static pressure sensor also depends on the type of terminal unit specified. For pressure-dependent types a sensor position close to the static pressure mid-point of the duct run ensures minimum pressure variation in the system. For 'pressure-independent' (variable-constant-volume) types a sensor location near the end of the index run allows maximum fan power savings with a minimum required pressure at the final terminal unit. The static pressure distribution in the system will also vary with a changing distribution of volume flow rates, and this will typically giving rise to a need for an element of *on-site adjustment* of the sensor location. For many systems a good starting point would be to provide an initial sensor location between two-thirds and three-quarters of the distance from the supply fan to the end of the main trunk duct.

Other references may be found to sensors positioned at points in the system defined in terms of the proportion of system volume flow rate at the sensor location under design conditions (50–70%), the proportion of system volume flow rate which has been fed off to terminal units upstream of the sensor under design conditions (*c*.70%), or the proportion of the total pressure loss, under system design conditions, accumulated between the root section of the network and the sensor location (*c*.75%). Yet in some systems the static pressure sensor has been installed in the fan discharge plenum[17].

In order to avoid unnecessarily high fan total pressure during part-load operation, the static pressure sensor may be positioned where the duct static pressure is lowest, i.e. near the furthest terminal unit. However, while this sensor location may be desirable from the point of view of reducing fan energy consumption, it carries with it a risk that the excessive influence of individual terminal units on the static pressure sensed (Fig. 8.12) may give rise to stability problems. The question of stability will also depend on the response characteristics of both the fan speed control (blade pitch control in the case of a VPIM axial fan) and the terminal unit local contol loop itself. Greater stability may be offered, albeit at the expense of increased energy consumption, by a sensor location in the first third of the main duct.

For supply networks with long main ducts a two-sensor *reset* arrangement may be preferred. In this a static pressure sensor (PS1 in Fig. 8.12) located close to the fan maintains adequate static pressure at high system volume flow rate. Its set-point is reset from a second sensor (PS2) positioned near the furthest terminal unit – in response to a reduced volume flow rate in that part of the network. In a variation of this arrangement the reset sensor uses air velocity (as a measure of volume flow rate), rather than static pressure. Again where the root section of the distribution network feeds several main ducts, each with similar peak total pressure requirements, the reset sensor arrangement may be duplicated for each main duct sub-system. Where the reset sensor measures air velocity, locating it at the furthest terminal unit will give rise to a measurement problem as the minimum zone flow rate is approached. In this event the velocity reset sensor will need to be moved further back along the index duct run, to where a sufficient number of terminal units are served to ensure an adequate minimum diversified volume flow rate for practical measurement. The anticipated pattern of cooling load variation in the zones served by these terminal units will need to be considered, since if all the units in the downstream group are likely to turn down together, the low-load measurement problem may still remain.

The type and sensitivity of the static pressure control loop must also be considered. Proportional-plus-integral control is essential, since a proportional offset from the static pressure set-point value results in excess fan total pressure being wastefully throttled off by the VAV terminal units. It should also be remembered that the duct static pressure control loop responds quickly. If commissioned for the relatively slow response of a temperature control loop, instability may result. It is also worth considering that in commercial HVAC applications the sensing of static pressure typically relies on differential pressure sensors, and uses a local internal ambient pressure as reference. Any variation in this reference pressure will be interpreted in an analogous (but opposite) manner to a variation in the duct static pressure, so it is important to avoid any location that may be subject to excessive pressure fluctuations. It has already been noted that VAV systems themselves generate variations in building pressurisation. Depending on the nature of the infiltration paths within a building, the combined effects of wind pressure and stack effect in high-rise buildings may affect sensors.

Supplementary static pressure control in main branch ducts is not commonly applied with variable-constant-volume ('pressure-independent') terminal units, but is typically an inherent requirement in systems using pressure-dependent VAV terminals. It is achieved by locating a pressure control damper in each main branch duct serving a group of pressure-dependent terminal units (Fig. 8.13). The damper is typically electrically actuated (although pneumatic systems continue to be available, and have their staunch adherents), and throttles off excess pressure to maintain a static pressure set-point at a specified sensor location. The requirements for this location are that the static pressure difference above ambient should be relatively low, and that no individual terminal unit should unduly influence the sensor. In most installations these requirements are likely to be adequately met by a sensor position between one-half to two-thirds of the way downstream along the main branch duct, although again a degree of site adjustment may be necessary to find the optimum position. Adequate downstream spacing should be allowed for the flow disturbance caused by the damper (which typically discharges a number of 'jets' from the blade

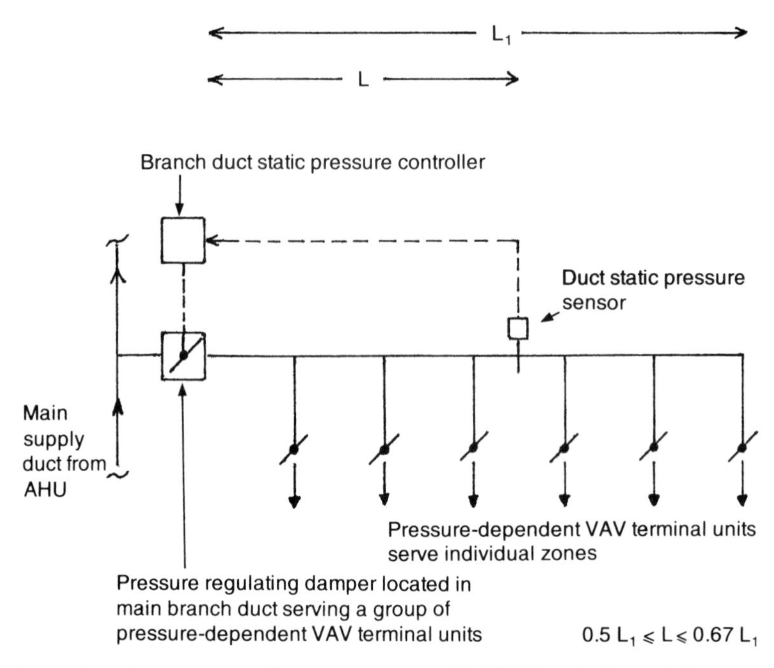

Fig. 8.13 Supplementary static pressure control in VAV main branch ducts.

pairs) to dampen out before the first terminal unit branch duct, and in any event prior to any measurement of duct pressures. Typical recommendations are a downstream spacing of 4–5 equivalent duct diameters, although strictly with opposed-blade dampers a minimum of ten times the blade (height) spacing may be acceptable. The sensor response must be compatible with the damper characteristic if the control loop amplification is not to be excessive (leading to instability). A steep damper characteristic may be compensated by a wide sensor range, and good control may be achieved by combining the inherently fast response of the control loop with a slow damper motor. With good duct design, the technique allows close control of the static pressures at inlet to the terminal units served by each main duct branch.

In contrast to the position of the index diffuser or grille in a constant-volume system, the index terminal unit in a VAV system is not fixed under all system operating conditions (Fig. 8.14). Throughout this book the assumption is typically made that the index terminal unit under system design conditions is the furthest from the fan. This is not *automatically* the case for the same reasons that the furthest diffuser or grille is not automatically the index in any air system, but it is a very reasonable first assumption. In any event the characateristic of the index terminal unit, whatever conditions are considered, is that it has the damper that is *nearest to the fully open position.* The location of the duct static pressure sensor must be such as to ensure that no terminal units are starved of air under any normal operating condition. A

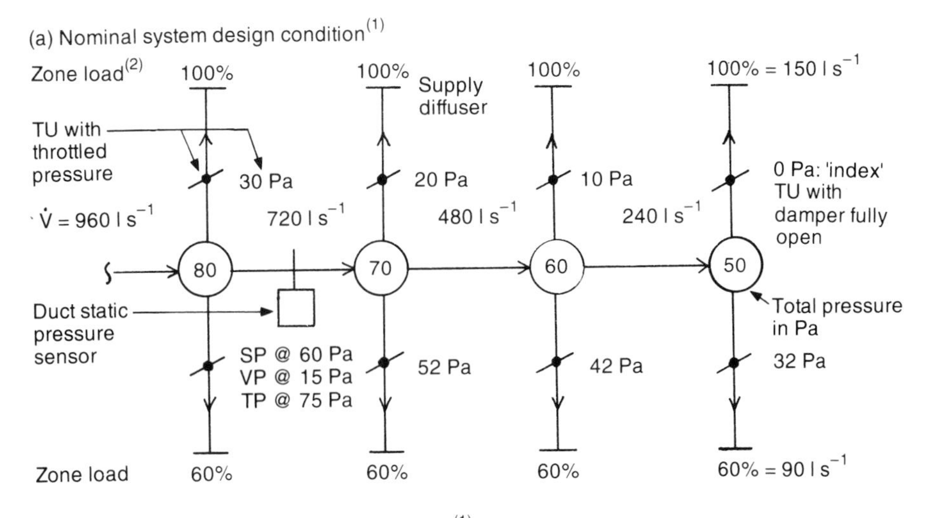

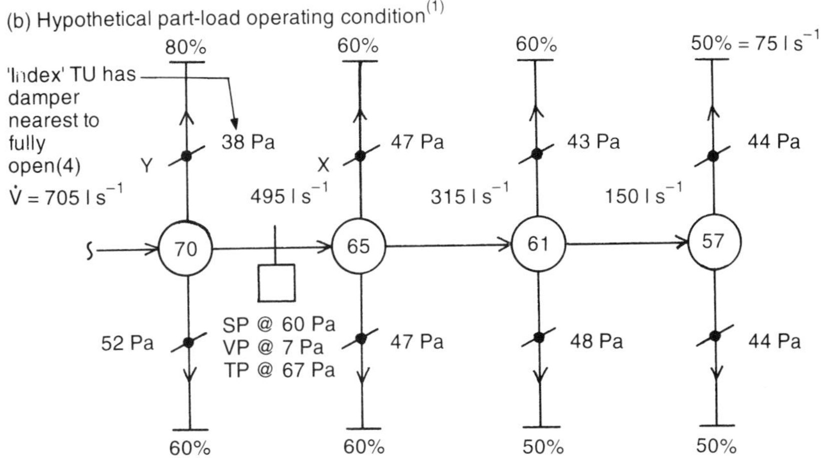

Notes:
(1) All values chosen for illustration only in this particular example. All terminal units assumed of same type & size, and in identical branch ducts. Real VAV systems are likely to be less symmetrical, and are certainly much more extensive. All pressure drops based on simple 'square law' relationship between ΔTP and V.
(2) Expressed as % of design value, assumed identical for all zones.
(3) Difference between minimum ΔTP required across branch duct with TU damper fully open and ΔTP actually available across branch.
(4) ΔTP available to each branch duct naturally increases towards the fan. If the increase between branches X and Y (5 Pa) $<$ the additional ΔTP that would be required, 50 Pa $\times\left[\left(\frac{80}{100}\right)^2 - \left(\frac{60}{100}\right)^2\right]$, for the increased $\dot{V}$ of TU 'Y', while maintaining the damper angle of TU 'X', then the damper in TU 'Y' must open relative to that in TU 'X' in order to sustain the increased $\dot{V}$ and maintain zone temperature.

Fig. 8.14 Variable position of the index VAV terminal unit.

terminal unit is 'starved' of air when there is insufficient static pressure available at its inlet spigot to pass the volume flow rate of conditioned air demanded by its zone temperature controller, even with the terminal unit damper fully open. In such circumstances both the terminal unit and the zone are effectively out of control.

The risk of any terminal unit being *starved* of air is typically greatest in two cases. The first is when a single duct static pressure sensor is used in a system featuring multiple main ducts, each having peak total pressure requirements of similar magnitude (as, for example, would be the case for identical floors in a multi-storey building), although the use of multiple duct sensors is recommended in this case. The second is when the duct sensor is located close to the furthest terminal unit (or whichever is the index unit under system design conditions). The risk of any terminal unit being starved of air at some time depends both on the nature and pattern of variation of the zone sensible cooling loads and the design of the ductwork system. In any event the potential seriousness of a shortfall in zone supply depends both on the nature of the application, the layout of the zones served, and the potential duration of the shortfall. Open-plan comfort cooling applications are inherently more 'robust' to shortfalls in duty than are cellular spaces in a process control environment, and occasional brief shortfalls are naturally more acceptable than long periods.

We may conclude that conventional static pressure control of the supply fan typically allows VAV terminal units to deliver the correct amount of cooling to each zone, but that the system could be more *energy-conserving*. By fixing part of the system pressure loss at a level based on the requirements of design conditions, it inherently requires throttling of excess fan total pressure under all part-load operating conditions. We have seen that these account for almost all the potential annual operating hours of VAV systems in typical commercial applications (and probably all of the actual operating hours in many instances). The throttling is carried out at the VAV terminal units, rather than at a main duct damper, but it still wastes fan total pressure, which wastes fan energy. As fan and variable-speed drive efficiency typically decrease under part-load conditions, each unit of throttled fan total pressure becomes more 'expensive' in wasted fan energy as system load factor falls – and as the degree of throttling required at the index terminal unit increases. Again operation at low system load factor typically represents a significant proportion of the annual operating hours.

8.8 Reset control of duct static pressure

While putting the duct static pressure sensor close to the furthest terminal unit minimises wasteful throttling, and with it fan energy consumption, it does not *remove* it (Fig. 8.12). Any further reduction requires the duct static pressure set-point to be varied (or reset) according to the actual needs of the supply network. Wasteful throttling of fan total pressure is eliminated when the duct static pressure is reduced to the point that the volume control damper of (only) one of the VAV teminal units connected to the system is *just fully open*. This will be the index unit under the particular operating conditions. Fan total pressure cannot be further reduced without at least this terminal unit being starved of air.

Reset of duct static pressure has the added benefit that the position of the control sensor is no longer critical, and it is probably most convenient to place it in the root section of the network (that immediately downstream of the fan), prior to the first branch take-off. However, the flow at discharge from the fan is strongly disturbed, and it will be wise to position the static pressure sensor a minimum of five equivalent duct diameters downstream of the fan discharge. Ten equivalent diameters would be preferable, if space permits. Although VPIM axial fans used in VAV applications will usually be fitted with downstream guide vanes to reduce the degree of outlet *swirl* imparted to the air stream, it is advisable to maximise the separation of the static pressure sensor where this type of fan is used. A control signal to reset duct static pressure may be used to control the fan directly, eliminating the duct static pressure control loop – although it has been suggested[18, 19] that its retention may be desirable from a stability perspective. Terminal regulated air volume is sometimes also used to describe the feedback scenario where the duct static pressure control loop has been eliminated[20–23].

Maximum benefit from reset of the duct static pressure control set-point accompanies the use of either a centrifugal fan with AC inverter drive or a variable-pitch-in-motion axial fan. However, both of these techniques for modulating fan duty permit a close approach to the ideal minimum fan power required for any given system operating condition. This ideal typically assumes a fan efficiency that remains constant, at its design peak value, for any system operating condition. The efficiency characteristics of inlet guide vane control for centrifugal fans may make reset of duct static pressure substantially less beneficial for this mode of fan duty modulation. However, at present the published evidence is inconclusive[11].

Several approaches are possible to reset control of duct static pressure (or direct control of fan speed or VPIM action). In all cases suitable *feedback* must be available from the VAV supply network. In theory some feedback is available from the behaviour of the volume flow rate in the system as a whole, or in parts of it (e.g. floors of a multi-storey building), to incremental changes in duct static pressure set-point (or fan speed). If the set-point is too low, a small upward reset results in an increase in system flow rate. If it is too high, system flow rate recovers from a small downward reset. However, practical control strategies depend upon feedback from individual VAV terminal units[5, 18, 19, 24, 25]. The capacity to handle a potentially very high volume of inter-device communication is essential for control systems employing such feedback control strategies. They are thus inevitably based on DDC technology.

Feedback from each terminal unit of the error between measured and set-point volume flow rates allows upward reset of duct static pressure to eliminate the errors (Fig. 8.15). As only *positive* errors (set-point – measured variable) are possible with variable-constant-volume (VCV) control action, this provides no information for downward reset to minimise duct static pressure. An additional *forcing function* is therefore required to automatically reset duct static pressure downwards under specified conditions, which include the absence of any positive requirement to reset it upwards, until the feedback shows a positive volume rate error for one of the terminal units. Only positive errors are possible with VCV control action since a terminal unit simply throttles off too much inlet pressure. Negative errors are therefore only transient. When a terminal unit damper is already fully open, however, any further drop in inlet pressure results in an error which will remain until either duct static pressure or fan speed is increased.

Rather than obtaining feedback of the actual values of flow rate error existing at each terminal unit, the feedback may simply identify (or 'flag' up) the *existence or absence* of a 'low-flow' alarm condition (Fig. 8.16). At the terminal unit the alarm is 'switched on' if a positive volume flow rate error is sensed when the damper is fully open. Recognition of an alarm condition at any terminal unit initiates an *upwards reset* of duct static pressure until the alarm is cancelled. However, the absence of any low-flow alarms does not distinguish whether the duct static pressure set-point value is only just sufficient to cancel the alarm or well in excess of this value. Minimisation of duct static pressure therefore requires that the absence of any alarms is itself the feedback which triggers a downwards reset until an alarm condition is recognised. The system may be operated with a number of alarms registered, the number allowed 'floating' between specified maximum and minimum numbers. However, this requires that the effect on zone conditions of operating in an alarm condition can be closely predicted, and is acceptable from a design and operating perspective. For example, zone ventilation rate must not be unduly compromised. Once a terminal unit damper is fully open, the zone it serves is no longer under full control, since the terminal unit cannot respond to any further increase in sensible cooling load. A two-state (on/off) 'low-flow' alarm gives no information as to how large the positive volume flow rate error can become if the alarm is allowed to persist. Once a terminal unit damper is fully open, the unit behaves as a fixed resistance. Under these conditions the loss of total pressure across the terminal unit bears a simple relationship to the volume flow rate through it to the zone served, which aids any analysis of the situation that may be required. The use

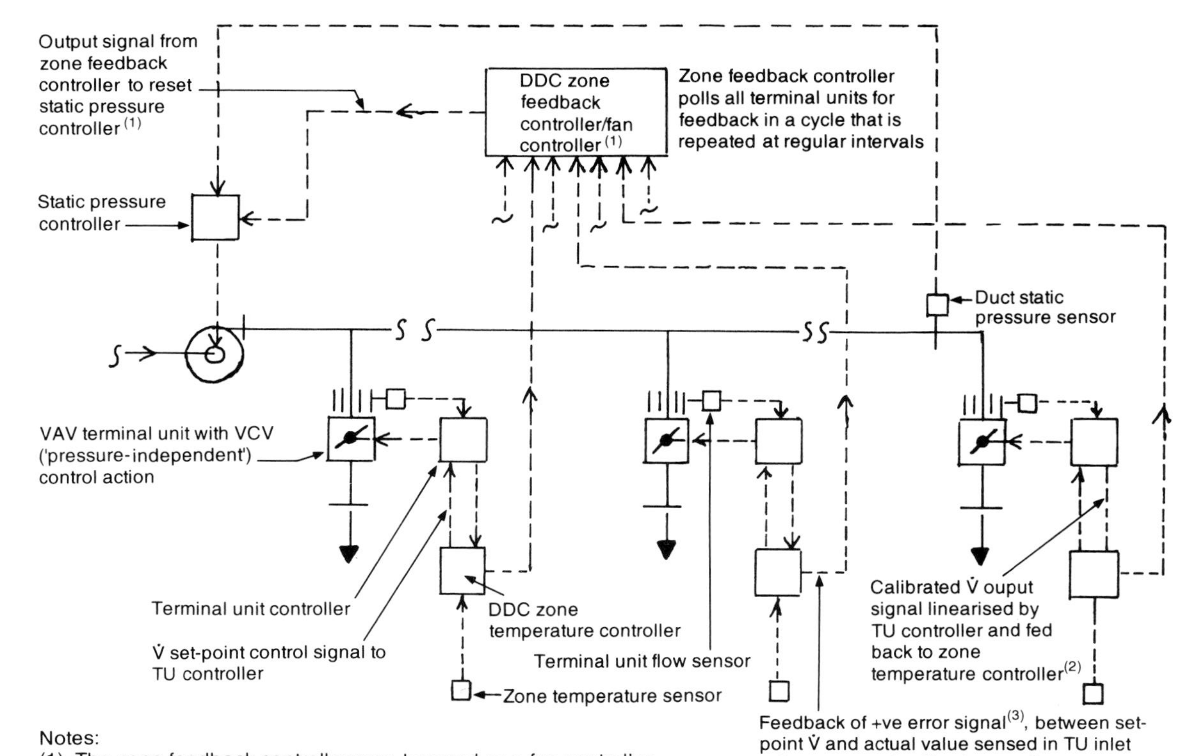

Notes:

(1) The zone feedback controller may be used as a fan controller to modulate fan speed directly without reference to a duct static pressure control loop.
(2) Feedback from terminal units may be direct to the zone feedback/fan controller.
(3) Only +ve error signals (sensed $\dot{V}$ < set-point value) are sustainable with variable-constant-volume ('pressure-independent') control action at the VAV terminal units. As a result a rule-based approach must be used to generate a 'forcing function' to reset duct static pressure (or fan speed) downwards in the *absence* of +ve errors.

Fig. 8.15 Reset of duct static pressure through feedback of volume flow rate error.

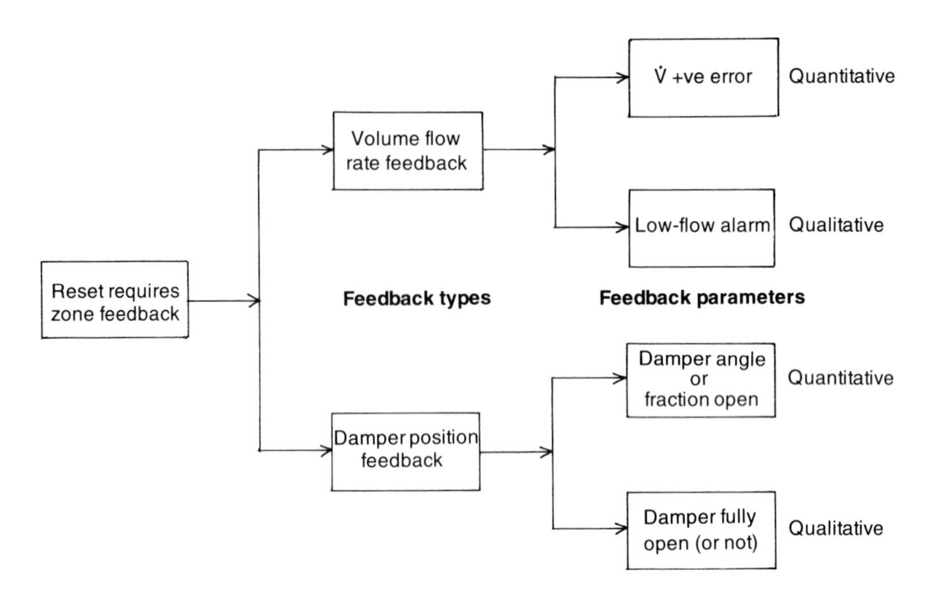

Fig. 8.16 Alternative reset strategies for duct static pressure using zone feedback.

of 'low-flow' alarms to identify terminal units with a positive volume flow rate error is also possible as a means of significantly reducing processing requirements.

Feedback from the terminal units may also be based on *damper position*. In this case the feedback may be qualitative or quantitative. Qualitative feedback provides information only on whether dampers are either fully open or 'not fully open', whereas quantitative feedback provides either actual damper angles or degrees of proportional opening/closure (Fig. 8.16). With qualitative or two-choice feedback the recognition of a fully open damper can be used to initiate an upwards reset of duct static pressure until this condition is eliminated. To minimise duct static pressure, the absence of a fully open damper must be used to force duct static pressure down until that condition is again established. Again the system may be allowed to 'float' between specified maximum and minimum numbers of dampers fully open. As with the use of low-flow alarms, the potential effects of this must be considered thoroughly. With quantitative feedback on damper angle, duct static pressure can be reset *both upwards and downwards* to maintain one damper at a specified set-point position – say 10° from fully open or *c*.90% open[25]. The penalty in additional duct static pressure for a small damper angle, rather than a damper just fully open, is small. However, terminal unit dampers are without exception strongly non-linear devices (Chapter 7). At small damper angles damper movement is at its most sensitive to changes in terminal unit inlet pressure, or the set-point value of volume flow rate. It should also be remembered that most terminal unit dampers go from fully open to fully closed over an angular range that may be only 45–60°. Damper movement is therefore likely to be too sensitive for control that tries to maintain any damper at approximately 90% open. Values of 75–80% open or 10–15° from fully open may be practical. In any event it appears unlikely (*ibid.*) that quantitative feedback on damper angle could be used without at least basic processing of the feedback data (determining a running average may be adequate) to smooth and slow down the reset of duct static pressure. At any time processing of the feedback data is only necessary for that group of terminal units that are close to the set-point damper angle.

Alternatively, a high-limit/low-limit 'floating' control approach may be adopted, according to simple rules[5]. One such application of rules would be as follows.

(1) Look for the most open terminal unit damper.
(2) If the most open damper is less than X° from fully open, then reset duct static pressure up until the damper closes by a further θ_1° (to $(X + \theta_1)^\circ$ from fully open).
(3) If the most open damper is more than $(X + \theta_1 + \theta_2)^\circ$ from fully open, then reset duct static pressure down until the damper opens by a further θ_2° (to $(X+\theta_1)^\circ$ from fully open).

The values of X, θ_1 and θ_2 should be defined according to the characteristics of the particular terminal unit used. Owing to the strongly non-linear characteristics of the terminal unit damper, and its reducing sensitivity with increasing angle from the fully open position, it is logical for θ_2 to be smaller than θ_1.

As noted, feedback control can potentially be applied *directly* to the VSD of a centrifugal fan or to the pitch change actuation of a VPIM axial fan, without the intermediary of duct static pressure control. In this case the scenario of a failure of the communication system linking the zone terminal units with the central controller must be considered. Reverting to static pressure control after a defined absence of zone feedback is an option. A sensor located in proximity to the supply fan can also perform *high-limit* control for the system. Where zone feedback is used to reset the duct static pressure control loop, the system may simply revert to the design set-point value in the event of communications failure.

Feedback from the individual VAV terminal units requires all the units served by a particular supply fan to communicate with a central controller, whether part of an overall control system or a dedicated fan controller. It is hard to conceive of such a system being possible without the use of direct digital control (DDC) techniques and equipment. In practical terms feedback may be achieved through 'polling' of the individual terminal unit controllers. This will inevitably introduce a 'polling delay' into the control system. The length of this delay will be minimised if distributed processing is used. Rapid polling of a large number of terminal units is likely to require the fan controller (or each controller in multiple-fan installations) to have its own dedicated local communications bus.

Unfortunately there is little firm published guidance for HVAC designers regarding practical criteria for duct static pressure (or direct fan speed) control strategies based on zone feedback.

As far as polling terminal unit controllers for volume flow rate errors is concerned, at least three methods of deriving a control signal have been identified[18, 19].

(1) The first option is simply to identify the *largest positive error* for any terminal unit. A control strategy based on this will ensure that the requirements of every zone are ultimately met, and has the advantages that it is simple and easy to understand. However, it also carries with it the risk that signal fluctuations may lead to instability, or that an undersized or poorly balanced terminal unit may dominate control of the system.

(2) A second option is to *average all positive flow rate errors*. This should result in a smoother control signal, and should also not allow an undersized terminal unit to dominate control of the system. The disadvantage is that any zone whose terminal unit is, in fact, undersized will likely overheat.

(3) A third option is to derive a *moving average of the largest positive flow rate error*, to yield a control signal that changes more slowly and smoothly. Stability may be expected to be improved. An undersized terminal unit will still tend to dominate system control, and the response of the supply fan may be slowed. However, this will depend on the number of samples averaged and the polling frequency, and it is possible to help matters by introducing 'dead band' control and providing a store of previously polled data for the controller to draw from.

Since the same pattern of 'low-flow' alarms or volume flow rate error signals may repeat at different duct static pressure settings, these are not amenable to a simple reset schedule of the kind employed to vary the flow temperature of a perimeter heating system according to outside air temperature. The reset increment must be added to (or subtracted from) the existing set-point value. The following three alternatives appear possible.

(1) The existing set-point is incremented by a fixed amount, repeated as necessary.

(2) The existing set-point value is increased or decreased at fixed rates.

(3) The reset increment or rate is scheduled according to the feedback data.

Fan speed will typically be increased and decreased at fixed constant rates, although these need not necessarily be the same. Whether operating through the intermediary of duct static pressure or not, the fan should in any event respond more slowly than the terminal unit actuators, so that these have time to respond to changing duct pressure without generating instability. However, the fan also needs to respond fast enough to effectively minimise duct static pressure and generate energy savings during most of the daily operating period in typical commercial applications. The fan controller itself may be based on traditional proportional-plus-integral (P + I) control, although the use of DDC permits the option of a rule-based approach. This opens up the possibility not only of making control actions conditional on the results of the current sampling cycle, but also of making them conditional on changes since the last sampling cycle.

8.9 Alternatives to duct static pressure control

Where pressure-dependent VAV terminal units are employed, the supply fan may be controlled according to the *system return temperature*, as measured in the main extract duct. This technique showed improved operating economy when compared with a combination of duct static pressure control and 'pressure-independent' terminal units in a simulation study[9]. However, good pressure balancing of all paths through the supply ductwork is important if all zones are to be properly served under this method of fan control. The system return temperature will not be unduly influenced by an individual zone, unless this generates a significant proportion of the total sensible cooling load. This is not usually the case in typical commercial applications.

On the negative side there is an inherent delay between air leaving the zones and reaching the return sensor. This results in a system which responds very slowly in comparison with a duct static pressure control loop. Slow response may allow local hot and cold spots to appear, leading to fan instability (hunting). The technique is best suited to applications where the range of load factor variation between zones is limited, and where the dominant pattern of change in zone load is a gradual one. Zone temperature controllers must be of the simple proportional-only type, and the technique is not suitable where return air light fittings are used. It is unclear how the technique (illustrated in Fig. 8.17) can be fully set up and accurately commissioned in the absence of a full range of system cooling load.

8.10 Limit control of VAV supply fans

Limit control of VAV supply fans is required to protect the system from excessive duct pressure. This may either shut the fan down when a static

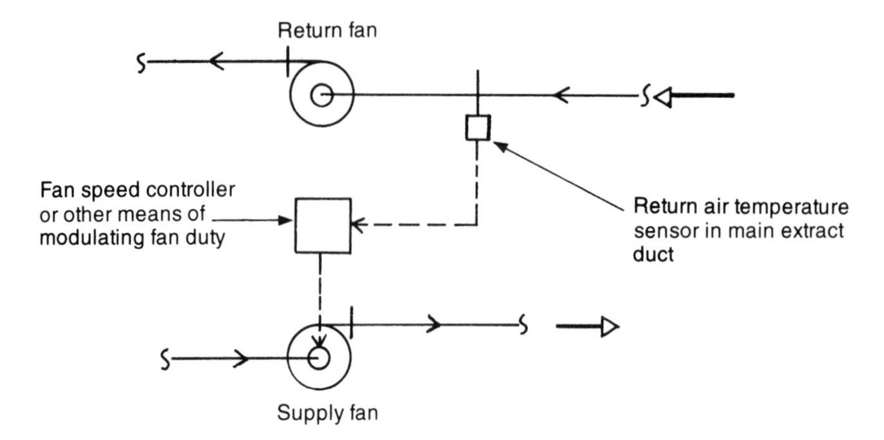

Fig. 8.17 VAV control from return air temperature.

pressure sensor at the fan discharge reaches a *safety* high-limit, or override the duct static pressure control loop at a *controlling* high limit. In the latter case a fire damper closing accidentally somewhere in the system would result in a drop in duct static pressure, leading to a demand for increased fan speed until overridden by the high limit function. If fan stability is likely to be a problem at high system turndown, the volume flow rate handled may be maintained at a low limit value, at low system load, by increasing the off-coil temperature in the central AHU.

8.11 Predicted energy consumption of VAV supply fans

In order to compare the energy performance of alternative system designs or control schemes, it is essential for HVAC engineers to be able to predict, *at least to a reasonable approximation*, the annual energy consumption of the fans serving an air conditioning system. For constant-volume operation this involves a straightforward application of the conventional expression:

$$E_a = [0.1 \times \dot{V}_{system} \times FTP \times H] \div \eta_{fan} \quad (8.1)$$

where

E_a = annual energy consumption (kWh)
$\dot{V}_{system}$ = system design volume flow rate ($m^3\,s^{-1}$)
FTP = fan total pressure (Pa)
H = annual operating period (h)
η_{fan} = overall efficiency of the fan, motor and drive combination (%).

By definition for variable-volume operation the system volume flow rate, fan total pressure and the overall efficiency of the fan, motor and drive combination vary according to the simultaneous cooling load on the system at any time. The prediction of annual fan energy consumption is thus more involved. Using a 'bin method' approach, annual energy consumption of a VAV supply fan becomes:

$$E_a = \dot{V}_{design} \times H \times 10^{-5} \times \sum [(z \times A \times FTP) \div \eta_{fan}]_{bin} \quad (8.2)$$

where

E_a = annual energy consumption (kWh)
$\dot{V}_{system}$ = system design volume flow rate ($m^3 s^{-1}$)
H = annual operating period (h)
z = the mid-point value of a 'bin' of system volume flow rates ($m^3 s^{-1}$)
A = the percentage of the annual operating period covered by the bin of system volume flow rate
FTP = a characteristic value of fan total pressure that may be applied to a bin of system volume flow rate (Pa)
η_{fan} = a characteristic value of the overall efficiency of the fan, motor and drive combination, which now includes the component efficiency of the method of fan duty modulation, that may be applied to a bin of system volume flow rate (%).

For many purposes it may be sufficient to use a typical annual operating profile of the form shown in Fig. 6.3 to establish the bin ranges and mid-point values. The difficulty is of course in adopting a suitable value of fan total pressure for each bin. With duct static pressure control simply treating the variation of fan total pressure with system volume flow rate according to the 'square law' is unacceptable. Since for a given system volume flow rate and overall fan/motor/drive efficiency fan power is proportional to fan total pressure, this approach underestimates energy consumption as system load factor reduces. The effect is progressive, being greatest at the minimum load factor, and is very significant because of the large proportion of the annual system operating hours typically spent at relatively low load factor. It is naturally most marked where conventional duct static pressure control to a fixed set-point value is employed. This may be seen by considering the example of a system with a design fan total pressure of 1500 Pa. At a load factor of 40% a simple 'square law' treatment of system pressure losses suggests a reduction to 240 Pa. If the 1500 Pa design pressure loss is now considered to comprise an element of 1200 Pa which does allow a 'square law' treatment, together with an element of only 300 Pa which is constant (20% of the total), then at 40% load factor the new prediction of fan total pressure is 492 Pa. A simple 'square law' treatment would therefore predict an hourly fan energy consumption under these conditions that is approximately half of what might be expected in practice.

The use of a duct static pressure reset control strategy naturally reduces the discrepancy, but cannot in any event fully remove it, since the system characteristic will not remain constant even where duct static pressure can be minimised under all operating conditions. Because of its ability to reposition all of the terminal unit dampers individually, the system does not have a unique operating configuration. Neither, therefore, is a strict application of the fan laws possible.

On the other hand, treating fan total pressure as constant will certainly significantly overestimate annual fan energy consumption, even where duct static pressure control is to a fixed set-point which is a relatively high proportion of design fan total pressure. In any event pressure losses across plant items in the central AHU do follow a 'square law' variation with system volume flow rate. Where a relatively basic approach of the kind illustrated above, in which fan total pressure is aportioned between constant and 'square law' elements, is inadequate, recourse must be made to more detailed calculations of system pressure losses. While manual and, certainly, spreadsheet-based techniques have a place in this increasing sophistication of treatment, the implication is clearly that detailed investigations of VAV systems probably require the availability of suitable simulation software. To be of any practical use this must include air-side simulation of the ductwork system, and specifically allow modelling of the characteristics of duct static pressure control.

The annual operating profile of Fig. 6.3 is worthy of further comment in regard to the distribution of annual fan energy consumption for control to a fixed duct static pressure set-point. The high proportion of annual operating hours in the lowest 'bins' covering ranges of system volume flow rate below 50% of design value typically account for *disproportionately less* fan energy consumption than the hours spent in the uppermost bins covering ranges of system volume flow rate above 80% of design value. The former may account for approximately 25% of annual fan energy consumption over approximately 40% of the annual operating hours, while the latter may represent a similar proportion of energy consumption over only some 15% of annual operating hours. However, up to half of the potential fan energy *savings* which may accrue from reset control strategies for duct static pressure are obtainable at system volume flow rates below approximately 50% of design value, while the operating envelope above approximately 80% of system design volume flow rate typically affords minimal savings. It is thus the low-load operating envelope that justifies the consideration of such reset control strategies.

For a VAV system under duct static pressure control (such as that shown schematically in Fig. 8.1) fan total pressure may be considered to be built up as follows:

$$FTP = SP_S + VP_S + \Delta TP_{F,S} + \Delta TP_{AHU} + \Delta TP_{UP} \tag{8.3}$$

where

FTP = fan total pressure (Pa)
SP_S = duct static pressure control set-point value at sensor location S (Pa)
VP_S = velocity pressure at duct static pressure sensor location S (Pa)
$\Delta TP_{F,S}$ = loss of total pressure in ducts and fittings, etc. between duct static pressure sensor S and fan discharge F (Pa)
ΔTP_{AHU} = loss of total pressure across equipment items in AHU (Pa),
ΔTP_{UP} = loss of total pressure in ducts and fittings, etc. upstream of AHU (Pa).

From this it is clear that minimising the value of VP_S under any operating condition will also minimise the value of FTP. At any point X in the supply ductwork the total pressure available is:

$$TP_X = SP_S + VP_S + \Delta TP_{X,S} \tag{8.4}$$

where

TP_X = total pressure available at point X (Pa)
SP_S = duct static pressure control set-point value at sensor location S (Pa)
VP_S = velocity pressure at duct static pressure sensor location S (Pa)
$\Delta TP_{X,S}$ = loss of total pressure in ducts and fittings, etc. between duct static pressure sensor S and point X (Pa).

If point X is at a branch take-off feeding the runout to a terminal unit, this is the pressure that is *available to balance* (and which must be balanced by):

(1) frictional and velocity pressure losses due to duct and fittings in the runout to the terminal unit;
(2) the pressure loss across the terminal unit itself and any associated secondary attenuation, including any throttling by the control damper, which naturally has a minimum value when the damper is fully open;
(3) frictional and velocity pressure losses due to duct and fittings downstream of the terminal unit;
(4) the pressure drop across the index unit of the supply diffusers served by the terminal unit;
(5) the velocity pressure at outlet from the index supply diffuser.

Because a balance *will always be achieved*, the difference is whether it is achieved at the *desired* volume flow rate to the zone (when TP_X, and hence SP_S, is adequate), or at some value below this (when SP_S is inadequate). The sign convention for $\Delta TP_{X,S}$ follows the natural sense of $(TP_X - TP_S)$, i.e. it is positive for terminal units upstream of the duct static pressure sensor and negative for units downstream. The duct static pressure set-point value required to ensure that any given terminal unit can supply the volume flow rate demanded by its zone therefore depends on the sensor location, on the layout and sizing of the main supply ductwork, on the selection and sizing of the terminal units and supply diffusers, and on the extent and sizing of both the runout to the terminal unit and the downstream ductwork from the unit. In particular sizing the branch runouts to terminal units too small can unnecessarily increase the duct static pressure set-point required. The effect of *overdesign* is also apparent, since a fixed duct static pressure set-point is naturally based on design volume flow rates to the zones. If these are never required in practice, fan total pressure and energy consumpion are penalised under all operating conditions. Not only is the fan oversized under design conditions, but duct static pressure control to a fixed set-point maintains fan total pressure at a higher than necessary value *throughout its operating life*, unless subsequently adjusted down. The excess pressure is simply throttled off by the terminal unit dampers, generating increased noise in the process.

Terminal units downstream of the duct static pressure sensor have their worst-case requirements for static pressure at the sensor under design conditions. For units upstream of the sensor the worst-case operating scenario is one in which the zone load factor is high while the load factors of the downstream zones are low. The former condition maximises the required value of the left-hand side of equation 8.4, while minimising the terms in VP_S and $\Delta TP_{X,S}$ on the right-hand side.

Control strategies are sometimes proposed which involve reset of the VAV *supply temperature*, either involving resetting this down when the outside air temperature is low enough or resetting it upwards with reducing system volume flow rate (possibly linked to an upward reset of chilled water flow temperature). While the aim is typically to reduce the energy consumption of the *water chiller*, there will inherently be associated implications for fan energy consumption. At a given system sensible cooling load system volume flow rate is $\dot{V}_{SYS1}$ (m^3 s^{-1}) at a supply air temperature differential of Δt_{S1} (K). If the supply air temperature differential becomes Δt_{S2} by resetting the discharge temperature off the central AHU, the new system volume flow rate is $\dot{V}_{SYS2}$. Assuming that changes in the specific heat of the supply airstream may be ignored,

$$\dot{V}_{SYS2} = \dot{V}_{SYS1} (\rho_1/\rho_2) (\Delta t_{S1}/\Delta t_{S2}) \tag{8.5}$$

where ρ_1, ρ_2 are the densities of the supply airstream 'before' and 'after' the change (kg m^{-3}).

For the supply fan in a system under duct static pressure control from a sensor at a position S, fan total pressure is given by Equation 8.3. If the supply air temperature differential is the same to every zone, resetting the discharge temperature from the central AHU changes all volume flow rates throughout the supply ductwork by the same proportion, i.e by $[(\rho_1/\rho_2)$

$(\Delta t_{S1}/\Delta t_{S2})]$ in Equation 8.5. The new fan total pressure is thus:

$$FTP_{S2} = SP_S + (VP_{S1} + \Delta TP_{F,S1} + \Delta TP_{AHU1} + \Delta TP_{UP1}) [(\rho_1/\rho_2)(\Delta t_{S1}/\Delta t_{S2})]^2 \quad (8.6)$$

where all terms have their meanings as previously defined and suffixes 1 and 2 refer again to the 'before' and 'after' conditions. This results in a *difference* in supply fan power under the two conditions of:

$$\Delta \dot{W}_{S2,1} = 0.1 \dot{V}_{SYS1} [(X \times FTP_{S2}/\eta_{S2}) - (FTP_{S1}/\eta_{S1})] \quad (8.7)$$

where

$\Delta \dot{W}_{S2,1}$ = difference in supply fan power (kW)
X = $(\rho_1/\rho_2)(\Delta t_{S1}/\Delta t_{S2})$
η_{S1}, η_{S2} = overall efficiencies of supply fan, motor, drive under conditions 1, 2 (%).

Any change in supply fan volume flow rate also results in a change in that of the return fan. Without duct static pressure control, the total pressure developed by this is proportional to the square of the change in volume flow rate. Where the volume flow rate is a constant proportion of that of the supply fan, the difference in return fan power between the two supply air temperature conditions is given by Equation 8.8.

$$\Delta \dot{W}_{E2,1} = 0.1 \; Y \; V_{SYS1} \; FTP_{E1} \; [(X^3/\eta_{E2}) - 1/\eta_{E1}] \quad (8.8)$$

where

V_{SYS1} = system supply volume flow rate under condition 1 ($m^3 s^{-1}$)
Y = the constant *ratio* of system supply air volume flow rate to system extract air volume flow rate, expressed as a decimal

and all other terms with suffix 'E' are the return fan equivalents of those previously defined for the supply fan.

Alternatively, where a constant *absolute difference* ($C\, m^3 s^{-1}$) is maintained between system supply and extract volume flow rates:

$$\Delta \dot{W}_{E2,1} = 0.1 (V_{SYS1} - C) FTP_{E1} [\{[(X V_{SYS1} - C)/(V_{SYS1} - C)]^3/\eta_{E2}\} - 1/\eta_{E1}] \quad (8.9)$$

where all terms are as previously defined.

8.12 The air supply system

An adequate ductwork design is an *essential* element in achieving successful operation and minimum fan energy use in a VAV system. It is generally agreed that a good duct system is one which is able to deliver air at *approximately* the same static pressure to all VAV terminal units, that is able to withstand the pressures encountered under any operating condition with a leakage that does not exceed approximately 5% of system volume flow rate, and does so without generating excessive noise. Low air leakage requires effective sealing of ducts and the connection between ductwork and the supply diffusers. It will naturally benefit capital cost, improve performance under both full and part-load conditions, and enhance the control of terminal units generally if the smallest duct and terminal units are used, with the minimum pressure at inlet to each. A lower requirement for throttling of excess pressure will inevitably reduce the generation of noise by the terminal unit damper, and may obviate the need for secondary acoustic attenuation (with further 'knock-on' effects for reduced inlet pressure and lower capital cost).

Traditional advice proposes that ducts should wherever possible serve zones with *unsynchronised* patterns of cooling load (typically through different orientation), thus taking maximum advantage of load diversity to minimise duct size and capital cost. This is generally consistent with an aim of keeping the duct layout as simple and symmetrical as possible, and will contribute to achieving a quiet system. However, 'duct looping' or the use of ring mains (Fig. 8.18) has acquired its fair share of proponents in recent years, particularly in the USA[25–27]. However, there are effectively two separate aspects of this.

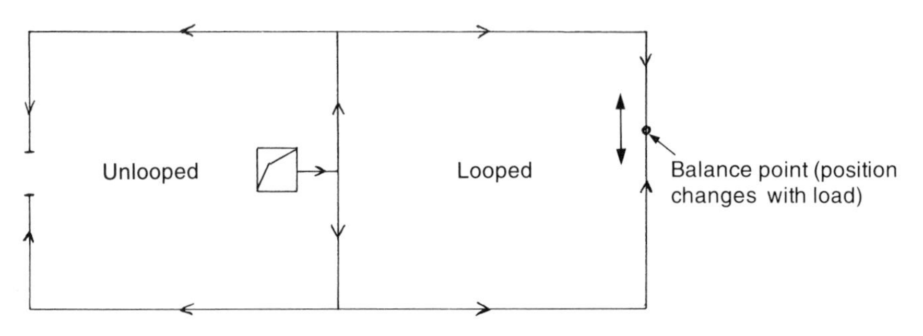

Fig. 8.18 The use of 'duct looping' or ring mains.

The first is the 'looped' or ring main configuration itself. The proposition here is that each 'side' of the loop may be sized at a lower volume flow rate capability, since a section which would otherwise experience a flow rate requirement above its design capacity may be helped out by flow from the 'other side' of the loop (through duct sections operating below their own design capacity). The aim is a simple one of reducing the capital cost and space requirements of the ductwork installation. While consideration of the potential benefits of this technique are certainly justified during the design stage, these will inevitably be dependent on the configuration of the layout and orientation of the building, the nature and pattern of the loads to be served, the type of duct construction and ductwork manufacturing and installation costs. With regard to the latter point, comparisons based on ductwork cost models appropriate elsewhere in the world may not be equally appropriate to UK practice, and vice versa.

The second aspect relates to whether or not it is sufficiently beneficial in overall terms to maintain a constant duct size over a particular section of the 'loop' or ring main, rather than the progressive reduction in cross-sectional area associated with conventional equal-friction-loss duct sizing. Naturally if size is not progressively reduced along a length of duct, while design volume flow rate capacity is, then the contribution to system pressure loss will be less. This, however, is not a unique characteristic of the 'looped' configuration, and may equally be applied to conventional duct layouts. It involves a consideration of optimum duct sizing. While this has been addressed in a number of alternative approaches for ducts in constant-volume systems, there appears at present to be no adequate or widely acknowledged treatment that takes into account the different operating characteristics for ducts in VAV systems. In any event the 'looped' configuration results in a balance point, at which the flow is zero. The position of this point is not fixed, being dependent on the load pattern. Its movement is accompanied by a local reversal of flow direction.

As far as basic ductwork design is concerned, there continues to be an oft-expressed preference for the *static regain* technique for main ducts and risers, with branch ducts and final runouts using the conventional *equal-friction-loss* method. However, the static regain technique is considered by some to be *fundamentally flawed*[28]. On grounds of fan energy consumption and moderating noise generation, modern VAV systems are likely to have a medium-pressure classification as far as main ducts and risers are concerned, with a low-pressure classification for the remaining bulk of the system. However, the ASHRAE Handbook notes that care should be taken to avoid leakage from ducts by treating all ductwork upstream of the terminal units as medium-pressure classification[16].

It is generally to be desired that an *approximate balance* of pressure drops should be achieved across main and sub-main branch ducts under design conditions. Where conditions deviate very significantly from this, the installation of a pressure regulating device close to the branch take-off from the main duct should be considered. An adequate length of straight duct should be provided for the final runout to each terminal unit, so that this is not influenced by non-uniform flow at its inlet spigot. Studies of the potential influence on 'pressure-independent' terminal unit flow sensors of the velocity profiles downstream of sheet metal HVAC duct fittings have shown that flow is unlikely to be either completely developed or fully symmetrical at ten duct diameters after a fitting[29]. The results of these studies show that for bends turning angle has the most pronounced effect on the results, vertical flow profiles showing near symmetry while horizontal profiles remain asymmetric. The most symmetry is shown by conical reducers, while conical expansions show pronounced 'jetting', with separation near the wall of the fitting. For circular-section branch take-offs from circular and rectangular-section main ducts, the angle of the branch take-off with the main duct, and the ratio of the main duct to branch duct flow rates influence flow profiles. Downstream profiles are nearly symmetric, while horizontal profiles are markedly asymmetric. Where calibration or checking of terminal unit flow sensors may be required following installation, it is still necessary to have adequate lengths of straight rigid duct to enable in-duct measurements to be made with sufficient accuracy.

8.13 Noise in VAV systems

Good acoustic design is more important for VAV systems, since the changing conditions in the air supply system (and to a lesser extent the air extract system) make the sound levels more noticeable.

As far as the system supply fan is concerned, typically the same factors which affect fan efficiency also affect fan noise levels. Hence more efficient fans should also be quieter fans. The acoustic reason for modulating the duty of the supply fan is to minimise duct static pressures under part-load conditions (by not allowing the fan to 'ride' its pressure–volume flow rate characteristic), and thus to minimise both fan noise and the noise generated by the throttling action of the terminal unit control dampers. VPIM axial fans are typically directly driven at relatively high speeds, and the high-frequency sound from them may typically be more simply treated using acoustic attenuators. However, duct-mounted acoustic attenuators themselves introduce additional pressure losses into the system. In their turn these increase the pressure generation required by the fan, which adds back some of the noise attenuated out.

An *oversized* supply fan may be efficient under design operating conditions, but is likely to have a more limited range of duty modulation. Separation of the air flow from the fan blades may result in increased noise generation under part-load conditions, and the minimum required volume flow rate may occur in an unstable region of its operating envelope, resulting in the occurrence of 'surge'. 'Surge' can occur with all fan types, and should wherever possible be avoided by selection, although the degree of the problem may be more or less acute. Forward-curved centrifugal fans may not have objectionable surge characteristics at low static pressure in the theoretical 'surge' region, while backward-curved centrifugal and axial are typically susceptible to more severe noise pulsations and vibration when 'surge' occurs. In any event a 'surge' condition will have an adverse effect on operation of the VAV terminal units, and would cause erratic readings of volume flow rate and pressure measuring instruments.

On the other hand, *undersized* fans can inherently show a wider range of duty modulation, but the high air velocity through the fan may result in high noise levels and inefficient operation at (or near) design operating conditions. Overall it is therefore best to avoid extremes of both oversizing and undersizing. Where the characteristics of the application mean that the supply fan will operate at very low system load factor for most of the time, it should be selected to minimize noise and energy consumption under these conditions. If only a low proportion of the annual operating hours are spent under low-load operating conditions, there is no point in penalising fan operation over the whole range of duty modulation to achieve optimum performance in just this one small area. The impact of air from each fan blade passing the discharge makes a noise at a frequency given by (*number of blades* × *RPM* ÷ 60). Since this noise is additive to the other sources of fan noise, it may be possible, by changing fan speed and shifting the *blade passage frequency* to another octave band, to reduce the overall peak sound level generated by the fan.

As far as techniques for the modulation of fan duty are concerned, the use of inlet IGVs typically increase noise level under design operating conditions, and may also increase it as the vanes close during part-load operation. (However, forward-curved centrifugal fans with IGVs become quieter during part-load operation.) Where VSDs are employed, fan noise decreases with reducing fan speed, although the contribution of high-frequency electromagnetic noise from the drive itself typically increases. Anti-vibration isolation must also be effective over a wide range of fan speed, and *resonant frequencies* need to be avoided. Since the VSD could stop at just that frequency, and subsequently remain there for some time, resonant frequencies need to be *positively avoided.* (A constant-volume fan may simply accelerate and decelerate through a resonant frequency with relative impunity.)

Sound reaches the occupied space via two routes. The first is airborne sound from the supply diffuser. The second is sound radiated from the terminal unit through the ceiling of the space. The route producing the greatest effect is the critical sound path. In a situation where the design noise rating (NR) or noise criteria (NC) level is exceeded, this is the path to treat – and is the only path that can have any effect on the room NR (NC) level. Typical sound sources in a VAV system are indicated in Fig. 8.19. The air distribution system itself can, however, provide a 'masking' or 'white noise' that may make occupants feel more at ease. Use of duct static pressure control strategies reduces the throttling required at the terminal units, thus reducing both radiant and airborne noise throughout most of the system's annual operating period.

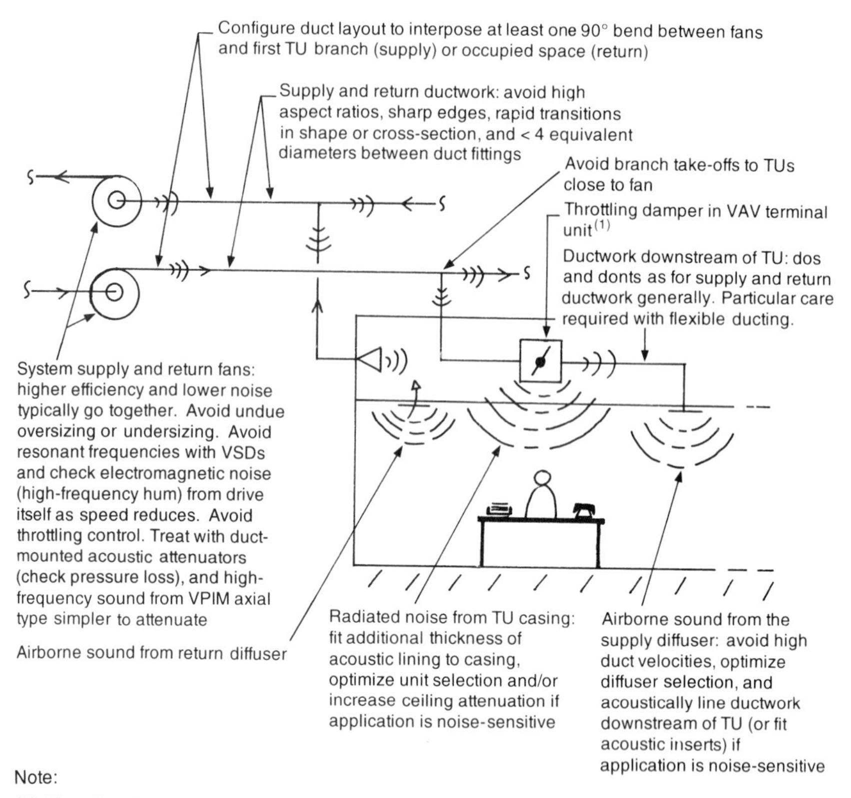

Fig. 8.19 Noise culprits in a typical VAV system.

As far as the prediction of NR (or NC) levels is concerned, several alternatives are available. First *mockups* may be used. However, these are typically too expensive for use in other than very high-profile projects. They also require a very detailed standard of work if true accuracy is to be achieved. The principal alternative is *octave band analysis*. This is accurate, time-consuming and tedious, and is therefore typically only ever applied to a sample of rooms, or specific known critical cases. Finally a method is available based on *sound power per unit volume flow rate*. This is reported to be very accurate, but unsuitable for manual calculations[30].

As far as duct design is concerned, the usual rules apply, i.e. avoid high aspect ratios, sharp-edged fittings and rapid transitions in either shape or cross-sectional area. All these increase regenerated noise. Branch take-offs should not be taken from the root section of the supply ductwork at less than three fan wheel diameters from the fan discharge, and no branch take-off to any actual terminal unit should be taken from the main duct before at least one 90° bend has been made downstream of the supply fan. (Similarly there should be at least one 90° bend in the return air duct between the plant room and the occupied space.) Wherever possible, allow at least 4–6 duct diameters between adjacent duct fittings. In particularly noise-sensitive applications it may be necessary to consider the acoustic lining of ductwork downstream of the terminal unit. Particular care should be exercised over the use and installation of flexible ducting used to connect terminal unit 'octopus boxes' to linear slot diffusers. Even acoustically lined flexible ducting, while attenuating, may *cause* the regeneration of noise if poorly installed (typically with sharp bends).

8.14 Commissioning of VAV systems

The idea is still widely accepted that VAV systems are inherently *self-balancing*, and that any design imperfections will be taken up by self-adjustment of the system. The ASHRAE Handbook states[16]:

> True VAV systems, except pressure-dependent systems, are virtually self-balancing, impaired only by inadequate static pressure control of volume regulation. The larger the network, the greater the need for these controls. This feature also makes it impossible to balance VAV systems with full-load supply air quantities greater than actual load requirements.

However, this self-adjustment is almost inevitably at the expense of increased energy consumption (and running costs) and poorer control, with an accompanying increased tendency towards system instability. It also makes it difficult to determine by how much system performance falls short of the potential indicated by design analysis, and what potential exists for improving it.

At an early stage in the application of VAV systems in the UK attention was drawn to the differences between the regulation (balancing) and commissioning of VAV and constant-volume systems[31]. It is inherently not possible to separate the *regulation* of VAV systems from their *design*, and in a poorly-designed system the required performance may be impossible to obtain during commissioning. The judgement that VAV as a self-regulating system was only likely to be possible where design was of a very high standard and all components functioned correctly is probably no less valid today. Since the control of conditions in the zones served cannot be considered independently of the performance of the air supply system, it was rightly concluded that the regulation of VAV systems should be carried out in conjunction with commissioning of the system control loops.

A VAV system is, in the end, a collection of air-side and water-side control loops threaded together by a variable-volume air stream. Each control loop has the potential to influence the flow rate of this airstream, and in turn to be influenced by it. The integration of system and controls design is thus particularly important for VAV systems. However, the traditional approach to HVAC design has tended to regard each control loop as independent of all others, both from the point of view of its design and its setting to work and commission on site. While the potential for interaction between these 'individual' control loops has long been recognised, the complexity of the effects involved in attempting to consider even a small group of control loops as a whole has lead to the persistence of the 'individual-loop' approach. Zone, duct static pressure, airflow tracking and outside air economiser cycle control loops that can be demonstrated to operate effectively in isolation during the commissioning process may exhibit degraded or unstable performance when operating interactively with other control loops. The introduction of DDC techniques, coupled to the use of detailed building and system performance simulation, may ultimately lead to a practical system control approach for widespread use.

Interaction between individual control loops naturally has implications for the level of thermal comfort in the spaces served, although to what extent specified comfort criteria can be achieved by a VAV system is dependent upon a broader range of factors which include:

(1) appropriate zoning;
(2) the performance characteristics and accuracy of the equipment and controls themselves:
(3) the location and accuracy of zone temperature sensors;
(4) the design of the room air distribution and the performance of the supply diffusers.

Compensation for more than minor deficiencies in these respects, often taken for granted of VAV systems, is not a reasonable expectation if *satisfactory control is to be achieved.* However, with realistic environmental design criteria it seems likely that transient shortfalls in the volume flow rate to individual zones are acceptable – and may even be imperceptible in their local effect.

There are five main stages in the commissioning of a VAV system[32]:

(1) Initial proving of the system supply and return fans.
(2) Testing of the supply system and its VAV terminal units.
(3) Setting up the duty of the supply fan and the control mechanism for duty modulation.
(4) Regulation of the return air system, including the return fan and its tracking (or other) control mechanism.
(5) Setting up of the outside air economy cycle dampers to ensure that the

system's outside air requirements are met over its full operating range. Checking that minimum outside air is drawn into the system when the supply and return fans are both set in turn to system design duty and minimum system volume flow rate, with dampers set for maximum recirculation.

Each VAV box and its downstream ductwork may be treated as a *separate system*. Where multiple supply diffusers are served from a single terminal unit, these should be proportionally balanced at the unit's high-limit volume flow rate. The effect of using a measuring hood at the diffuser is compensated by variable-constant-volume ('pressure-independent') terminal units, although a calibration traverse will still be required to obtain the hood factor.

Calibration of terminal units, or a check on the manufacturer's works calibration, should be made for two reasons. First, settings may have been affected during transit to site. Secondly duct velocity profiles may be ideal for the manufacturer's works calibration, but not so as installed, leading to sensing errors. The basic procedures for variable-constant-volume ('pressure-independent') terminal units include the following.

(1) Set terminal unit controller to high-limit volume flow rate, verifying that the unit is 'in control', with damper between fully closed and fully open.
(2) Similarly set terminal unit controller to low-limit volume flow rate, verifying that the control damper moves to a new position and subsequently stabilises there. Take a reading of the 'index' supply diffuser downstream of the terminal unit. If it is not within 10% of the required value, the terminal unit controller should be adjusted accordingly.

Where FATs are employed, the unit's fan performance should be checked at both maximum and minimum primary supply volume flow rates. With 'series' FATs this requires measurement of the discharge volume flow rate, while for 'parallel' types it requires the measurement of the volume flow rate of return air induced from the ceiling void.

The supply fan should be capable of achieving 105% of design duty for commissioning purposes, so that *with minimum duct leakage* it can deliver the design volume flow rate to the terminal units. The system should then be set up to represent design load diversity. It should be borne in mind that no balancing procedure will produce repeatable data unless the changes in system load are simulated by using either the same configuration of zone temperature controller settings or the same terminal units fixed at their high- and low-limit volume flow rates. In practice it is simpler to set groups of terminal units to their low-limit volume flow rate settings, or to shut off branch ducts fully where a suitable fire damper is available which can be closed (the remaining terminal units remaining at their high-limit flow rate settings), rather than to try to organise a random pattern of reduced flow rate settings among the terminal units.

Having identified the index terminal unit under design conditions (probably the furthest terminal unit from the fan), all terminal units on this branch are set to their high-limit volume flow rate. If the static pressure measured at inlet to the index terminal unit is less than its design value, the duct static pressure set-point is increased until the measured static pressure at inlet to the index terminal unit reaches this value. For terminal units of the pressure-dependent type, the local pressure reducing damper may first be adjusted.

If it is unclear where the optimum position is for the duct static pressure control sensor, provision should be made in the contract specification of works for several sensors to be installed in the *most likely* positions. Static pressures should then be recorded at each of these during commissioning, with one subsequently being designated as the chosen control sensor position. Where conventional duct static pressure control is employed (fixed set-point value), it is prudent to set the controller a little higher than required to satisfy branch static pressures – by a small amount for simple layouts and a greater amount for more complex layouts[32]. Ideally the controller should be adjusted until the index terminal unit is approximately 95% open. This is then the minimum static pressure required to operate the terminal units. The most searching test for the supply fan and system performance is obtained by setting the terminal units closest to the fan to their low-limit volume flow rates and forcing air towards the furthest units, which then have the minimum duct pressure.

References

1. Daly, B.B. (1978) *Woods Practical Guide to Fan Engineering*, 3rd edn, Woods of Colchester Ltd.
2. Jones, W.P. (1994) *Air Conditioning Engineering*, 4th edn, Edward Arnold, London.
3. Kell, J.R. and Martin, P.L. (1979) *Faber and Kell's Heating and Air Conditioning of Buildings*, 6th edn (revised), Architectural Press, London.
4. Larson, E.D. and Nilsson, L.J. (1991) Electricity use and efficiency in pumping and air-handling systems. *ASHRAE Transactions*, **97**, Part 2, 363–377.
5. Building Research and Energy Conservation Support Unit (BRECSU) (1996) *Variable Flow Control*, General Information Report 41, BRECSU, Garston/Watford.
6. Building Research and Energy Conservation Support Unit (BRECSU) (1993) *Energy Efficiency in Offices. A Technical Guide for Owners and Single Tenants*, BRECSU, Garston, Energy Efficiency Office Best Practice Programme, Energy Consumption Guide 19.

7. Woods Ballard, W.R. (1984) Comparative energy savings in fan systems, *Australian Refrigeration, Air Conditioning and Heating*, **38**, May, 18–26.
8. Spitler, J.D., Hittle, D.C., Pedersen, C.O. and Johnson, D.L. (1986) Fan electricity consumption for variable air volume. *ASHRAE Transactions*, **92**, Part 2B, 5–18.
9. Johnson, C.M. (1987) Comparison of variable pitch fans and variable speed fans in a variable air volume system. In *Proceedings of the CIBSE Technical Conference*, held at Brunel University, London, 1–2 June 1987. Chartered Institution of Building Services Engineers, 120–149. Also *Building Services Engineering Research and Technology* (1988), **9**, 89–98.
10. Noon, C. (1987) Energy savings from variable duty fans. *Heating and Air Conditioning Journal*, **56**, March, 17–20.
11. Englander, S.L. and Norford, L.K. (1992) Saving fan energy in VAV systems – Part 1: Analysis of a variable-speed drive retrofit. *ASHRAE Transactions*, **98**, Part 1, 3–18.
12. Lorenzetti, D. and Norford, L.K. (1992) Measured energy consumption of variable-air-volume fans under inlet-vane and variable-speed drive control. *ASHRAE Transactions*, **98**, Part 2, 371–9.
13. Chartered Institution of Building Services Engineers (CIBSE) (1985) *Automatic Controls and their Implications for Systems Design*, CIBSE, London.
14. Chartered Institution of Building Services Engineers (CIBSE) (1986) *Ventilation and Air Conditioning: Systems, Equipment and Control*, CIBSE Guide, Section B3, CIBSE, London.
15. American Society of Heating, Refrigerating and Air Conditioning Engineers (ASHRAE) (1987) All air systems. Chapter 2 in *ASHRAE 1987 Handbook: HVAC Systems and Applications*, ASHRAE, Atlanta.
16. American Society of Heating, Refrigerating and Air Conditioning Engineers (ASHRAE) (1987) Automatic control. Chapter 51 in *ASHRAE 1987 Handbook: HVAC Systems and Applications*, ASHRAE, Atlanta.
17. Arnold, D. (1986) VAV and the economy cycle. *Building Services CIBSE Journal*, **8**, May, 68–69.
18. Englander, S.L. and Norford, L.K. (1991) VAV-system simulation, Part 2: Supply fan control for static pressure minimization using DDC zone feedback. In *Proceedings of the Third International Conference on System Simulation in Buildings*, held in Liege, 2–5 December 1990, Laboratory of Thermodynamics at the University of Liege and the International Building Performance Simulation Association, Inc., 581–606.
19. Englander, S.L. and Norford, L.K. (1992) Saving fan energy in VAV systems – Part 2: Supply fan control for static pressure minimization using DDC zone feedback. *ASHRAE Transactions*, **98**, Part 1, 19–32.
20. Hartman, T. (1989) TRAV: A new HVAC concept. *Heating, Piping and Air Conditioning*, **61**, July, 69–73.
21. Hartman, T. (1992) Promising control innovations with new generation DDC. *Heating, Piping and Air Conditioning*, **64**, December, 55–59.
22. Hartman, T. (1993) *Direct Digital Controls for HVAC Systems*, McGraw-Hill, New York, 110–120.

23. Hartman, T. (1993) Terminal regulated air volume systems. *ASHRAE Transactions*, **99**, Part 1, 791–800.
24. Goswami, D. (1986) VAV fan static pressure control with DDC. *Heating, Piping and Air Conditioning*, **58**, December, 113–117.
25. Mayhew, F.W. (1993) Application of direct digital temperature control systems for maximum system efficiency. *ASHRAE Transactions*, **99**, Part 1, 801–807.
26. Coe, P.E., Jr. (1983) The economics of VAV duct looping. *Heating, Piping and Air Conditioning*, **55**, August, 61–64.
27. Khoo, I., Levermore, G.J. and Letherman, K.M. (1996) Findings of a UMIST VAV research project. In *Variable Flow Control*, General Information Report 41, BRECSU, Garston/Watford.
28. Tsal, R.J. and Behls, H.F. (1988) Fallacy of the static regain duct design method. *ASHRAE Transactions*, **94**, Part 2, 76–89.
29. Griggs, E.I., Swim, W.B. and Yoon, H.G. (1990) Duct velocity profiles and the placement of air control sensors. *ASHRAE Transactions*, **96**, Part 1, 523–541.
30. Brandenburg, J. and Dudley, J. (1982) What you must do for a quieter VAV system. *Specifying Engineer*, **47**, April, 62–68.
31. Holmes, M.J. (1977) Air systems balancing – regulating variable flow rate systems. In *Proceedings of the Symposium on Testing and Commissioning of Building Services Installations*, held at the Institute of Marine Engineers, London, 3 November 1977, Chartered Institution of Building Services (CIBS), G1–9, GD1–2.
32. Building Services Research and Information Association (1991) *Commissioning of VAV Systems in Buildings*, BSRIA, Bracknell, Application Guide 1/91.

Chapter 9
Return Fans in VAV Systems

9.1 The need for control of VAV return fans

The principal requirement for controlling the duty of VAV return fans stems from considerations of building *pressurisation*. It should be remembered that the HVAC designer aims at achieving a slight positive pressurisation of the building as a whole by providing an excess of system supply airflow over extract. If return fan duty remains constant as supply fan duty decreases under part-load-conditions, more air will be extracted from the building than is supplied. The initial slight positive building pressurisation under design conditions will be reduced through neutral to negative. Unfiltered outside air will be drawn into the building through infiltration, both adding to sensible and latent loads on the space and contributing airborne contaminants. A decreasing negative pressure at the junction of the return and outside air ducts may impact the intake rate of outside air into the system. If the duct pressure at this point becomes positive, the outside air intake duct may rather provide a relief path to atmosphere exhausting return air. Hence if a constant level of positive building pressurisation is to be maintained, and the characteristics of the leakage paths for supply air from the building also remain constant, a fixed differential must be maintained between system supply and return volume flow rates as the supply fan duty modulates up and down. Since the supply fan must be free to respond to the system's load requirements without considering the effect on building pressurisation, it is the return fan's duty that must be synchronised with the supply.

The differential required between supply and return air flows is defined not only by the surplus of supply over extract required for building pressurisation but by the amount of the supply air that is ultimately exhausted from the building via local mechanical extract systems. The former depends on the building's air leakage characteristics, while the latter will tend to be continuous at constant volume. However, there is a current trend towards the intermittent use of local mechanical extract systems under the control of passive infra-red (PIR) type occupancy detectors. The principal such systems encountered in commercial applications are toilet extracts. This trend may need to be considered in particular applications.

9.2 Return fan duty control

The techniques available for the control of return fan duty are identical to those discussed for the supply fan. Modern design preference again favours the use of either a centrifugal fan with AC inverter variable-speed drive or a VPIM axial fan, although formerly both eddy current couplings and inlet guide vanes were commonly used with centrifugal fans. For the purpose of the present discussion a variable-speed centrifugal fan with AC inverter drive will typically be assumed.

9.3 The principle of fan tracking control

Control of the return fan to achieve the required system supply–return volume flow rate differential is conventionally achieved using a technique known as 'fan tracking' (or 'airflow tracking'). *Simple* fan tracking uses a common input signal to both fan speed controllers (Fig. 9.1). The return fan speed controller accepts an input signal from the supply fan duct static pressure control loop (or from zone feedback control where this technique is employed), with the signal being offset (or 'lagged') to maintain the required supply–return flow differential. However, this is an *open-loop* control technique, and is only likely to be acceptable within strict limitations. The supply and return fans must have similar pressure–volume flow rate and flow modulation characteristics, and the minimum flow rate of the return fan

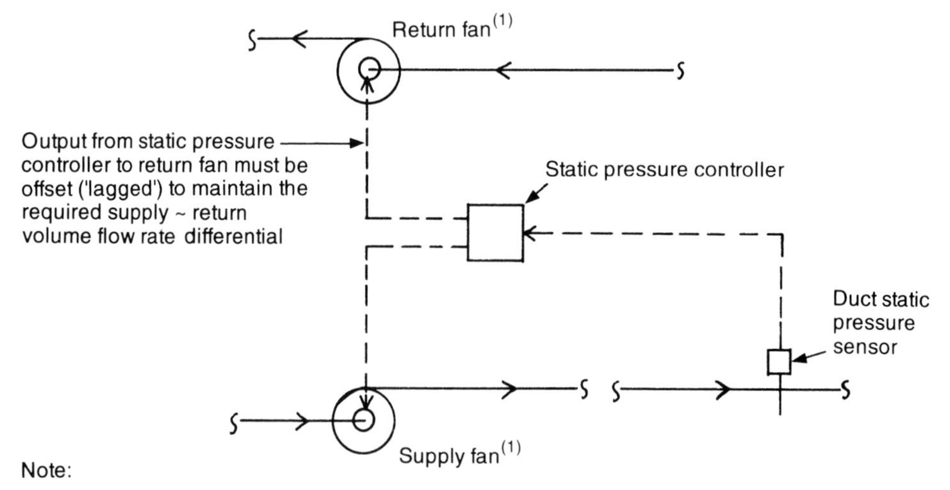

Fig. 9.1 Fan tracking control of VAV return fans.

should not be less than *c*.50% of the design value[1]. With conventional analogue control techniques it is unlikely that a constant flow rate differential can be maintained between supply and return fans using this method. Since the flow rate of local exhaust systems remains constant, the level of positive building pressurisation may be expected to decrease as the system turns down. In principle, DDC techniques permit the incorporation of a suitable compensating control algorithm within the software control strategy. The technique is likely to require a lengthy setting-up process on site, and recalibration of the return fan control system will be required in the event of any subsequent changes to the supply and extract systems themselves.

In *volumetric* fan tracking (Fig. 9.2) volume flow rates or mean duct velocities are determined from velocity pressures sensed in the root sections of both main supply and extract ducts. In principle the volume flow rate or mean velocity in the supply duct is adjusted for the required differential between supply and return, and becomes the set-point value for return fan duty. The volume flow rate or mean velocity in the extract duct is compared with this, and the error is used to generate a control signal that modulates either the speed of the return fan (centrifugal) or its VPIM action (axial fan) in a corrective manner. If mean velocity is used *directly* as a set-point, rather

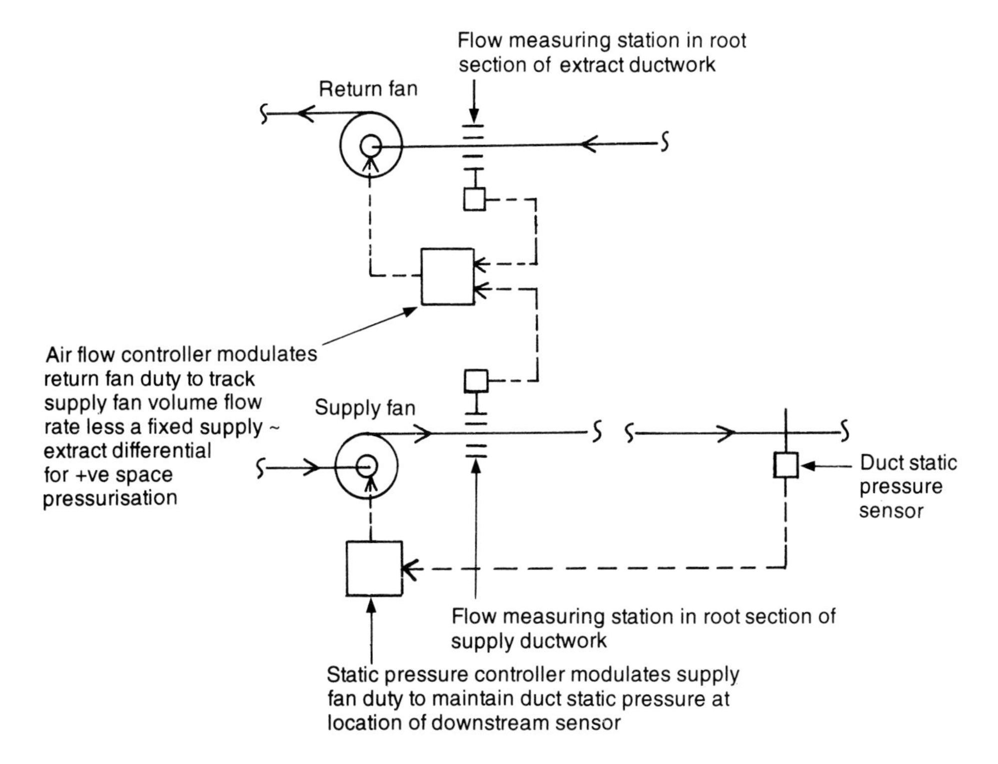

Fig. 9.2 Volumetric fan tracking control of VAV return fans.

than volume flow rate, it must be corrected for any difference between the cross-sectional areas of the supply and extract ducts at the in-duct measuring points (or flow measuring stations).

As has been seen in Chapter 5, reliance on volumetric fan tracking alone to ensure minimum building ventilation rate with a VAV system is likely to be optimistic, and is not common design practice in the UK. While control of minimum outside air intake rate and building pressurisation may not be considered in total isolation of each other, they are independent objectives which have each to be achieved to an acceptable level irrespective of the other. The situation should be avoided whereby the required supply–return fan differential is greater than the design minimum outside air intake rate, since this will increase the outside air fraction on to the central plant and adversely affect its predicted performance under design conditions.

With a fixed flow differential between supply and return fans, static pressure within the spaces served will vary if the exhaust and leakage air flows change. The control of local exhaust systems needs to be considered, and an assessment made of its likely overall impact on building pressurisation. Building pressure will quickly become negative if sufficient outside air make-up is not provided to replace that either exhausted or exfiltrating. To some extent negative pressurisation will cause local exhaust fans to 'ride their fan curves'. While this may reduce their duty slightly, it will not eliminate it. The addition or deletion of local exhaust systems should also be considered for its implications on building pressurisation and supply–return fan differential. In the short term, changes in air leakage are primarily a result of opening and closing of doors or windows. In the long term, building air leakage may increase as a result of normal wear and tear and weathering effects. Unless this effect can be periodically reassessed, and the supply–return fan differential revised accordingly, it will result in a reduction in the level of building pressurisation that can be maintained over the life of the installation.

9.4 The importance of air flow measurements

Achieving accurate in-duct measurements of volume flow rates or mean duct velocities is thus a vital element of airflow tracking. Measuring stations should be located well downstream of the supply fan discharge, and well upstream of the return fan. In-duct measurement of velocity pressure may use single-point or multi-point averaging techniques (inset, Fig. 9.2). Because of the square-law relationship between velocity pressure and volume flow rate, accurate measurement of velocity pressure typically requires either the use of in-duct flow grids or pressure probes that provide an enhanced velocity pressure signal.

Although typically simpler in form, the principle of operation of an *enhanced* velocity pressure (EVP) probe is the same as that of the flow cross in the inlet spigot of the variable-constant-volume VAV box described in Chapter 6. While a number of proprietary designs of pressure sensing probe are available, in each case their purpose is to maintain a measurable differential pressure signal for improved accuracy under the low-flow conditions that will exist at high system turndown. If minimum zone volume flow rate is set at 30% of design, and the system has a design load diversity of 80%, minimum system supply flow rate is 37.5% of design. The existence of a supply–return flow differential results in the return fan having a greater turndown than the supply fan. If, in the present case, building pressurisation requires a surplus of supply over return of only 10% of design volume flow rate, the minimum return volume flow will be only 27.5% of design. These values yield unenhanced velocity pressures of approximately 14% and 8% of their respective values under design conditions. With a mean duct velocity in the range 10–15 $m\,s^{-1}$ for the supply duct under design conditions, and perhaps 7.5–10 $m\,s^{-1}$ for the extract duct, the corresponding ranges of unenhanced velocity pressures are respectively 60–135 Pa (supply duct) and 34–60 (extract duct) under design conditions, and 8–19 Pa (supply duct) and 3–5 Pa (extract duct) under maximum turndown. Velocity pressure on the duct centreline is again greater than these values, and manufacturers of EVP probes typically claim to provide a pressure signal at least several times greater than the unenhanced value.

With the circular-section ducts an alternative option is to sense at a point across the duct radius where the axial velocity has the value of the mean duct velocity. For a fully-developed flow profile this remains constant at 0.242 times the duct radius from the inner wall of the duct. However, if there are any doubts about flow symmetry, two such measuring points should be used on diametrically opposite sides of the same circumference. Single-point sensing of any kind requires a location where the flow is substantially undisturbed, either by the influence of the fan itself or by duct fittings upstream or downstream of the sensing point. This is frequently difficult to achieve in tight plant room spaces and, under non-uniform flow conditions, single-point sensing leads to considerable errors.

In the absence of an acceptable location for single-point sensing of duct velocity pressure, an averaging technique may be preferred. At its most comprehensive, this is based on the use of a full flow grid. As this provides total pressure and reduced static pressure outputs that are each the average of a number of values sensed at the points of a grid fixed within the duct, a flow grid is more robust to duct flow profiles that are non-symmetrical, although it will typically still be adversely affected by strong turbulence. A flow grid can thus cope with *reduced* separation from duct fittings, although adequate separation from the turbulent air stream at a fan discharge must still be maintained. A flow grid is inherently a more expensive option than single-point sensing. A compromise between single-point sensing and a full flow grid is offered, in terms of both cost and performance, by the use of multi-point sensing probes. In its simplest form such a probe comprises an EVP probe that traverses the full width of the duct to allow the averaging of pressures sensed at a number of points across the traverse. A more comprehensive installation might comprise two pressure sensing probes traversing the full width of the duct.

Whether single-point or averaging pressure probes are employed, the pressure outputs from the devices in both main supply and extract ducts must each be passed to their own velocity pressure transmitter. This is a differential pressure sensor, which transduces the velocity or enhanced velocity pressures and amplifies them to a control signal which is proportional to velocity pressure. Typically the control signal will be an electrical output in the range 0–10 VDC, for input to analogue or digital electronic control loop, although an output of pneumatic control pressures in the range 3–15 psi remains an alternative. Since simple transduced outputs from the differential pressure sensors would be linear only with respect to the velocity pressures or enhanced velocity pressures sensed, they cannot be *directly* manipulated or compared for the control of the return fan. The signals must be *linearised* with respect to the volume flow rates (or mean duct velocities) at the measuring station, a process that has come to be known as 'square root extraction'. This function may be incorporated within the air flow controller or be carried out by the velocity pressure transmitter itself, which then becomes more correctly a flow transmitter.

Again, irrespective of whether single-point or averaging pressure sensing techniques are used, initially the device must be accurately calibrated against system volume flow rates measured using standard in-duct techniques, such as multi-point duct traverses using a pitot-static tube. Single-point and multi-point pressure sensing probes need accurate on-site calibration.

In-duct flow grids may be calibrated by the manufacturer off-site, for installation within specified constraints. Outside these constraints additional on-site calibration is necessary. Successful on-site calibration requires full and convenient access to the measuring stations, and between these and the fan duty controls. Calibration must not be undertaken with dirty filters in the system.

Again the alternative to measurement of velocity pressure is to measure duct air velocity using sensors based on the so-called 'hot wire' principle. In practice it may be impossible to achieve adequate separation of the sensor from upstream and downstream duct fittings, and large errors can result from measurements with a single-point velocity sensor in a non-symmetrical or non-uniform flow. *Averaging* techniques are, however, possible using an in-duct grid of hot-wire sensors, with averaging of the sensor outputs and subsequent linearisation of the average output signal. In any event, on-site calibration is required.

Where there is insufficient separation between the supply and return fans and the first branches in the distribution systems, or where a tortuous duct layout does not permit an acceptable installation for pressure sensing probes or velocity sensors, several measuring points may be used further downstream. The derived volume flow rates are summed and manipulated to give a set-point for control of the return fan. However, this approach is probably only practical using the software capabilities afforded by DDC techniques.

The measuring points for airflow tracking control should always be located in the main supply and extract ducts, and never in the air handling unit itself. Pressure sensing probes or velocity sensors must be kept free of contamination by dust or moisture. In particular, care should be taken that probes or sensors are not contaminated or damaged on start-up of the system fans by dirt or debris left in the ductwork during installation. Where steam is injected into the main supply duct for humidification, the position of the steam lances should be considered when choosing the location of the supply duct measuring point.

A modern development may have the potential to remove the problems that are often experienced in achieving accurate in-duct measurements for airflow tracking control. Centrifugal fans are available which offer a differential pressure output calibrated against volume flow rate handled. The technique is typically achieved using calibrated-orifice inlet cones having experimentally derived inlet loss factors, which naturally vary according to fan diameter (Fig. 9.3). Fan duty (in $m^3 s^{-1}$ or $m^3 h^{-1}$) is then obtained from an expression of the form:

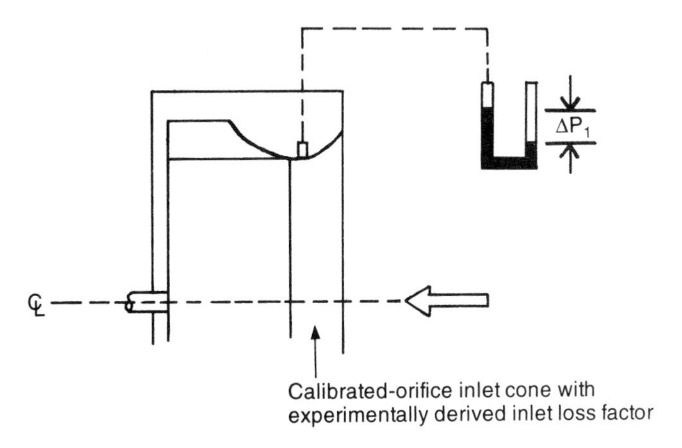

Fig. 9.3 Calibrated volumetric output for VAV fans.

$$\dot{V} = k\,[2\Delta P_1/\rho_1]^{0.5} \tag{9.1}$$

where

k = an experimentally determined loss coefficient
ΔP_1 = the effective pressure at the fan inlet cone (Pa)
ρ_1 = the air density at the fan inlet ($kg\,m^{-3}$).

In principle the technique of *calibrated volumetric output* (CBO) has the potential for a high level of accuracy.

9.5 The air flow control loop

The control loop for volumetric fan tracking is shown in schematic form in Fig. 9.4. The flow measuring stations comprise the primary (in-duct) velocity pressure sensing elements, which have already been considered in the preceding section. The air flow controller has three functions:

(1) to compare the controlled variable with the reference input, and to derive from this an error. Naturally the controlled variable is return fan duty (in the form of the volume flow rate in the main extract duct), while the reference input is derived from the supply fan duty (in the form of the volume flow rate in the main supply duct);
(2) to apply the control algorithm to the derived error;
(3) to provide an output signal to the final control element that will decrease the magnitude of the error, and achieve the required return fan duty. The final control element is either the return fan VSD (centrifugal fan) or its VPIM actuation (axial fan).

The *VP transmitter* is perhaps the most vital component of the control loop[2]. It transduces the sensed velocity pressure (or enhanced VP) signal from the in-duct sensor and amplifies it for output to the control loop. The device cannot make this output any better than sensed by the in-duct sensor, but it can destroy the loop's ability to control by distorting or corrupting the pressure signals sensed. The transmitter may incorporate the 'square root extraction', in which case it is probably more correctly termed a *flow transmitter*. Reference accuracy of such a device may be quoted as a percentage of full span, where the span represents the difference between the maximum and minimum velocity pressure that it is capable of measuring. If the device is zero-based, the span is equal to the maximum measurable velocity pressure. However, the reference accuracy is a characteristic of the *device*, and must therefore be adjusted for its actual operating (input) span in the *particular installation*.

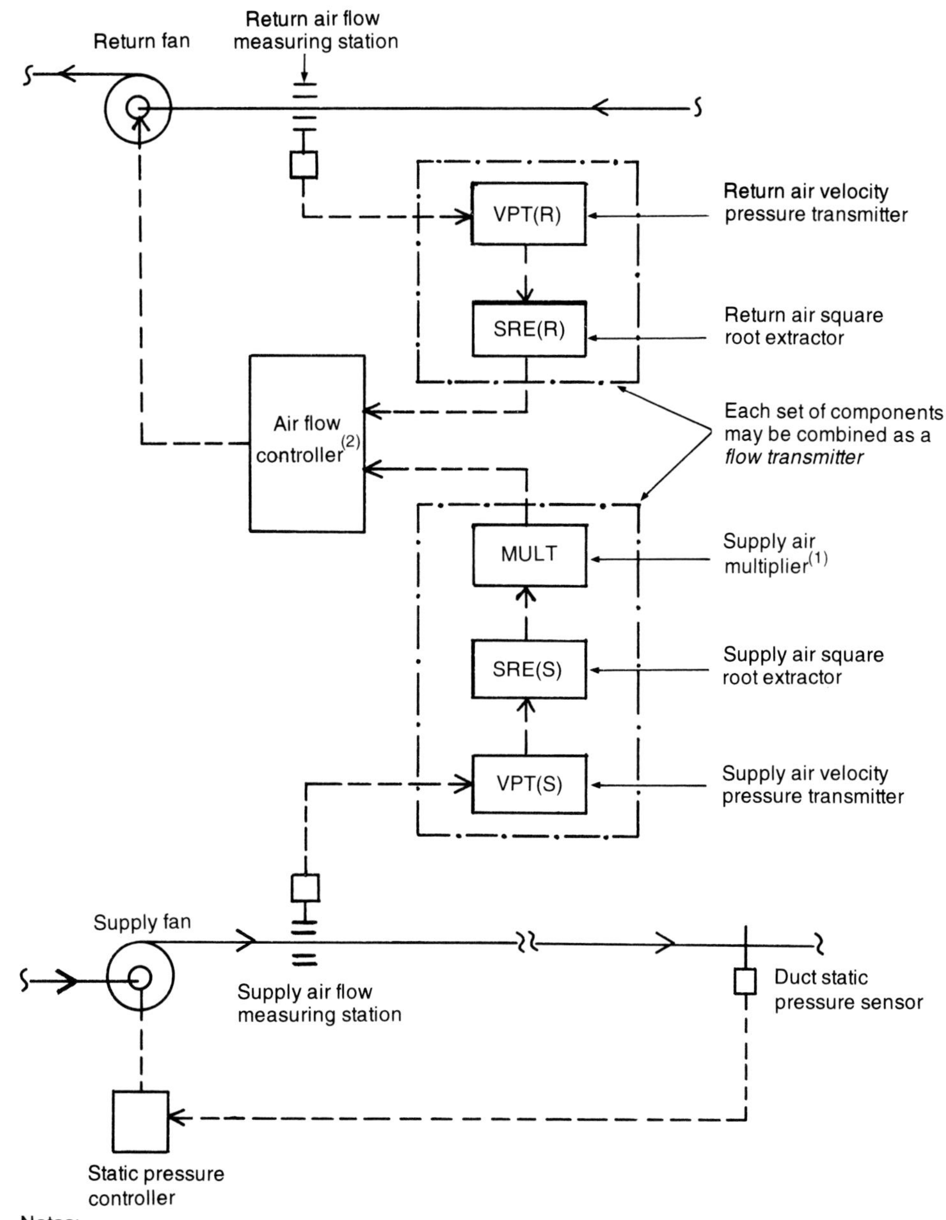

Notes:

(1) A constant factor which refers the output signal from the supply-side square root extractor to a common duct cross-sectional area and VP transmitter span with the return side square root extractor output.

(2) The air flow controller receives the ouput from both flow transmitters. The supply-side signal is offset to compensate for the desired supply ~ extract volume flow rate differential, and forms the $\dot{V}$ set-point value for the return fan. Comparison with the output of the return flow transmitter generates an error signal from which a control output is derived and passed to the return fan VSD controller (or other form of duty modulation). The algorithm for the control output must be P + I to avoid offset, although full 3-term control (P + I + D) may be advisable to avoid potential instability.

Fig. 9.4 Schematic of the control loop for volumetric fan tracking.

$$\textit{Accuracy at operating span} = \textit{Reference accuracy} \text{ (\% full span)} \times \textit{Full span} \div \textit{Operating span} \quad (9.2)$$

This is the accuracy of the device itself, still expressed in terms of its full span, and the *absolute* accuracy must be considered at both design volume flow rate and (particularly) maximum turndown for the fan served. After all, a turndown of 3:1 in volume flow rate corresponds to a turndown of 9:1 in velocity pressure. There is also a need to consider the effect of this accuracy at the maximum turndown in the operating span. Temperature effects may also be present, since the transmitter will be made from different materials, which are unlikely to experience identical rates of thermal expansion and contraction, although temperature compensation or self-calibration facilities may be built into some types. Finite construction tolerances will also have an effect on the accuracy of the device. Unless these effects are included within the reference accuracy, they will also need to be adjusted for the actual operating span of the device in the particular installation. Again the implications for absolute accuracy at design volume flow rate and maximum turndown must be considered.

It would naturally represent the ideal solution simply to size the duct at the location of a VP transmitter to give exactly the maximum velocity pressure capability of the device selected. However, the duct must allow an acceptable velocity pressure to be achieved at maximum turndown, and unless a custom-made transmitter can be afforded, selection of the device must conform to commercially available product ranges. If the maximum velocity pressure to be measured exceeds the maximum capability of the VP transmitter, control will not be possible for part of the operating span. For a zero-based type with an output range of 0–10 *VDC* the transduced output of the basic VP transmitter is given by the expression:

$$\textit{Output (VDC)} = [\textit{VP sensed} \text{ (Pa)} \div \textit{Span} \text{ (Pa)}] \times 10 \quad (9.3)$$

This is proportional to *velocity pressure*, and 'square root extraction' is necessary to gain a direct measure of air velocity, and hence volume flow rate, at the flow measuring stations.

Again for an output range of 0–10 VDC from the VP transmitter, a new output from the square root extractor may be defined as:

$$\sqrt{\textit{Output}} \text{ (VDC)} = [\textit{VP Transmitter output} \times 10]^{0.5} \quad (9.4)$$

This is now proportional to mean duct velocity, and hence volume flow rate. However, unless the duct areas at both supply and extract flow measuring stations are identical, and the spans of both VP transmitters are also identical, the square-root-extracted outputs of the supply and return flow transmitters still cannot be compared (because they are referred to different volume flow rates). Hence a multiplier function is introduced, the purpose of which is to ensure that the reference input to the air flow controller would be the same as the measured-value input from the return air square root extractor for identical volume flow rates at the flow measuring stations. By substituting the outputs of the supply and return VP transmitters from equation 9.3 into equation 9.4 and using the multiplier to equate them for identical flow rates at the FMSs:

$$\text{Return}\sqrt{\ }\text{Output} = \text{Multiplier} \times \text{Supply}\sqrt{\ }\text{Output} \tag{9.5}$$

where

$$\text{Multiplier} = \left[\frac{\text{Return VP}}{\text{Supply VP}}\right]^{0.5} \times \left[\frac{\text{Span of supply VP transmitter}}{\text{Span of return VP transmitter}}\right]^{0.5} \tag{9.6}$$

Since velocity pressure is proportional to (Volume flow rate ÷ Area of FMS)^2, and the area of the respective FMSs are naturally constant, a final expression for the multiplier may be derived as:

$$\text{Multiplier} = \left[\frac{\text{Area of supply FMS}}{\text{Area of return FMS}}\right] \times \left[\frac{\text{Span of supply VP transmitter}}{\text{Span of return VP transmitter}}\right]^{0.5} \tag{9.7}$$

The multiplier function may be carried out within the flow transmitter itself, and it may be possible to achieve a value of unity by a choice of appropriate VP transmitters. However, the latter option may have associate disadvantages in that a custom-made instrument may be required (with implications for cost and availability). Operating accuracy may also suffer if a larger span instrument cannot be fully utilised.

The signal output from the multiplier is still not the end of the story, since it is typical in VAV systems to employ a supply return volume flow rate differential between the system fans. In order for the measured-value input to the air flow controller to have the same value as the reference (set-point) input when the specified supply–return flow differential is just maintained, a suitable offset must be applied to the output of the return air square root extractor. Since the flow rate differential is conventionally intended to be constant, this offset may be derived by considering the difference between the supply-side reference input to the air flow controller at supply volume flow rates equal, respectively, to the system design supply and return air volume flow rates. The required offset is added to the return-side input to the air flow controller, which typically performs this function in modern volumetric fan tracking systems.

Naturally if in-duct sensing is based rather on the use of velocity sensors than velocity pressure, the function of 'square root extraction' becomes unnecessary. Simple proportional control is unacceptable for volumetric fan tracking, because of the offset that is inherent in it. A Proportional-plus-Integral (P + I) control algorithm will ultimately eliminate offset, but will result in the persistence of oscillation in the meantime. The response of the supply fan to changes in system requirements is basically slow, and this may generate instability when combined with P + I action alone. Use of a Proportional-plus-Integral-plus-Derivative (P + I + D) algorithm will increase the rate of response and reduce the oscillatory nature of the response that may arise with P + I action alone, although rapid response in itself is not actually required by the basically slow-acting nature of changes in supply fan duty. An alternative was suggested of using a P + I algorithm in the controller's forward loop and a D-only algorithm in the feedback loop ('feedback derivative'). It was suggested that this would make the output of the air flow controller compatible with the dynamics of the airflow system and the response of the return fan, and capable of achieving the set-point of return air duty without overshoot[3]. However, it is not known whether this in fact proved successful or not (indeed whether it was ever implemented in practice), and it was naturally put forward for an earlier generation of fan duty control equipment.

In any event the action of the airflow controller, and hence of the control algorithms used, is very significant as far as both the accuracy of volumetric fan tracking and its stability are concerned. Modern DDC techniques are generally preferable to conventional analogue electronics, both because of their improved accuracy and the ease of incorporating 'square root extraction', volume flow rate calculations and supply–return differential manipulation within the software-based control algorithms.

9.6 Additional control requirements

Since the return fan inherently has a lower design duty than the supply fan in a VAV system, where a morning warm-up cycle is employed, the supply fan duty should have a high limit equal to the return fan design duty for the duration of the cycle. To ensure matching of supply and return fan duties in this phase of operation, it has been further suggested that the return fan should be controlled to maintain zero static pressure at the *neutral pressure point* that will exist between the return fan discharge and the supply fan suction when both fans are handling the high-limit duty on full recirculation. The advantage claimed for this technique is its lack of dependency on the accuracy of volume flow rate measurement in the main supply and extract ducts.

9.7 Alternatives to fan tracking

Since the primary objective of controlling the return fan is to withdraw just enough air from the occupied spaces to maintain them at a slight positive pressure relative to ambient conditions, a logical alternative to fan tracking would seem to be control direct from this *differential pressure* (Fig. 9.5). Indeed, at first sight this has the advantage of avoiding the generally unsatisfactory nature of simple fan tracking, and both the relatively complex control loop required for volumetric fan tracking and the potentially problematic in-duct velocity pressure (or velocity) measurements associated with this. However, measurement of building pressurisation introduces its own particular set of problems, which are no less demanding to solve:

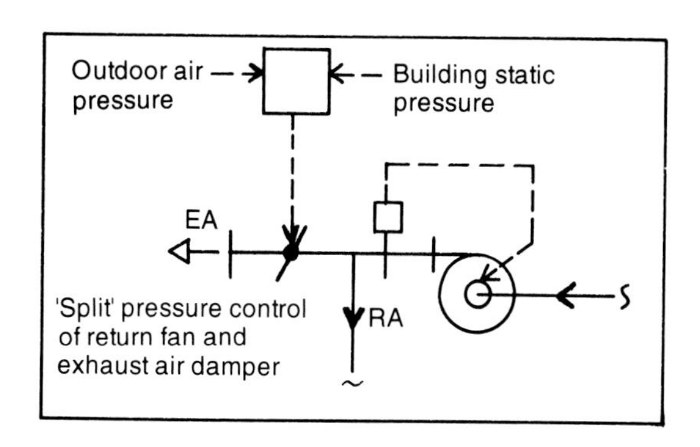

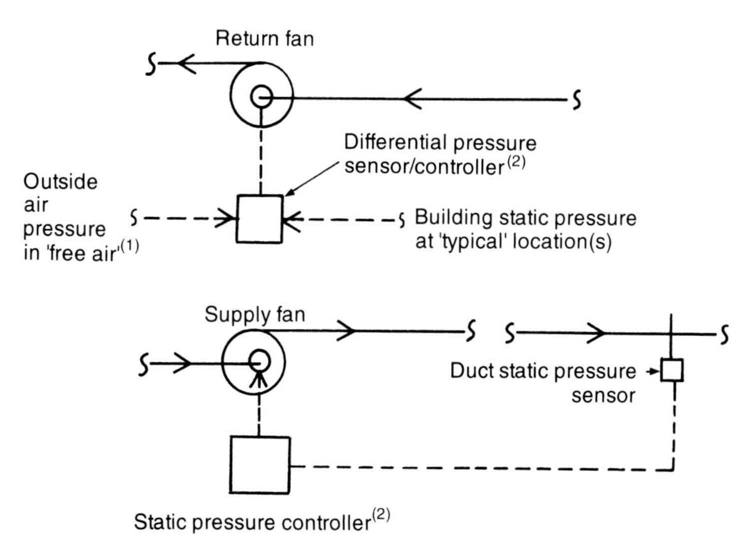

Notes:

(1) Shielded from direct wind pressure and the influence of wind-generated effects.

(2) Modern VSD controllers typically incorporate control algorithms to accept input from sensors directly for common forms of control.

Fig. 9.5 Control of VAV return fan from building differential pressure.

(1) The level of pressurisation required is typically small (25–50 Pa), if problems with door opening are to be avoided, and may be difficult to measure accurately with standard commercially available sensors.
(2) The level of pressurisation may not remain constant throughout the building. This is particularly true of large multi-storey buildings (frequent commercial applications), where wind and stack effects are significant. The variation in pressurisation level throughout such buildings may be much larger than the control value.
(3) Since it is the internal–external differential pressure that is the control variable, a suitable location is required for measurement of pressure outside the building. This should be in 'free air', reliably shielded both from direct wind pressure (a wind speed of 30 mph has a velocity pressure of approximately 108 Pa) and the influence of wind-generated effects, protected against the elements and the risk of vandalism, and readily accessible for periodic inspection and maintenance. Finding such a location may be difficult.
(4) Careful selection of the internal sensing point is required. To reduce the risk of the return fan 'hunting', the sensor location must not be affected either by external conditions (particularly gusts of wind), by the sudden changes of pressure which may result when doors or windows are opened and closed, or by stack effects within the building. It should thus not be within (or immediately adjacent to) horizontal or vertical access routes within the building or vertical services risers. Where DDC controls are used, the sensor output signal may be processed to filter or damp out transient pressure changes, and improve stability.

Basic control of the return fan to a set building differential pressure may be modified by the addition of a static pressure sensor in the discharge duct from the fan. This allows override control of the return fan VSD (or other form of duty modulation) to maintain a slightly positive static pressure there at all times. Taking the idea one stage further, this static pressure set-point may be made more substantial and the basic control input from the building differential pressure transferred from the return fan to the exhaust air damper directly (inset to Fig. 9.5).

Of course a sensor location in the inlet side of the return fan is of no use, since this would only generate constant-volume operation. Attempting to maintain building pressurisation in a 'leaky' building may result in an excessive reduction in return fan duty, leading to the supply fan attempting to draw the corresponding amount of make-up air from outside, either through the outside air damper in its fixed minimum position or even through the exhaust air discharge. On the other hand, if reliance has been placed on the outside air make-up for pressurisation control providing an adequate minimum building ventilation rate, a lower than expected leakage will result either in inadequate ventilation or excessive pressurisation. However, the technique may be applicable to industrial air conditioning applications where there is a requirement for pressure control between adjacent areas. Where a morning warm-up cycle is employed, the set-point for building pressurisation should be reset to zero for the duration of the cycle.

A *hybrid* of volumetric fan tracking and building differential pressure control is possible, in which the fixed supply–return fan differential is reset to maintain building pressurisation, either to take account of extreme short-term effects (although transient conditions may be neglected) or long-term changes in building air leakage characteristics.

Depending upon the layout of individual systems and the relative isolation of floors and zones, active pressurisation control may be used to limit (or encourage) the flow of air between spaces, and to compensate for seasonal stack effects or dynamic wind loading conditions. Return fan control according to pressure differential generally may be applicable to *industrial* air conditioning applications where there is a requirement for pressure control between adjacent areas.

9.8 Alternatives to the use of a return fan

Strictly-speaking a return fan is not necessarily required to extract air from the zones served and return it to the central AHU. This can be achieved with a *single fan* on the supply side of the system, since the return ductwork is connected to the suction side of this fan (Fig. 9.6). If the loss of total pressure along the index run of the return ductwork is greater than that along the outside air intake duct, a greater pressure development will be required from the single fan. However, a single fan cannot by itself exhaust air from the building. Local exhaust systems and exfiltration through positive pressurisation may be able to account for exhaust requirements during operation with minimum outside air. However, if an outside air economy cycle is to be employed, such a system configuration will only be practical where the increased exhaust requirements under full outside air operation can be discharged, directly from the spaces served, through pressure relief dampers in the walls or roof of the building fabric itself. This requirement effectively limits the configuration to single-storey buildings (perhaps extending to some low-rise commercial applications). The pressure relief dampers are set to maintain a design static pressure level in the spaces. It is recommended that simple counterbalanced dampers are used, rather than motorised types, to avoid backflow from wind pressure.

A more widely applicable alternative to the use of a return fan is offered by a relief or exhaust fan (Fig. 9.7). Although not widely used in UK design practice, this technique has been strongly supported in the USA. Assuming

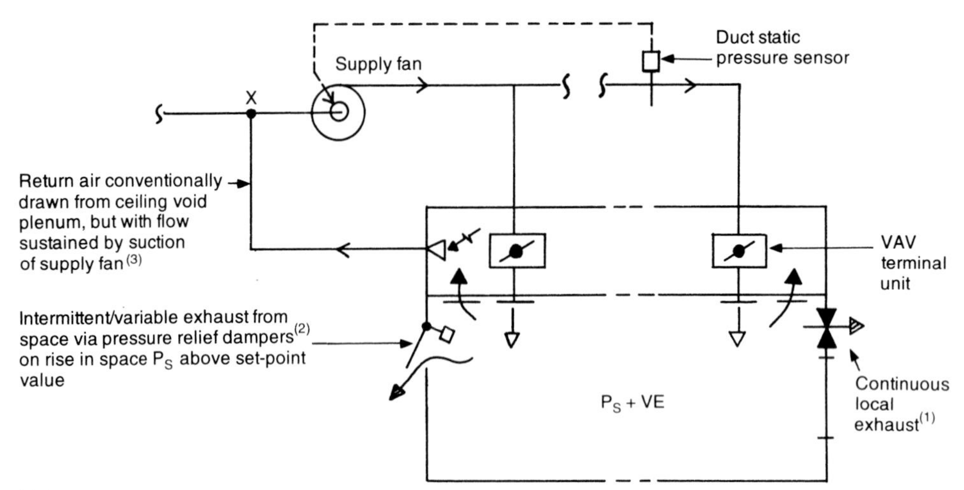

Notes:

(1) Commonly via toilets or (industrial applications) exhaust hoods.

(2) Simple mechanical (counterbalanced) type preferred to avoid backflow from wind pressure.

(3) FTP and –ve P_S at X may be increased.

Fig. 9.6 Single-fan VAV system with direct pressure relief.

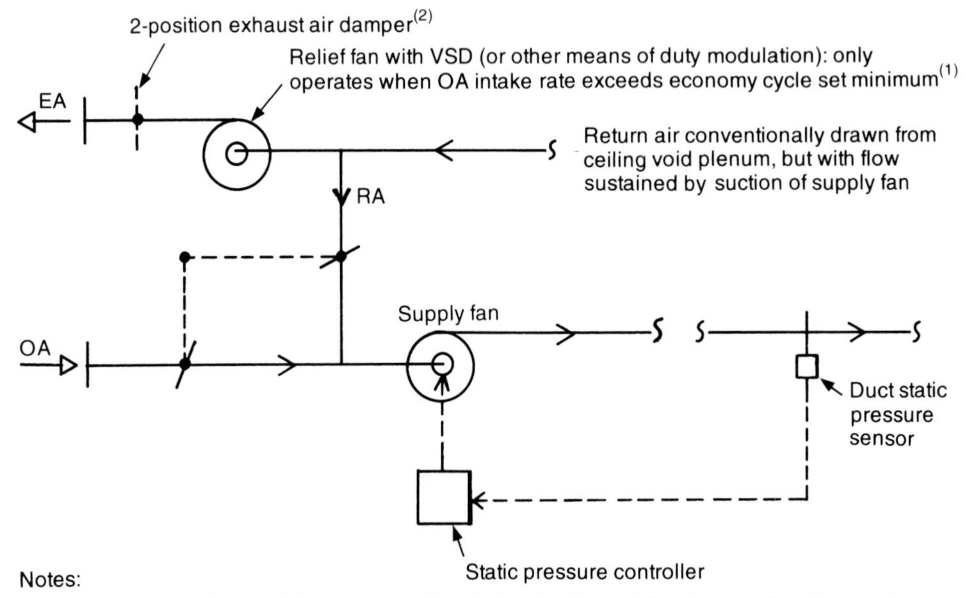

Notes:

(1) When exhaust from building is catered for by local exhaust (via toilets, etc) and general exfiltration resulting from +ve building pressurisation. Assumes that minimum OA requirement for building ventilation can be reconciled with this.

(2) Fully open with fan on, at full shut-off when fan off to prevent OA intake through EA discharge duct.

Fig. 9.7 VAV system with relief fan.

that the outside air intake rate for minimum building ventilation can be reconciled with the make-up air requirement for positive building pressurisation, the relief fan does not run when the system operates on minimum outside air. Air is extracted from the zones and drawn through the extract/return ductwork back to the central AHU using the suction available at the inlet to the supply fan. The minimum outside air intake requirements are exhausted from the building either via local exhaust systems which draw their make-up air from the VAV supply or by exfiltration through a slight positive pressurisation of the building. The exhaust air damper should be fully closed to prevent outside air being drawn in through the exhaust air louvre.

The duty of the relief fan should be controllable through a variable-speed drive (VSD) or some other means. When operation with more than minimum outside air is required, the exhaust air damper moves to fully open and the relief fan runs to exhaust the excess air (only that amount above the minimum outside air intake rate) from the building.

Control of relief fan duty may be to maintain a set-point of building pressurisation, or the fan may be synchronised with the intake rate in the outside air duct using a volumetric tracking technique (Fig. 9.8). A further option had been proposed where a separate minimum outside air duct is employed, enabling the maximum outside air damper to modulate to full shutoff. In this the relief fan is controlled to maintain a constant static pressure ratio across the return air damper, which it is proposed will ensure matching of the flow rates through the maximum outside air damper and the relief fan[4]. While the static pressure upstream of the return air damper no longer has to be kept positive with respect to ambient where a relief fan is employed, a positive pressure difference must still be maintained across the damper (from return to supply) if the relief fan is not to draw air from the outside air intake.

As far as energy consumption is concerned, it has been suggested that between 4 and 8% of total annual system energy consumption for cooling may be saved by use of a relief fan in lieu of a return fan in VAV systems employing outside air economy cycles[5–7]. This is based on the relief fan not running during operation with minimum outside air, on it handling less air than the corresponding return fan when variable air is required (only that actually exhausted from the building), and the heat gain from it not appearing as cooling load on the central plant. These estimates are all based on analyses for various climatic regions in the USA, and may therefore not be directly applicable to the UK's climate. In particular operation with minimum outside air accounts for a relatively small proportion of the potential annual operating hours in the UK climate. There is also a need to consider any increased energy consumption by the system supply fan, if its pressure development must be increased. It may also be necessary to consider

(a) Control from building differential pressure

Differential pressure sensor/controller

Outside air pressure in 'free air'[(1)]

Building static pressure at 'typical' location(s)

2-position exhaust air damper

Relief fan with VSD or other means of duty modulation

EA

RA

OA

Supply fan

(b) Control by volumetric tracking of OA intake rate[(2)]

Relief fan with VSD or other means of duty modulation

Flow measuring station in EA discharge duct

EA

2-position exhaust air damper

RA

Air flow controller

Flow measuring station in OA intake duct

Supply fan

OA

(c) Control from pressure ratio across the return air damper

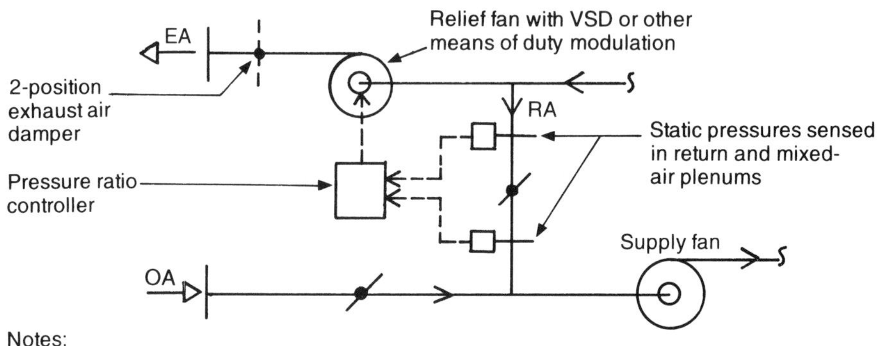

Notes:

(1) Shielded from direct wind pressure and the influence of wind-generated effects.

(2) Simplified schematic for clarity. Similar considerations apply as for volumetric tracking control of a return fan (fig. 9.4).

Fig. 9.8 Control options for the relief fan in a VAV system.

the effects of reduced efficiency of the relief fan VSD, if it is going to run at greater turndown than for the corresponding return fan.

9.9 Predicted energy consumption of VAV return fans

Where only control of overall system extract is employed, the annual energy consumption of a return fan may be predicted by treating it as a particular example of the situation for the supply fan (Equation 8.2) in which fan total pressure bears a simple 'square law' relationship to extract volume flow rate. In any event the return fan may typically account for only about 25% of the energy consumption of the supply fan. There are four reasons for this:

(1) The supply–return air differential results in the return fan handling a lower volume flow rate than the supply fan under all system operating conditions.
(2) Without duct static pressure control, the extract network behaves as a *more-or-less fixed* airflow resistance, exhibiting a square-law relationship between system total pressure loss and volume flow rate over the whole range of operation.
(3) The index run of the return air ductwork has no terminal unit, with its association pressure loss.
(4) The extent of the return air ductwork is typically abbreviated by using ceiling voids as common return air plenums, reducing the design system total pressure in comparison with the supply ductwork.

The implications for annual fan energy consumption of using an exhaust fan in lieu of a return fan have already been noted.

References

1. American Society of Heating, Refrigerating and Air Conditioning Engineers (ASHRAE) (1987) Automatic control. Chapter 51 in *ASHRAE 1987 Handbook: HVAC Systems and Applications*, ASHRAE, Atlanta.
2. Smith, R.B. (1990) Importance of flow transmitter selection for return fan control in VAV systems. *ASHRAE Transactions*, **96**, Part 1, 1218–1223.
3. Hill, A. (1981) Air flow synchronization controls in VAV systems. *Heating, Piping and Air Conditioning*, **53**, August, 47–53.
4. Avery, G. (1989) Updating the VAV outside air economiser controls. *ASHRAE Journal*, **31**, April, 14–16.
5. Avery, G. (1984) VAV economizer cycle: don't use a return fan. *Heating, Piping and Air Conditioning*, **56**, August, 91–94.

6. Avery, G. (1986) VAV: designing and controlling an outside air economiser cycle. *ASHRAE Journal*, **28**, December, 26–30.
7. Kalasinsky, C.C. (1988) The economics of relief fans vs return fans in variable volume systems with economizer cycles. *ASHRAE Transactions*, **94**, Part 1, 1467–1476.

Chapter 10
VAV and Alternative System Comparisons

10.1 Comparing the relative merits of alternative HVAC systems

Before asking the question *how does VAV compare* with the alternative systems design approaches available to present-day HVAC designers and clients, we should perhaps first ask *how the comparison should be made*.

In commercial HVAC applications the relative merits of different design solutions are, in the end, typically weighed in terms of their capital or initial investment cost, running cost, and space requirements. Space requirements may be interpreted both in terms of the loss of rentable area and the sufficiency of installation space, as in the available depth of ceiling voids or a raised floor, or the available cross-section of a service riser. The relative importance of each of these factors varies according to the application (for example, whether a new building or a refurbishment), the type of client (whether owner-occupier or developer), and their socio-economic perspectives.

Among many construction industry professionals there is an increased awareness of the need to service the HVAC requirements of buildings with a minimum of environmental harm (although a positive environmental friendliness is perhaps still some way off), and a growing number of large client organisations are finding the environmental 'high ground' either a fashionable place to be for public relations or a lucrative selling point for their products or services. However, while the relative environmental performance of different design solutions to a particular application may come to acquire an importance independent of simple capital and running cost implication, it is presently still the rare few designers who enjoy the luxury of a client prepared to take more than a very small step away from the 'bottom line'.

Whether it concerns cost, energy use or environmental impact, comparisons between HVAC systems ought properly to follow a life-cycle approach. The absence of a suitable and freely available database for estimating either the overall *embodied energy* (as analagous to capital cost) or environmental impact of a complete HVAC system prevents the practical adoption of these yardsticks for comparing the life cycle performance of systems. The basis for comparison between systems thus still remains primarily an *economic* one. While life cycle costing is a well-developed technique, the reality is that in many modern commercial applications (especially for office buildings in financial centres) the annual running cost for HVAC systems may represent only *c*. 5% or less of the cost of staff employment (salaries themselves, plus employment overheads). Many markets permit high running cost and energy use to simply be shrugged off on to the ultimate end user of the facilities. Unsurprisingly, under such conditions it is often difficult to persuade clients to place enough importance on running costs and energy use, and its resultant environmental impact.

There are several factors why meaningful generalisations about the realistic relative performances of different HVAC systems may be difficult to make. Design time and cost do not typically permit thorough comparison of competing design solutions for large numbers of real buildings. Neither is the time available after the completion of design work to go back and work out what the cost would have been with alternative system types. Real data may sometimes not be published because of commercial sensitivity. Hence the database on system comparisons available to those without extensive personal experience of varied competing systems tends to be quite narrow. Those comparisons that exist in published form have typically been made for a particular building, whether a real application or a model serving simply as a tool for system comparison. The results of extrapolation to other applications typically require considerable care and knowledge in their interpretation.

The correlation of capital cost with building and HVAC design parameters is not a simple one, since the options available to both architects and building services engineers are so varied. Cost per square metre has been widely used for office buildings. Although not necessarily accurate in absolute terms, it will tend to put competing design solutions on an even relative footing. Cost per unit of installed refrigeration capacity, or per unit of supply air volume flow rate may not do so. Comparisons of running cost and energy consumption require assumptions to be made about system operating profiles and the performance of system controls. Finally, whether system comparisons are actually valid in reality depends both upon the correct matching of system and application, and upon good design. In this respect we should be reminded that a system which is designed well, installed well, commissioned well, maintained well and well operated is always likely to prove superior, regardless of type. It is estimated that where air conditioning is necessary, up to 30% of typical annual energy cost may be saved by ensuring[1]:

(1) appropriate system selection;
(2) energy-conscious design;
(3) good system control;
(4) effective system commissioning;
(5) effective maintenance of system.

10.2 System comparisons for VAV

Early in their development the cost and energy performance of VAV systems would have been compared principally against those of perimeter induction, fan coil, and constant-volume terminal reheat systems. To a lesser extent a comparison would have been made with dual-duct systems, terminal heat recovery units and chilled ceiling systems. Since the 1980s the comparison for commercial multi-zone air conditioning applications at the medium-to-large end of the market has become principally one between VAV and fan coil systems. Industry statistics suggest that there is effectively no new-build market for perimeter induction systems in the UK[2,3].

The introduction of variable-refrigerant-flow (VRF) systems has introduced a further competitor for some applications that would previously have been virtually the sole province of either VAV or fan coil systems. It is not yet clear to what extent VRF systems are likely to gain acceptance in this role by HVAC designers and clients in the UK. Industry opinion has been divided on the question of whether market share gained by variable-refrigerant-flow systems would be at the expense of either VAV or fan coil systems, or both equally. A complicating factor is the consideration of health and safety issues involved in the piping of even low mass flow rates of refrigerant in proximity to normally occupied spaces. What effect the intended phasing out of production world-wide of the present hydrochlorofluorocarbon (HCFC) refrigerants, following on from the earlier phasing out of the chlorofluorocarbon (CFC) types, will have on designers' attitudes to VRV systems is as yet unclear. The HCFC R22 is currently the refrigerant predominantly used in VRF systems, as it is indeed for a wide range of other HVAC refrigeration purposes. Renewed interest in the application of proprietary types of chilled ceiling and chilled beam systems will naturally generate renewed comparison with VAV systems.

10.3 Factors affecting cost and performance

A thorough-going and detailed technical comparison of the various types of air-conditioning system available to the HVAC designer would justify a lengthy volume in its own right. However, while *not exhaustive*, the following list summarises the principal factors that ought commonly to be considered when making a comparison of cost and performance between system types, or between the main sub-types of a particular system.

Acoutics	performance; ease of attenuation; implications of constant or fluctuating noise levels.
Comfort	control of temperature; humidity; room air movement; IAQ; variables influenced directly, indirectly or allowed to 'float'; quality of control and likely deviations of indirectly influenced or 'floating' variables.
Controls	requirements; extent; type; accuracy; degree of integration.
Design load diversity	system sizing to peak simultaneous zone sensible cooling loads or sum of peak zone loads.
Fans	type; duty required; techniques and equipment available for modulation and control (and their characteristics relative to the application, e.g. generation of harmonics); energy consumption.
Flexibility	to accommodate changes in use of the areas served; whether or not this incurs a penalty in performance or operating cost.
'Free cooling'	availability and basis (air-side or water-side economy cycles).
Humidity control	capabilities (close control or high/low limit).
Maintenance	nature and extent; frequency; localised or distributed plant and equipment.
Room air distribution	risk of 'dumping' or stratification; low or high room air movement (stagnation or draughts).
Space heating	full heating/cooling flexibility or separate cooling and heating.
Space requirements	for plant; depth of ceiling/floor voids and cross-section of service ducts (requirements vs. availability); loss of rentable space.
Ventilation	constant or varying; control for minimum level.
Zoning	requirements and flexibility; control and energy use.

Table 10.1 shows a generalised comparison of these factors for the principal types of air conditioning system in current use for new or replacement systems.

The variable-flow-rate systems, VAV and VRF (variable-refrigerant-flow), allow full advantage to be taken of available design load diversity. So

Table 10.1 The principal factors affecting comparisons of cost and performance for various air conditioning systems

Factor	Single-duct VAV	Fan coils with ducted outside air	Reversible heat pumps with ducted outside air	Chilled ceilings with ducted outside air	Variable refrigerant volume (VRF)
Acoustics	Consider fluctuating noise from fans, TUs[1], diffusers	For good performance select on low speed only	Compressor cycles in zone; consider TU performance carefully	Conventional consideration for ducted OA	Fan, diffuser noise; consider TU performance carefully
Control of temperature	Zonal	Zonal	Zonal, but close control not possible	Zonal	Zonal
Control of humidity	Not zonal; consider worst-case rise under part-load	By dehumidification of OA Check capacity vs. ventilation needs	TUs dehumidify Close control not possible	By dehumidification of OA Check capacity vs. ventilation needs	TUs dehumidify Close control not possible
Controls	Added complexity of fan modulation/ synchronisation, TU controls Risk of interaction	Large number of zone control valves for ch.w., LPHW to fan coils	Central plant and ducted OA basic; TUs cycle on/ off (complex internally)	Large number zone control valves for ch.w.	TU controls (complex internally); refrigeration circuit controls complex (typically proprietary)
Design load diversity	Full advantage for plant/ system sizing	None	Full advantage for plant/ system sizing	None	Full advantage for plant/ system sizing
Fan and pump power	High potential fan power savings; select for high efficiency over most operating hours Terminal fan power significant if FAT VAV, but possible cold air distribution	CAV OA system fans much smaller than VAV, but fan coil fans make up for this (and lower efficiency) Ch.w. secondary + LPHW pumping to fan coils	CAV OA system fans much smaller than VAV, but TU fans and compressors significant; also water loop pumping	CAV OA system fans much smaller than VAV, but high ch.w. pumping power	No air system fans unless ducted. OA for ventilation, but TU fans significant; variable-speed compressors at central refrigeration plant

Continued

Table 10.1 Continued

Factor	Single-duct VAV	Fan coils with ducted outside air	Reversible heat pumps with ducted outside air	Chilled ceilings with ducted outside air	Variable refrigerant volume (VRF)
Flexibility	Very good both addition of zones (subject system capacity OK) and rearrange existing (especially modular designs)	Some for addition of zones; better for rearrangement of existing	Flexible for addition of zones (if system capacity OK) and rearrange existing, subject to TU space and piping constraints	Strictly limited unless considered at design	Flexible for addition of zones (if system capacity OK) and rearrange existing, subject to refrigerant pipework constraints
Free cooling/economy cycle	Available on air side	Limited to water side if cooling tower OK	Not available	Limited to water side if cooling tower OK	Not available
Specific maintenance requirements	System fan control and TUs; FATs if FAT VAV	Fan coils (including filter changing) and zone ch.w., LPHW control valves	TUs (specialist, plus housekeeping/changing filters)	Zone ch.w. control valves	TUs (specialist, plus housekeeping/changing filters)
Reheat for zone control	Not unless risk of serious overcooling or high humidity	Sequence control of cooling/heating at fan coil	No	No	No
Risk of simultaneous cooling and heating	Yes, if combined with LPHW perimeter heating; use local emitter control and 'dead band' techniques to minimize risk	Use 'dead band' in sequence control of cooling/heating at fan coil to minimize risk	Individual TUs either cooling or heating	Yes, if combined with LPHW perimeter heating; use local emitter control and 'dead band' techniques to minimise risk	Individual TUs either cooling or heating
Room air distribution	Risks of dumping and stratification need careful attention to design; risk of low air movement (especially if system oversized)	Conventional design considerations for ducted OA and fan coil discharge	Conventional design considerations for ducted OA; typically poor from TUs at low level (tho' CV units possible)	Conventional design considerations for ducted OA	Depends on TU type; integral discharge typically poor; conventional design considerations apply if remote diffusers

Space heating	Full heating/cooling flexibility available at zones. Consider options carefully at design. VAV cooling/LPHW perimeter heating is common	Full heating/cooling flexibility available all zones with four-pipe system. Two-pipe changeover type not suited to UK climate	Full heating/cooling flexibility available all zones	Requires separate space heating system (typically LPHW to radiators, natural convectors)	Full heating/cooling flexibility available all zones with heat pump TUs
Space requirements	Lowest all-air for PR[2], ducts; ducts minimised if cold air distribution (but FATs, ice storage); no zone space	Much less than all-air for PR, ducts (none if local OA); no zone space if fan coils in CV	Much less than VAV for PR, ducts (none if local OA); high zone space at low level unless TUs in CV (uncommon in UK)	Much less than all-air for PR, ducts; no zone space	Minimal PR (none if roof plant) and for refrigerant pipework; no zone space if TUs in CV
Ventilation	Varies with zone, system load; consider carefully at design	Constant minimum OA to all zones with CAV ducted or local OA	Constant minimum OA to all zones with CAV ducted or local OA	Constant minimum OA to all zones with CAV ducted or local OA	Constant minimum OA to all zones if CAV ducted or local OA
Zoning	Zone as little as possible for maximum design load diversity (never by orientation)	Not required, but may be convenient	Zone as little as possible for maximum design load diversity	Not required, but may be convenient	Zone as little as possible for maximum design load diversity

Notes

(1) Terminal unit

(2) Plant room

also does a system of reversed-cycle heat pumps connected to a common water loop (rejecting heat from room air to the water circuit during cooling, and pumping heat from the water circuit to room air during heating). The variable-temperature systems, constant-volume terminal reheat and fan coils, allow no design load diversity.

However, VAV systems have additional equipment and control requirements associated with the system fans – including electrical or mechanical means of modulating fan duty, and control loops both linking the modulation of supply fan duty to feedback from the system and suitably synchronising the modulation of return fan duty with this. Fan selection in constant-volume systems aims to achieve peak efficiency at the design operating point. Selection of VAV fans should aim to achieve near-optimum efficiency over a range of duty covering a significant proportion of the annual operating period.

A fan coil system with ducted primary (outside) air requires much lower design duty for the central system fans (perhaps only 20% or less of the VAV system's requirement, as far as the supply fans are concerned), and has fewer pressure loss components in the supply ductwork. Pressure losses downstream of the fan coils are made good by the units' own fans. Under design conditions the energy consumption of the central system supply and extract fans will be much less than those in an equivalent VAV system. While the position naturally improves somewhat with increasing turndown of the VAV system fans, the overall annual energy consumption of the central system fans is still likely to be greater than in the fan coil system by a factor of the order of 2. However, to this must be added the energy consumed by the large number of zone fan coils themselves. The fans in these units may be of relatively low overall efficiency. Chilled water must also be piped – and pumped – throughout the building to the fan coils.

An all-air design approach, such as embodied in VAV systems, inevitably has a greater space requirement for ducted air distribution. Any attempt at economising on this by reducing duct size will result in increased fan energy consumption *over the whole operating life of the system*. While this is true for any all-air system, in absolute terms the effect will be less for a VAV system than for a constant-volume one, since the former spends a high proportion of its annual operating period at relatively low system load factor. Furthermore the VAV approach may be combined with low-temperature air supply (cold air distribution) to allow the size and cost of the distribution ductwork from the central AHU to be minimised. There is a 'knock-on' effect from this, since the smaller ductwork associated with cold air distribution may allow a reduction in the floor-to-floor height of a building. This generates cost savings in the structure and fabric of the building, and in many other building components – not least of which include the mechanical, electrical and public health engineering services. In parts of the world where height restrictions apply to high-rise construction, the ability of VAV with cold air distribution to maximise the number of floors that can be fitted within a specific height restriction may in itself constitute sufficient justification for adopting this approach. There is a further benefit to be derived from the smaller AHU and primary-side ductwork where plant room space is at a premium.

There is of course also a capital cost premiuim associated with the change from conventional throttling VAV terminal units to fan-assisted terminals (FATs) of the series-fan type that are conventionally used with such systems, and with the plant associated with production of the 'cold air' (which will typically involve ice storage techniques). Furthermore, while the energy consumption of the primary supply fan may be reduced by a factor of the order of one-third, this is typically more than compensated by the energy consumption of the series-type FATs themselves. However, the cost of this fan energy consumption may frequently be passed on by the building owner or operator to the tenants. Indeed, in shell-and-core developments the provision of the FATs themselves becomes the responsibility of the tenant. In both cases this may thus be seen as a benefit to the developer and building owner/operator.

The use of intermittent-operation parallel-fan type FATs only when necessary to maintain proper room air distribution, or combined with a supply air temperature reset strategy, is one possible alternative means of removing the energy use penalty, relative to conventional VAV systems, of using 'series' FATs with cold air distribution. The use of induction-type VAV terminal units is another, while the successful supply of cold air *directly* to the spaces served is considered by some to be practical if supply diffusers are carefully selected and again a *supply temperature reset* strategy is employed. A further variant of the latter approach involves the use of high-induction linear diffusers, which are partially (and continuously) fed with supply air bled from upstream of the VAV terminal units. This supply air bleed provides a relatively high-velocity, and hence highly inductive, jet of air at the diffuser. This is instrumental in achieving the high level of room air induction necessary to permit low-temperature air to be supplied directly to the space.

Whilst it is typically advantageous to zone perimeter induction and constant-volume all-air systems generally according to the orientation of the building facades, the *opposite* is true for VAV systems. Space and access requirements may generally be handled more efficiently for a single air handling plant than for several smaller ones. Similarly, a single duct is more efficient in its use of space than are two smaller ones. For example, a 500-mm diameter sheet metal duct has a volume flow rate capacity of approximately $1.3\,m^3\,s^{-1}$ at a rate of frictional pressure loss of $1\,Pa\,m^{-1}$. At double the daimeter, a 1000-mm duct has a flow rate capacity of approximately $8\,m^3\,s^{-1}$ at the same pressure loss rate.

Room air distribution design techniques and equipment selections differ between VAV and constant-volume systems. The risk of 'dumping' must be considered when fixed-geometry diffusers are used with a VAV zone supply. While removing this risk, the use of variable-geometry diffusers or fan-assisted terminals adds to the capital cost of the installation. FATs also increase maintenance and running costs and energy use, and require consideration of design NR levels. The risk of *stratification* must be considered if it is proposed to use a VAV system for space heating at maximum turndown of the terminal units. Again, the use of FATs overcomes this, but at the cost premiums noted.

Where terminal units incorporating fans or compressors are to be installed either in false ceiling voids or within the space served itself, the acoustic performance of the units must be carefully considered. For typical installations, manufacturers' acoustic performance data may be over-optimistic in its assessment of the benefits of room effect and the attenuation offered by a suspended ceiling. The acceptability of the fluctuations in noise level which are likely to be experienced with a VAV system must also be considered. Different methods of modulating fan duty may have different acoustic charactersitics.

The fan coil system cannot take advantage of the 'free cooling' available to a VAV system using an outside-air economy cycle. The primary cooling coil in the central AHU handles 100% outside air even under summer design conditions, and refrigeration plant may be required to run down to relatively low outside air temperatures. *Water-side* economy cycles are possible, but require the availability of a *cooling tower*. The latter are the subject of stringent health and safety requirements. In view of the potential risks and onerous costs associated with operating cooling towers (inspection and maintenance, water treatment and the water use itself), the modern trend is towards the use of *dry air cooling* techniques, enhanced where necessary by limited adiabatic (water spray) action.

The fan coil system again has no capacity to utilise recirculation as a form of heat recovery, although mechanical heat recovery techniques may be used – since the AHU will handle 100% outside air even under winter design conditions. However, mechanical heat recovery is inherently less efficient than recirculation, and increases the capital cost of the system. Either mechanical cooling or heating is typically required, or both. Comments made regarding fan coil systems are generally equally valid for other air/water systems.

The primary air supply to fan coils is able to provide a constant supply of ventilation air to all zones under all operating conditions, albeit this is limited to a constant minimum level. A similar consideration applies to all constant-volume air supply systems. For a large part of the annual operating period, a VAV system using an outside air economy cycle also supplies 100% outside air, and at much larger volume flow rates than a fan coil system in the same application.

However, when operating with minimum outside air, the VAV system may need to introduce a greater volume flow rate of outside air into the system (than the primary supply to the fan coils) to ensure adequate ventilation of all zones.

10.4 The sensitivity of cost/performance comparisons

Within the overall comparison between systems, the relative importance of the various individual component packages typically varies. This may affect the *sensitivity* of the comparison to potential design changes, the vagaries of tendering procedures, and fluctuations in the economy. Air distribution systems are a case in point, i.e. ductwork and associated thermal insulation, diffusers and grilles. For a typical medium-to-large office block this component may represent up to 50% of the capital cost of a VAV system with perimeter heating, but little more than half of this for a fan coil system with ducted primary air. The capital cost of VAV systems generally is thus more sensitive to the cost of ductwork than that of fan coil systems, whether through contractors' overhead and profit margins, labour costs or increases in the price of steel. Additionally, the way in which each is designed can significantly affect both relative proportions and absolute costs.

Finally, the quality of the environmental control that can be achieved, and the flexibility of the system to adapt to changes in end-user requirements may differ. Factors such as these may be hard to quantify, and are typically not adequately reflected by the traditional comparative measures of capital and running costs, or even in terms of energy use. One exception to this concerns the amount of outside air that must be supplied to provide adequate ventilation of the spaces served. Where ventilation performance varies between systems, opinion may vary as to what constitutes 'adequate' ventilation. However, the influence of different outside air supply rates can generally be fully quantified in terms of cost and energy use.

10.5 Comparative capital costs for VAV systems

Table 10.2 summarises data from several sources on the comparative capital costs of various HVAC systems. The comparison is presented in terms of a relative cost ratio which is the ratio of system cost to that of a VAV system with perimeter heating. The data due to Jones have been converted from their original form, in which the cost ratio was presented relative to that of a two-pipe perimeter induction system[4]. As originated, it was considered that to reflect average system costs, 'high' and 'low' values for each system type covering a range of 30–40% each side of the average, and to be applicable to

Table 10.2 Comparative capital costs for various air conditioning systems

System	Relative cost ratio[1]							
	Jones (1981)	Hayward (1987)	Handel, Lederer and Roth (1992)		Scott and Hunt (1995)[4]			EEO[5]
			Control case 1[2]	Control case 2[3]	$35\,W\,m^{-2}$	$60\,W\,m^{-2}$	$90\,W\,m^{-2}$	
VAV with perimeter heating	1.0	1.0	1.0	1.0 (0.96)[6]	1.0	1.0 (1.13)[7]	1.0 (1.27)[7]	1.0
Perimeter induction (2-pipe)	0.91	1.05	—	—	—	—	—	—
Perimeter induction (4-pipe)	1.05	—	—	—	—	—	—	—
Ceiling-mounted induction units	—	—	0.76	0.79	—	—	—	—
Four-pipe fan coils with ducted OA	1.0	0.97	—	—	1.03	0.97	0.92	0.94
Reversible heat pumps with ducted OA	—	0.95	—	—	—	—	—	0.72
Chilled ceiling with mixing ventilation	—	—	—	—	—	0.88	1.02	—
Chilled ceiling with displacement ventilation	—	—	1.21	1.13	—	—	—	—
Ditto with LPHW perimeter heating	—	—	—	—	—	1.03	1.13	—
Ditto with electric panel heating	—	—	—	—	—	0.95	1.06	1.16
Dual-duct	1.23	1.20	—	—	—	—	—	—
Low-velocity constant-volume all-air	1.23	—	—	—	—	—	—	1.11

Notes

(1) Ratio of cost relative to that of a VAV system with LPHW perimeter heating (VAV with terminal reheat for data due to Scott and Hunt).
(2) Room temperature set point scheduled vs. outside air temperature.
(3) With 3K 'dead band' superimposed on Case 1 schedule.
(4) For cooling load levels of 35, 60 and $90\,W\,m^{-2}$.
(5) Source: Building Research Energy Conservation Support Unit (BRECSU) (1993) Energy Efficiency Office Good Practice Guide 71. *Selecting Air Conditioning Systems. A Guide for Building Clients and their Advisers.* Garston/Watford: BRECSU (Datum cost is £180 m^{-2} for VAV with perimeter heating.)
(6) Relative to Control case 1.
(7) Relative to VAV with perimeter heating at cooling load of $35\,W\,m^{-2}$.

office blocks with treated floor areas in the range 1000–10 000 m^2 and refrigeration loads in the range 110–150 W m^{-2}. The data due to Hayward is unspecified, being based on the contributed experience of practising HVAC designers, and a two-pipe perimeter induction system has been assumed[5].

In the *two-pipe* perimeter induction system the induction units in the zones have a single heat exchanger with flow and return pipe connections (two pipes). In the *changeover* mode of operation all units are supplied with either secondary chilled water to do sensible cooling or LPHW for space heating. In *non-changeover* operation the heat exchanger can only do sensible cooling, and any requirement for space heating is met by heating the primary air supply to the units. In *four-pipe* systems the perimeter induction units each have both heating and cooling coils. Each is served by its own flow and return pipes (four pipes in all), effectively doubling the quantity/cost of distribution pipework to the units and its thermal insulation and increasing the cost of the control system. As with a four-pipe fan coil system, heating and cooling coils are controlled *in sequence*, and full heating and cooling capacity is always available.

The data due to Handel *et al.*[6] compares the capital costs of three different approaches to air conditioning an office building of four floors. The total area of the building is given as 5000 m^2, with a ratio of window surface area to outside wall area of 0.5. The exact orientation of the building is unclear, but may be north–south (principal axis east–west), since a 'diversity factor' of 0.95 is noted for simultaneous peaks of cooling. This raises an important point, since the capital cost of a VAV system will be least when design load diversity is greatest. Thus the most advantageous orientation for a building served by a VAV system is when the major axis is north–south, giving principal glazed facades facing east and west. Handel *et al.* quote a sensible heat load of 60 W m^{-2}, intending this to represent standard office accommodation with a medium level of gain from office equipment (suggested as 10 W m^{-2}) and a good level of shading.

The three approaches compared were a VAV system with perimeter heating, an air–water system using *ceiling-mounted* induction units, and a chilled ceiling system combined with displacement ventilation. In UK practice the use of induction units would have been synonymous with terminal units mounted at low level around the perimeter of the building. The choice of a chilled ceiling system is interesting in view of the strong resurgence of interest in this system and recent developments of the principle. Unfortunately little detail is given regarding system design. The VAV approach involved two roof-top AHUs handling a total design supply volume flow rate of 95 000 $m^3 h^{-1}$. No reason was given for the use of two VAV systems. Again from a capital cost perspective it is important to take the maximum possible advantage of the potential design load diversity available to VAV systems. Both the primary supply volume flow rate of the induction system and the displacement ventilation associated with the chilled ceiling were 30 000 $m^3 h^{-1}$.

The induction system was used as the datum for relative costs, and the data in Table 10.2 have been converted from this.

The two sets of data presented by Handel *et al.* refer to alternative control strategies. In the first the set-point value of room air temperature is scheduled against outside air temperature, rising linearly from 22°C at an outside dry-bulb temperature of 22°C, to 26°C at an outside dry-bulb of 32°C. In the second a *dead band* of 3 K is superimposed on to the schedule, allowing room air temperature to 'float' between 22 and 25°C up to an outside dry-bulb of 26°C, the limits of this dead band increasing linearly to between 24 and 27°C at an outside dry-bulb of 32°C. Introduction of the dead band was of particular benefit to the VAV system, with the total design supply air volume flow rate for the two AHUs reducing to 75 000 $m^3 h^{-1}$. Capital cost of the induction system was unaltered, whilst that of the VAV system was reduced by 4% and that of the chilled ceiling by 11%. This again raises the question of the sensitivity of overall capital cost to changes in individual elements of the design, since the 4% change inferred in the cost of the VAV approach has resulted from a reduction of 21% in the system volume flow rate.

The data due to Scott and Hunt similarly derives from a comparison between VAV, fan coils and chilled ceilings for a model building representing the refurbishment of a 1970s' six-storey office block[7]. Dimensions of this model building are given as 65 m by 12 m (net office floor area approximately 4000 m^2), refurbished to current Building Regulations, with principal east/west facades glazed to 50% of their area, with internal blinds and external shading via a horizontal overhang.

For the capital cost comparison in this study three levels of design cooling load were considered:

(1) 35 $W m^{-2}$ for a 'low-tech' office with energy-efficient lighting;
(2) 60 $W m^{-2}$ for a 'medium-tech' office;
(3) 90 $W m^{-2}$ for a 'high-tech' office with a high density of IT equipment.

As with all other data in Table 10.2, Scott and Hunt's have been converted to be relative to VAV as unity. The study revealed the following points.

(1) At low design cooling load there is little to choose between any of the options. There may be a distinct capital cost advantage where displacement ventilation can deal with the cooling load on its own, although the question of flexibility in zone control may need to be addressed.
(2) A wide variance was noted in the potential cost of a chilled ceiling installation (a factor of approximately two between the highest and

lowest cost), raising the question of sensitivity of comparisons. Scott and Hunt reported that their study assumed an economic selection towards the lower end of the cost range. Chilled ceiling cost depended not only upon material type but on manufacturing and installation methods. Performance was also affected – for example, aluminium offers good conductivity, but requires water treatment to prevent corrosion, while capillary systems require segregation of primary and secondary water circuits across a heat exchanger. The cost of the chilled ceiling also depends upon the manner in which heating and ventilation are provided. Air distribution in particular influences the ceiling's performance and cooling capacity.

(3) At medium design cooling load there was again little to choose between the options.

(4) At high design cooling load the cost of a chilled ceiling increases sharply, as the active cooling area increases, although an all-air approach may have difficulty in introducing sufficient air into the conditioned space, in an acceptable manner, under design load conditions.

10.6 Comparative running costs and energy consumption for VAV systems

Unless absorption-cycle refrigeration plant is used, the provision of space cooling primarily consumes electrical energy. The latter's status as the *premium-cost* fuel makes it a prime target for the reduction of HVAC system running costs. Because of its high primary energy ratio, reductions in electrical energy consumption at point of use also have greater environmental impact, unit for unit, than reductions in natural gas or fuel oil.

As part of its best practice programme, the UK Energy Efficiency Office (EEO) has published guidance on energy consumption and costs in owner-occupied and single-tenant office buildings[8]. Four types of office building were considered, two heated and naturally ventilated and two air conditioned. The EEO guidance is based on an analysis of data for a large number of office buildings, and selected data for the air-conditioned types are summarised in Table 10.3. Air conditioning systems are assumed to be of VAV, fan coil or perimeter induction type.

The first of the air conditioned building types represented a standard commercial application, comprising open-plan office accommodation operated 'nine-to-five' and without dedicated computer suite or staff restaurant. The second type represented a prestige building, such as a national or regional head office, with a dedicated computer suite and sizeable staff restaurant, a generally high level of facilities and IT equipment, and a

Table 10.3 Energy consumption and costs in air conditioned offices

End use	Annual delivered energy in $kWh\,m^{-2}$ treated area[1]				Annual energy cost in £ m^{-2} treated area[2]			
	'Standard' commercial application		'Prestige' commercial application		'Standard' commercial application		'Prestige' commercial application	
	Typical	Good practice	Typical	Good practice	Typical	Good practice	Typical	Good practice
Heating and hot water (fossil fuel)	222 (58%)	100 (51%)	259 (57%)	124 (51%)	2.67 (23%)	1.20 (19%)	3.11 (24%)	1.48 (20%)
Refrigeration	33 (9%)	17 (9%)	41 (9%)	24 (10%)	1.83 (16%)	0.92 (14%)	2.06 (16%)	1.18 (16%)
Fans, pumps, controls	61 (16%)	39 (20%)	71 (16%)	47 (19.5%)	3.36 (29%)	2.14 (33.5%)	3.53 (28%)	2.35 (32%)
Lights	67 (17%)	39 (20%)	82 (18%)	47 (19.5%)	3.67 (32%)	2.14 (33.5%)	4.12 (32%)	2.35 (32%)
Totals								
Fossil fuel	222	100	259	124	2.67	1.2	3.11	1.48
Electricity	161	95	194	118	8.86	5.2	9.71	5.88
Overall	383	195	453	242	11.53	6.4	12.82	7.36
Reductions by good practice[3]								
Fossil fuel	—	55%	—	52%	—	55%	—	52%
Electricity	—	41%	—	39%	—	41%	—	39%
Overall	—	49%	—	47%	—	44%	—	43%

Notes
(1) Figures in brackets show value as percentage of overall total.
(2) 1990–91 values, typically 1.2 p kWh^{-1} for fossil fuel and 5.5/5.0 p kWh^{-1} for electricity in 'Standard'/'Prestige' commercial applications respectively.
(3) Relative to the corresponding 'typical' values.

more flexible pattern of operation to allow extended hours of use. Annual energy consumption and costs for the provision of heating and hot water and lighting are included for comparison, but consumption and cost data for office equipment, computer suite and catering use have been excluded.

Consumption and cost data is given per unit of *treated* area, defined as the total building area measured inside the external walls, but less plant rooms and other areas not *directly* heated or cooled (including stores, car parking and roof spaces). Treated area was expected to be in the region of 90% of gross area for the standard application, assuming a reasonably space-efficient building, and in the region of 85% for the prestige offices. The potential difference in energy consumption between buildings due to varying levels of energy efficiency is catered for in part by the separation of the EEO data into two categories. The first category represents a 'typical' building exhibiting performance near the middle of the range suggested for UK office stock as a whole. The second represents a building considered to be typical of 'good practice' in building and engineering services design and management, as based on past case studies of energy-efficient offices (although standards of energy efficiency will subsequently have moved on). Cost guidance allows a lower unit cost of electricity for larger and more intensively serviced buildings, on the assumption that a higher voltage supply and probable improved load factor gives more bargaining power in negotiations with electricity suppliers.

Since energy consumption by control systems themselves (not for control) is typically insignificant, it can be seen that for the standard air conditioned office building annual energy consumption by fans and pumps accounts for approximately 40% of annual electricity use (38% for 'typical', 41% for 'good practice'), approximately 20% of total energy use including fossil fuel for heating and hot water generation (16% for 'typical', 20% for 'good practice') and approximately 33% of total energy costs (29% for 'typical', 33% for 'good practice'). This is comparable to use for lighting, and approximately double that for refrigeration. The latter comes as a surprise to many people, since the refrigeration plant almost inevitably has the greatest current rating of any element of an air conditioning system.

For the good-practice examples of both the standard and prestige types of office buildings the annual electricity costs for air conditioning were down by approximately 40%, with that due to fans, pumps and controls down by approximately 35%. For comparison, for both types of building annual energy costs for space and water heating and lighting were down by approximately 55% and 42% respectively. Standards for energy efficient design of buildings and their engineering services are continuously evolving, and it is reasonable to suppose that present-day 'good practice' would improve on these reductions. On the basis that moving to a heated and naturally ventilated environment removes both the need for refrigeration and all but a now negligibly small element of fan and pump consumption, the energy use and operating cost premium associated with air conditioning is clear to see.

Table 10.4 shows the results of an early comparison, by Gosling and Williams, of annual electrical energy consumption between a VAV system with LPHW perimeter heating, four-pipe fan coils with ducted outside air and a two-pipe non-changeover perimeter induction system for air conditioning of an identical office building[9]. In the 'two-coil, four-pipe' system each fan coil has separate heating and cooling coils supplied with LPHW and secondary chilled water respectively. Heating and cooling is controlled in sequence, and full heating and cooling capacity is always available. Changeover systems are not suited to the lengthy spring and autumn characteristic of the UK climate, with their variable weather. All estimates of component energy consumptions for the three systems have been reduced to percentages of the estimated total annual electricity use by the VAV system (in kWh). The study did not compare the estimated energy requirements, from natural gas or fuel oil, for the heating plants associated with the three system types.

The basis of the comparison is an office block of ten floors, each approximately 70 m long by 18 m wide (1260 m^2 area) and 3 m high.

Table 10.4 Annual electrical energy consumption for alternative air conditioning systems: VAV with (wet) perimeter heating, four-pipe fan coils and non-changeover perimeter induction systems

	Electrical energy consumption as a percentage of the total for VAV with perimeter heating		
Plant/equipment	VAV with perimeter heating[1]	Two-pipe perimeter induction[2]	Four-pipe fan coils[3]
Supply fan	27.79	22.35	14.53
Return fan	6.95	—	—
Terminal unit fans	—	—	57.40
Water chillers	40.70	61.84	61.84
Chilled water primary pumps	7.28	16.76	23.47
Chilled water secondary pumps	—	12.85	—
Condenser water pumps	4.49	5.31	5.31
Cooling tower fans	7.28	8.62	8.62
LPHW pumps	5.51	3.67	8.45
System	100.0%	131.4%	179.6%
Routine inspection and maintenance requirements			
Total number of terminal unit inspection or maintenance actions/year	960	4800	2880

Notes

(1) LPHW – radiators or finned-tube natural convector.

(2) Non-changeover system.

(3) With outside air ducted from a central air handling plant.

Construction was assumed to include double glazing with light-coloured Venetian blinds and insulation levels average for the period (since improved). Each floor was divided into 96 modules of 1.5 m by 7 m (parallel to the major axis), based on the minimum controllable area for flexibility of partitioning. Estimates of cooling and heating loads were based on a location in London with an ambient design condition of 27.8°C dry-bulb and 20°C wet-bulb, and with a ventilation requirement of 12 litre s^{-1} per person for 1400 occupants. Inside design conditions were assumed to be 22.2°C dry-bulb temperature with 50% relative humidity for both induction and fan coil systems, but with a slightly reduced relative humidity of 45% for the VAV system. No indication was given of the orientation of the building. Naturally there have been significant changes, both in HVAC technology and in the economic climate, in the intervening eighteen years since Gosling and Williams' study. However, these are likely to have rendered their results more conservative for a current comparison than they would have appeared at the time. Further consideration of this study is therefore justified, as it illustrates well the fundamental characteristics of VAV systems.

The water chillers for all three design options were selected to provide constant condenser water flow rate and cooling tower size. Energy consumption for each plant was estimated on the basis of average monthly wet-bulb temperature at noon (for cooling and dehumidification of outside air), monthly average dry-bulb temperature (for sensible heat transmission through the building fabric), and average monthly sunshine hours (for solar gain). Heat gains due to occupancy and lighting were assumed constant throughout the year. The VAV system employed an outside-air economy cycle and return fan.

Both design refrigeration capacity and power requirements were slightly higher for the VAV system than for both perimeter induction and fan coil systems – by *c*.3% and 6% respectively. However, operation of the outside-air economy cycle results in a much reduced energy consumption for the VAV system chiller(s) over an assumed cooling season between April and October. In the fan coil system chilled water is shown as distributed directly to the fan coils at the same temperature as to the main cooling coil in the central AHU. Under such circumstances the fan coil units will inevitably perform some dehumidification. It is frequently the case that a primary/secondary circuit arrangement is employed for the chilled water system, in which the fan coils are fed from the secondary side with chilled water at *c*.11°C, and an element of electrical energy consumption for secondary chilled water pumping is involved. Since Gosling and Williams' study, concerns over the health risk from poorly designed and improperly maintained cooling tower installations has led many HVAC designers to favour the use of either air-cooled water chillers or adiabatic/dry air coolers in chiller condenser water circuits. Heat can typically be rejected more efficiently using latent cooling techniques than by means of sensible cooling, and the overall effect of the move away from cooling towers may be expected to benefit the VAV system, as this requires least use of mechanical refrigeration.

For the perimeter induction and fan coil systems all requirements for extract from the zones were assumed to be satisfied by a combination of the building's toilet extract ventilation system(s) and exfiltration due to building pressurisation. Comparisons of fan energy consumption are based on fan total pressures for the main supply fans in the central AHUs of 1.3 kPA (VAV), 1.1 kPa (fan coils) and 1.87 kPa (perimeter induction) respectively. This shows clearly the reason for the big difference between the estimated energy consumption of the main supply fans in the perimeter induction and fan coil systems, despite an identical design volume flow rate. The higher fan total pressure will also make it a potentially much more expensive fan in capital cost terms. Working back using Gosling and Williams' data, overall efficiencies for the fans at design duty are implied as being of the order of 70% for both VAV and fan coil systems, and *c.*78% for the perimeter induction system (it being essential that a high-efficiency fan be selected where high fan total pressure is required).

As effectively there is no longer a new-build market for perimeter induction system designs, further comments will in general be restricted to the fan coil and VAV systems. The same loss of total pressure under design conditions was assumed for the supply ductwork in both systems. The assumption in this is that the physical extent of the air supply systems is little different whether it is VAV terminal units or fan coils that are being supplied, and are in each case simply sized up or down to the same basic design rate of frictional pressure loss. The difference in fan total pressure between VAV and fan coil systems is down to the additional pressure loss across the VAV terminal unit, plus differences in component pressure losses in the central AHU. The fan coil (ceiling void units are assumed) makes good its own pressure losses, together with that of any downstream ductwork and the supply air diffuser, which could be of comparable linear type in both cases.

The net effect of the slightly different component configurations and losses in the central AHUs is small, and may be specific to this particular study. However, the allowance for pressure loss across the index VAV terminal unit, which for the present purpose may simply be equated to the terminal unit furthest from the supply fan, merits comment. At 238 Pa this represented *c.*18% of design fan total pressure. The authors' stated affiliation suggests that this would have been based on the use of a system-powered type, which requires a minimum static pressure to be available at the terminal unit (independent of volume flow rate) to ensure a satisfactory control function. A minimum requirement of between 100 and 150 Pa would be reasonable for a modern system-powered type discharging the supply airstream directly into the zone served through an integral linear slot diffuser. For a modern electrically actuated type an allowance of between 50 and 100 Pa for the minimum (damper fully open) pressure drop across the terminal unit at zone design volume flow rate would be typical, assuming a basic cooling-only unit without secondary attenuator. For any particular design or size of terminal unit the actual value depends on the design volume flow rate, and lower values may be possible. To this would need to be added pressure losses in the ductwork downstream of the unit and across the index supply diffuser.

A minimum system supply air volume flow rate of 41% of the design value was taken for the VAV system during the period November to March. This period accounted for approximately 41% of the total VAV system annual operating hours, and energy consumption by the supply fan estimated at approximately 30% of the total for both supply and return fans for the whole year. This is very significant since we have seen in Chapter 8 that a large proportion of the potential savings in fan energy consumption that are possible through improved control strategies occur in this area of VAV system operation. Variable-volume control of both VAV supply and return fans was achieved by the use of inlet guide vanes. One most important change that has occurred since the time of Gosling and Williams' study is the general availability for application to air conditioning fans of variable-speed drive technology based on solid-state AC inverters. In a current study this would typically be the preferred choice for centrifugal fans, and would be expected to yield improved absorbed power–volume flow rate characteristics.

The data lumps together fan power under part-load operation for the VAV supply and return fans. The absorbed power of the return fan motor at system design volume flow rate appears to have been taken simply as 25% of that of the supply fan. Assuming similar total efficiencies, this implies a design fan total pressure for the return fan of between approximately one-quarter and one-third of that of the supply fan, depending on any excess of system supply over extract. This would result from a design pressure loss in the extract ductwork of between approximately one-half and two-thirds of that in the supply ductwork (excluding the index VAV terminal unit). The combined power absorbed by the supply and return fan motors under part-load conditions is treated as a percentage of the design value. It seems clear that this percentage was based on an absorbed power–volume flow rate characteristic for a fan connected to a system which behaves as a fixed resistance. This *underestimates* fan total pressure (and absorbed power) of the supply fan at low system load factor where duct static pressure control is employed. Although not explicitly stated, it is reasonable to assume that a duct static pressure control loop would indeed have been employed for this application. Four points are worth noting in relation to this.

(1) Despite holding the duct static pressure constant at some point, the overall pressure loss through the supply ductwork will still reduce as

system load factor falls. Reductions in the pressure loss between the static pressure sensor and the fan discharge will cause fan total pressure to decrease. Even with the sensor located close to the actual fan discharge itself, the reduction in velocity pressure with system load factor will be reflected in some (albeit small) decrease in fan total pressure. A reduction in mean duct velocity from $15\,m\,s^{-1}$ at system design volume flow rate to $6\,m\,s^{-1}$ at 40% of this results in a reduction in velocity pressure of 113.4 Pa, or 8.7% of the design fan total pressure.

(2) At 505 Pa, the design pressure losses across the components of the central AHU and primary acoustic attenuator (the outside air inlet louvre is excluded here) slightly exceed those in the supply ducting. Since this element of the system does operate as a fixed resistance, a reduction in system load factor to 40% would result in a reduction in fan total pressure of 424 Pa (*c*. 33% of its design value).

(3) Return fans in VAV systems are typically not under static pressure control, and the absorbed power–volume flow rate characteristic is that when connected to a system which behaves as a fixed resistance.

(4) In the intervening period since their study it has been widely realised that to achieve the full potential for reducing the fan energy consumption inherent in VAV systems requires that the supply fan be controlled in such a way as to maintain both the minimum volume flow rate and the minimum fan total pressure required by the existing thermal or ventilation requirements of the system. In Table 10.4 the energy consumed by the system supply fan represents approximately 28% of the overall electrical energy consumption for the system. The electrical energy consumption of the system is four times more sensitive to reductions in the energy consumption of the supply fan than that of the return, and more sensitive than for any other item of equipment except (in this case) the chillers, which were responsible for approximately 41% of system annual electrical consumption.

In the fan coil system the additional energy consumption by the terminal unit fans themselves completely swamps the reduced consumption of a central AHU handling just enough outside air to satisfy the ventilation requirement. The potential for 'free cooling' with an all-air system is also clearly evident in the significantly lower chiller energy consumption of the VAV system. Since Gosling and Williams' study VAV systems employing fan-assisted terminals (FATs) have been applied in the UK. Such systems might well be considered as alternative options to a conventional VAV system employing the single-duct throttling type of terminal unit. Fan-assisted terminals incorporate a heater battery not dissimilar to that of a fan coil unit, and a broadly similar type of forward-curved centrifugal fan with the same permanent-split-capacitor type of motor. Absorbed power ratings of the motors would be broadly comparable. Such a VAV system will have a very significant energy consumption for the terminal unit fans, which in the series-fan type of terminal unit run continuously at constant volume. The overall effect of this on the electrical energy consumption of such a system depends on the potential to take advantage of two factors. First, constant-volume/variable-temperature operation on the zone side of the terminal unit permits the use of *low-temperature* VAV, or VAV with *cold air distribution*. The reduced supply temperature from the central AHU proportionately reduces the system volume flow rate at all load factors, and hence the energy consumption of the main supply fan. Second, a smaller refrigeration plant may run overnight on reduced-tariff electricity, building up a store of ice which can subsequently be used during the day to produce the lower chilled water temperatures required for cold air distribution. The situation is made more complex both because the coefficient of performance of the refrigeration plant will be lower in an ice-making role and because of the range of possible strategies for chiller sizing and chiller/store operating priority. Whilst caution should be exercised in making comparisons out of context, it is worth noting that reducing the total fan energy consumption by one-third, and the unit price of electricity consumed by the water chillers by two-thirds, would still not pay back the cost of energy consumption at the FATs equivalent to that at the fan coils. Reducing running costs without reducing actual energy use also has no *environmental* impact.

The total installed electrical power was given as 538.5 kW for the VAV system (100%), 563.5 kW for the fan coil system (105%), and 472.5 kW for the perimeter induction system (88%). During the peak heating months when UK electrical maximum demand charges are severe, the fan power of the VAV system reduces with system load factor. In the fan coil system fan power remains constant throughout the year. There may also be circumstances in which the availability of sufficient capacity from the electrical supply authority may be a limiting factor in the design of an air conditioning system. This is rarely the case in the UK, but may be encountered in other countries. Within Europe, Italy has experienced problems in this respect.

It is also instructive to consider briefly the relative requirements of the three systems for routine inspection and maintenance. Numerically the largest item concerns the terminal units in each system. Each system features a single terminal unit per zone: 960 in all for each system. The requirements of the VAV terminals are minimal, involving at most an annual inspection. The recirculation air filters in each fan coil are assumed to require replacement twice yearly for each unit, with an inspection of the fan motors annually. Cleaning of the lint screen filters in the induction units (for floor dirt, units typically being mounted at low level around the perimeter of the building) may be required up to four times annually for each unit, with the induction nozzles themselves being cleaned at least annually – although more frequent nozzle cleaning has been recommended[10]. If the labour man–hour allowances per unit for these basic tasks are not drastically different, the simple summary appended to Table 10.4 provides some comparison of the routine inspection and maintenance requirements of each system.

Table 10.5 summarises comparative data on the annual energy consumptions and running costs of different HVAC systems. The data are derived from several sources, including the study by Gosling and Williams described above[9]. The data due to Jones were presented for a hypothetical office block served alternatively by VAV, two-pipe (non-changeover) and four-pipe perimeter induction systems[4]. The 14-storey building was located in London, and had its major axis orientated north–south, the optimum for a VAV building. Electrical, thermal, and total energy consumption are quoted. The difference in specific thermal energy consumption of two-pipe and four-pipe perimeter induction systems derives from the local thermostatic control of the secondary heating and cooling coils in the four-pipe induction units, which allows advantage to be taken of casual heat gains from people, lighting, office equipment and winter sunshine (although the useful effect of the latter is difficult to quantify in estimating annual energy consumption). The four-pipe fan coil system also achieves this, as does the 'wet' perimeter heating installation associated with the VAV system if the heat emitters have local thermostatic control (e.g. thermostatic radiator valves). Perimeter heating in which the flow temperature is compensated or scheduled against outside air temperature without any local thermostatic 'trimming' control is unable to take advantage of such casual heat gains, resulting in cancellation of local overheating by active cooling. The data due to Hayward are unspecified, being based on the contributed experience of practising HVAC designers, and a two-pipe perimeter induction system has been assumed[5]. The data due to Handel *et al.* are for the four-storey office building described in Section 10.5,[6] while those due to Scott and Hunt are similarly derived from their study of a six-storey application. In the latter case the study applies to a single design internal cooling load of 45 W m^{-2} for equipment, lighting and people. The results have again been converted to be relative to VAV as unity. In this case the load diversity of the VAV system has been 'penalised' by the use of external solar shading and an even load distribution throughout the building (as noted by Scott and Hunt). While fan energy consumption was naturally reduced with the chilled ceiling, the use of a low chilled water flow–return differential (reported as only 2 K) resulted in increased energy consumption for pumping.

10.7 Other aspects of comparison for VAV systems

In making a true comparison between alternative air conditioning system types it is important to consider the *sensitivity of the design approach to*

Table 10.5 Comparative energy consumption, running costs, and environmental impact of various air conditioning systems

System	Relative energy consumption ratio[1]								Relative cost ratio[2]	CO_2 emissions[4]
	Gosling and Williams (1987)	Jones (1981)			Handel, Lederer and Roth (1992)		Scott and Hunt (1995)	EEO[3]	Hayward (1987)	(kgm^{-2}/yr^{-1})
	Electric	Electric	Thermal	Total	Electric (Control case 1)[5]	Electric (Control case 2)[6]	Total	Total		
VAV with LPHW perimeter heating	1.0	1.0	1.0	1.0	1.0	1.0 (0.78)[7]	1.0	1.0	1.0	40
(Datum energy use, $GJ\,m^{-2}$)	(0.19)	(0.12)	(1.1)	(1.27)	(0.5)	(0.39)	—	—	—	n/a
Perimeter induction (two-pipe)[8]	1.32	1.83	0.97	1.05	—	—	—	—	1.55	—
Perimeter induction (four-pipe)	—	1.83	0.72	0.83	—	—	—	1.33	—	50
Ceiling-mounted induction units	—	—	—	—	0.73	0.84	—	—	—	—
Four-pipe fan coils (ducted OA)	1.84	—	—	—	—	—	0.9	1.33	2.18	50
Reversible heat pumps (ducted OA)	—	—	—	—	—	—	—	1.33	1.33	55
Chilled ceiling (displacement ventilation)	—	—	—	—	0.78	0.89	0.7	—	—	—
Dual-duct	—	—	—	—	—	—	—	1.42	3.78	55

Notes

(1) Relative to VAV with LPHW perimeter heating.
(2) Assumed overall cooling and heating.
(3) Source: Building Research Energy Conservation Support Unit (BRECSU) (1993) Energy Efficiency Office Good Practice Guide 71. *Selecting Air Conditioning Systems. A Guide for Building Clients and their Advisers.* Garston/Watford: BRECSU. For comparison CO_2 emission for Central Heating with Mechanical Ventilation is given as $30\,kg\,m^{-2}\,y^{-1}$, that for Central Heating with Local Mechanical Ventilation only at $17\,kgm^{-2}\,y^{-1}$.
(4) Source: As (3). (Datum cost is £2.40 $m^{-2}\,y^{-1}$ for VAV with perimeter heating.)
(5) Primary energy; room temperature set point scheduled vs. outside air temperature.
(6) Primary energy; 3K 'dead band' superimposed on Case 1 schedule.
(7) Relative to Case 1.
(8) Non-changeover system.

changes in building design or end use. Both the capital and running costs of an all-air design approach, such as a VAV system, are more sensitive to changes in design loads than those of an air–water system. The value of flexibility to adapt to changes in end use of the conditioned space varies according to the particular application. Typically it is high on the client's list of priorities in a speculative office development, for which an end user must be found either during the project or following its completion.

Individual VAV terminal units are available to serve zone requirements ranging from less than 50 litre s^{-1} to greater than 1500 litre s^{-1}, or from approximately 0.5 to 15 kW, assuming a supply temperature differential of 8 K. Terminal units initially slaved under a master controller may subsequently be given the status of independent zones by the addition of suitable zone temperature controllers, giving some measure of zoning flexibility.

Maximum flexibility for partitioning of a space is afforded by a large number of small control zones, although there is an inevitable cost penalty associated with the aditional controls, ductwork and/or pipework and equipment required. This differs not only between different system design approaches, but according to the particular approach adopted (and even the type of equipment selected) for a given system.

HVAC designers are interested in whether any worthwhile improvement in thermal comfort and/or air quality can be expected to result, in a real working environment in real use, from selection of one system design approach rather than another. The problem then still remains of how such an improvement may be quantified against the conventional comparative measures of capital and running costs or energy consumption. For example, a design approach using reversed-cycle terminal heat pumps limited to simple on/off control will not achieve the quality of space temperature control that can be achieved with modulating control of either VAV terminal units, fan coils or induction units.

The thermal comfort and air quality likely to be achieved were compared for air conditioning the four-storey office block described in Section 10.5 using the design approaches of VAV plus perimeter heating, an air-water system featuring ceiling-mounted induction units, and chilled ceilings with displacement ventilation[6]. The study found that thermal comfort was relatively high both for the induction system and for the chilled ceilings with displacement ventilation. However, the latter was suggested as prone to causing draughts when the supply temperature rise was greater than 2 K and the room air temperature at a height of 0.1 m dropped below 21–22°C. As far as the VAV system was concerned, the comfort level was reasonable, and air quality was acceptable due to a higher supply air volume flow rate.

The *robustness* of different systems design approaches to misuse, incorrect installation or control, and poor maintenance may on occasion have to be considered. Whilst ideally such situations should not arise, in practice they may influence energy consumption and comfort achieved more than the differences between different system design approaches. The effects of overdesign could also be added to this list.

As far as routine inspection and maintenance is concerned, chilled ceilings in particular benefit from fewer mechanical components requiring maintenance. However, the results of chilled water leakage within the ceiling void are potentially more serious, and inevitably costly.

10.8 The application and design of VAV systems: a summary

We must conclude that VAV systems are *but one* of the tools at the disposal of an HVAC designer. Since HVAC designers, as a group, are not totally uninfluenced by technical 'fashions', VAV systems have undoubtedly been used in unsuitable applications, and the chips are always down for such systems. As is the case with all other air conditioning techniques, the VAV approach has its strengths and its weaknesses. However, if well designed, well engineered, well installed and well commissioned *in a suitable application*, a VAV system can provide acceptable standards of both thermal comfort and indoor air quality in a flexible and energy-efficient manner.

References

1. Building Research Energy Conservation Support Unit (BRECSU) (1993) *Selecting Air Conditioning Systems. A Guide for Building Clients and Their Advisers*, Energy Efficiency Office Good Practice Guide No 71, BRECSU, Garston/Watford.
2. Building Services Research and Information Association (1990) Product profile – terminal units (2), *BSRIA Statistics Bulletin*, **15**, June, i–iv.
3. Building Services Research and Information Association (1997) *Statistics Bulletin*, **22**, March, 13.
4. Jones, W.P. (1980) Energy consumption by variable air volume systems, *Building Services and Environmental Engineer*, **2**, July, 16–19.
5. Hayward, R.H. (1988) *Rules of Thumb. Examples for the Design of Air Systems*, Building Services Research and Information Association, Bracknell, Technical Note 5/88.
6. Handel, C., Lederer, S. and Roth, H.W. (1992) Energy consumption and comfort of modern air-conditioning systems for office buildings. In *Ventilation for Energy Efficiency and Optimum Indoor Air Quality: Proceedings of the 13th AIVC Conference*, held in Nice 15–18 September, 1992, Air Infiltration and Ventilation Centre, Coventry, 423–30.
7. Scott, R. and Hunt, J. (1995) A comparative commercial appraisal of chilled ceilings and displacement ventilation with conventional AC systems. *H & V Engineer*, **68**, 727, 5–6.

8. Building Research Energy Conservation Support Unit (BRECSU) (1993) *Energy Efficiency in Offices. A Technical Guide for Owners and Single Tenants*, BRECSU, Garston, Energy Efficiency Office Best Practice Programme, Energy Consumption Guide 19.
9. Gosling, C.T. and Williams, T.M. (1978) The design and application of VAV systems, Part 5: Energy comparison of air conditioning systems. *Heating and Air Conditioning Journal*, **48**, August, 20–25.
10. Jones, W.P. (1981) *Air Conditioning Applications and Design*, Edward Arnold, London.

A Bibliography of VAV Systems

In the UK the standard technical and design references for HVAC engineers concerned with the application of VAV systems are provided by the current editions of the Guide published by the Chartered Institution of Building Services Engineers (CIBSE) and the Handbooks of the American Society of Heating, Refrigerating and Air Conditioning Engineers (ASHRAE). In addition the *CIBSE Applications Manual: Automatic Controls and their Implications for Systems Design* (1985) also contains pertinent information on the control of VAV systems. By the very nature of their broad scope, the design guides of CIBSE and ASHRAE cover all areas of VAV systems engineering in relatively general terms.

Beyond these standard sources of reference, however, a substantial body of literature exists on various aspects of the design, control and performance of VAV systems. This primarily takes the form of journal articles and conference papers. These tend to be divided between those which take a broad overview of system or equipment design, control or performance, and those that treat a specific aspect in more detail. Both approaches have their merits, according to the state of the art in the subject matter and the audience to which the article or paper is addressed. A distinction is also evident between sources that report personal or corporate experience (with particular design practice or system applications) and those that offer insight through analysis of system behaviour. The approaches are complementary.

The body of literature identified represents a very large resource of corporate and individual knowledge and experience, and one of great potential value to HVAC engineers. Naturally some information becomes dated with time, and as the state of the art in VAV systems engineering advances, but much fundamental information and considered advice retains a timeless validity.

Alcorn, L.H. (1988) Decoupling supply and return fans for increased stability of VAV systems. *ASHRAE Transactions*, **94**, Part 1, 1484–1492.

Aldridge, K.G. (1972) Linear air distribution: principles and practice. *Building Services Engineer*, **40**, November, 167–174.

Alexander, J., Aldridge, R., and O'Sullivan, D. (1993) Wireless zone sensors. *Heating, Piping and Air Conditioning*, **65**, May 37–39.

Allary, L. (1987) Systemes de regulation pour climatisation de confort (Control systems for comfort air conditioning). *Chaud Froid Plomberie*, **4**, February, 69–75.

Alley, R.L. (1988) Selecting and sizing outside and return air dampers for VAV economiser systems. *ASHRAE Transactions*, **94**, Part 1, 1457–1466.

American Society of Heating, Refrigerating and Air Conditioning Engineers (ASHRAE) (1987) All-air systems. Chapter 2 in *ASHRAE 1987 Handbook: HVAC Systems and Applications*, ASHRAE, Atlanta.

American Society of Heating, Refrigerating and Air Conditioning Engineers (ASHRAE) (1987) Automatic control. Chapter 51 in *ASHRAE 1987 Handbook: HVAC Systems and Applications*, ASHRAE, Atlanta.

American Society of Heating, Refrigerating and Air Conditioning Engineers (ASHRAE) (1987) Testing, adjusting, and balancing. Chapter 57 in *ASHRAE 1987 Handbook: HVAC Systems and Applications*, ASHRAE, Atlanta.

American Society of Heating, Refrigerating and Air Conditioning Engineers (ASHRAE) (1988) Air-diffusing equipment. Chapter 2 in *ASHRAE 1988 Handbook: Equipment*, ASHRAE, Atlanta.

American Society of Heating, Refrigerating and Air Conditioning Engineers (ASHRAE) (1989) *Ventilation for acceptable indoor air quality*. ANSI/ASHRAE Standard 62-1989, ASHRAE, Atlanta.

Appleby, P. (1990) (FAT VAV): What Price Air Quality? *Building Services CIBSE Journal*, **12**, March, 41–42.

Arnold, D. (1986) VAV and economy cycle. *Building Services CIBSE Journal*, **8**, May, 68–69.

Atkinson, G.V. (1986) VAV system volume control using electronic strategies. *ASHRAE Transactions*, **92**, Part 2B, 46–56.

Atkinson, G.V. (1987) Rooftop variable air volume control. *Consulting Specifying Engineer*, **1**, March, 93–98.

Australian Refrigeration, Air Conditioning and Heating (1984) Perth Education Department Offices, **38**, June, 21–32.

Australian Refrigeration, Air Conditioning and Heating (1985) ACI House, 200 Queen Street, Melbourne, **39**, January, 21–25.

Avery, G. (1984) VAV economizer cycle: don't use a return fan. *Heating, Piping and Air Conditioning*, **56**, August, 91–94.

Avery, G. (1986) VAV: designing and controlling an outside air economiser cycle. *ASHRAE Journal*, **28**, December, 26–30.

Avery, G. (1989) Updating the VAV outside air economiser controls. *ASHRAE Journal*, **31**, April, 14–16.

Avery, G. (1989) The myth of pressure-independent VAV terminals. *ASHRAE Journal*, **31**, August, 28–30.

Avery, G. (1992) The instability of VAV systems. *Heating, Piping and Air Conditioning*, **64**, February, 47–50.

Barbetta, C. (1983) La regolazione dei ventilatori negli impianti a portata variabile (Fan control in variable air volume systems). *Condizionamento dell'Aria*, **27**, October, 1035–1043.

Barbetta, C. (1990) Fans and variable air volume. *Condizionamento dell'Aria*, **34**, April, 637–645 (in Italian).

Beck, P.E. (1990) Applying direct digital control to VAV systems. *Consulting Specifying Engineer*, **7**, 31–34.

Boeche, A. (1982) L'impianto a portata costante-portata variabile. *Condizionamento dell'Aria*, **26**, November, 959–963.

Boeche, A. (1991) Remarks on VAV installations with specific reference to double-duct systems. *Condizionamento dell'Aria*, **35**, March, 345–351 (in Italian).

Boeche, A. and Pasquale, L. (1980) L'impianto a doppio canale a portata variabile – portata constante. *Condizionamento dell'Aria*, **24**, August, 601–609.

Boldrini, S. (1990) VAV systems: an installation example. *Condizionamento dell'Aria*, **34**, April, 653–657 (in Italian).

Bonotti, M. and Marchetti, F. (1986) Terminali per impianti VAV: Regolazioni in campo e/o pretaratura in fabbrica (Terminal devices for VAV systems: on-site control and/or preadjustment at the factory). *Installatore Italiano*, **37**, March, 607–610.

Boyd, N. (1973) A new air conditioning system: the variable temperature variable volume system. *The Building Services Engineer*, **41**, May, 48–49.

Brandenburg, J. and Dudley, J. (1982) What you must do for a quieter VAV system. *Specifying Engineer*, **47**, April, 62–68.

Brehm, H.P. (1968) Variable volume system coupled with air floor in office building design. *Heating, Piping and Air Conditioning*, **40**, April, 87–92.

British Standards Institution (1986) *BS 4979: 1986, Aerodynamic Testing of Constant and Variable Dual or Single Duct Boxes, Single Duct Units and Induction Boxes for Air Distribution Systems*, British Standards Institution, London.

Bright, N.J. (1982) Microprocessor control applied to variable air volume air conditioning system, M.Phil. thesis, Brunel University.

Bringmann, A. and Hottinger, B. (1989) Air conditioning with variable air volume: a contribution to well-being at the workplace. *Sulzer Technical Review*, **71** (4), 15–18.

Brittain, J.R.J. (1997) *Oversized Air Handling Plant – A Guide to Reduce the Energy Consumption of Oversized Constant or Variable Air Volume Air Handling Plant*, BSRIA, Bracknell, BSRIA Guidance Note 11/97.

Brothers, P.W. and Warren, M.L. (1986) Fan energy use in variable air volume systems. *ASHRAE Transactions*, **92**, Part 2B, 19–29.

Brown, C.E. (1977) Retrofitting dual duct systems with VAV components. *Building Systems Design*, **74**, June/July, 24–34.

Brown, M.R. (1989) New technology at No.1 Spring Street, Melbourne. In *Proceedings of the CIBSE National Conference*, held at the University of Warwick, 16–18 April 1989, Chartered Institution of Building Services Engineers, London, 5–19.

Building Research and Energy Conservation Support Unit (BRECSU) (1993) *Selecting Air Conditioning Systems. A Guide for Building Clients and their Advisers*, Energy Efficiency Office Good Practice Guide No.71, BRECSU, Garston/Watford.

Building Research and Energy Conservation Support Unit (BRECSU) (1996) *Variable Flow Control*, General Information Report 41, BRECSU, Garston/Watford.

Building Services CIBSE Journal (1984) High Spec., **6**, February, 30–32.

Building Services CIBSE Journal (1990) FAT VAV: What is it?, **12**, March, 33–34.

Building Services and Environmental Engineer (1980) Variable air temperature and volume system at Watson House, **2**, April, 21–23.

Building Services Research and Information Association (1991) *Commissioning of VAV systems in Buildings*, BSRIA, Bracknell, Application Guide 1/91.

Building Systems Design (1973) Police get New York's first self-controlled VAV system, **70**, February, 5–7.

Cassidy, V.M. (1982) Saving energy with VAV. *Specifying Engineer*, **48**, August, 104–105.

Chartered Institution of Building Services Engineers (CIBSE) (1985) *CIBSE Applications Manual: Automatic Controls and their Implications for Systems Design*, CIBSE, London.

Chartered Institution of Building Services Engineers (CIBSE) (1986) *CIBSE Guide, Section B3, Ventilation and Air Conditioning: Systems, Equipment and Control*, CIBSE, London.

Chen, S.Y.S, Yu, H.C. and Hwang, D.D.W. (1992) Ventilation analysis for a VAV system, *Heating, Piping and Air Conditioning*, **64**, April, 36–41.

Clark, R. (1988) *A procedure for commissioning variable air volume systems*, Building Services Research and Information Association, Bracknell, Technical Memorandum 2/88.

Clima Commerce International (1980) *Systemubersicht: Variable Volumenstrom-systeme* (System Review: Variable Volume Systems), **14**, 4 July, 30–40.

Coad, W.J. (1984) DX Problems with VAV. *Heating, Piping and Air Conditioning*, **56**, January, 134–139.

Coad, W.J. (1984) Ideal terminal control and distribution system: I. *Heating, Piping and Air Conditioning*, **56**, October, 127–128.

Coad, W.J. (1984) Ideal terminal control and distribution system: II. *Heating, Piping and Air Conditioning*, **56**, November, 161–164.

Coad, W.J. (1996) Indoor air quality: a design parameter. *ASHRAE Journal*, **38**, June, 39–47.

Coe, P.E., Jr. (1983) The economics of VAV duct looping. *Heating, Piping and Air Conditioning*, **55**, August, 61–64.

Cohen, T. (1994) Providing constant ventilation in variable air volume systems. *ASHRAE Journal*, **36**, May, 38–40.

Colledge, B. (1993) Regaining control of VAV systems. *Premises and Facilities Management*, June, 31–33.

Collins, W.J., Jr. (1966) Choose variable volume air system for tower office building's interior. *Heating, Piping and Air Conditioning*, **38**, February, 85–88.

Coogan, J.J. (1994) Experience with commissioning VAV laboratories. *ASHRAE Transactions*, **100**, Part 1, 1635–1640.

Crane, J. (1990) (FAT VAV) The cool alternative. *Building Services CIBSE Journal*, **12**, March, 43–44.

Daly, B.B. (1978) *Woods Practical Guide to Fan Engineering*, 3rd edn, Woods of Colchester Ltd.

Daryanani, S. *et al.* (1966) Variable Air Volume Air Conditioning. *Air Conditioning, Heating and Ventilating*, **63**, March, 56–78.

Daryanani, S. *et al.* (1972) Variable air volume systems and hardware: system hardware. *Building Systems Design*, **69**, August, 13–16.

Daryanani, S. *et al.* (1974) Design engineer's guide to variable-air-volume systems. *Actual Specifying Engineer*, July, 68–76.

Day, T.L. (1974) VAV air distribution. *ASHRAE Transactions*, **80**, Part 1, 486–490.

Dean, F.J., Jr. (1970) How to avoid pitfalls in variable volume design. *Heating, Piping and Air Conditioning*, **42**, May, 74–79.

Dean, R.H., Dean, F.J., Jr. and Ratzenberger, J. (1985) Importance of duct design for VAV systems. *Heating, Piping and Air Conditioning*, **57**, August, 91–94, 101–104.

Dean, R.H. and Ratzenberger, J. (1985) Computer simulation of VAV system operation. *Heating, Piping and Air Conditioning*, **57**, September, 115–127.

Dean, R.H. and Ratzenberger, J. (1985) Stability of VAV terminal unit controls. *Heating, Piping and Air Conditioning*, **57**, October, 79–90.

Delp, W.W., Sauer, H.J., Jr., Howell, R.H. and Subbarao, B. (1993) Control of outside air and building pressurization in VAV systems. *ASHRAE Transactions*, **99**, Part 1, 565–589.

Dirkes, J.V. (1992) Duct protection for variable speed fans. *ASHRAE Journal*, **34**, November, 40–43.

Dowling, K.F. (1996) Airing the facts on VAV systems. *Consulting-Specifying Engineer*, **19**, January, 42–50.

Duffin, J. (1990) (FAT VAV) A question of control. *Building Services CIBSE Journal*, **12**, March, 38–40.

Elleson, J.S. (1991) High-quality air conditioning with cold air distribution. *ASHRAE Transactions*, **97**, Part 1, 839–842.

Elleson, J.S. (1993) Energy use of fan-powered mixing boxes with cold air distribution. *ASHRAE Transactions*, **99**, Part 1, 1349–1358.

Elovitz, D.M. (1995) Minimum outside air control methods for VAV systems. *ASHRAE Transactions*, **101**, Part 2, 613–618.

Energy Management (1989) First UK underfloor moduline for Charter Place, Uxbridge, No.12, March, 11.

Englander, S. (1990) Ventilation control for energy conservation: digitally controlled terminal boxes and variable speed drives. Master's thesis, Princeton University.

Englander, S.L. and Norford, L.K. (1991) VAV system simulation, Part 1: Development and experimental validation of a DDC terminal box model. In *Proceedings of the 3rd International Conference on System Simulation in Buildings*, held in Liege, 2–5 December 1990, Laboratory of Thermodynamics at the University of Liege and the International Building Performance Simulation Association, Inc., 553–580.

Englander, S. and Norford, L.K. (1991) VAV system simulation, Part 2: Supply fan control for static pressure minimization using DDC zone feedback. *Ibid.*, 581–606.

Englander, S. and Norford, L.K. (1992) Saving fan energy in VAV systems – Part 1: Analysis of a variable-speed drive retrofit. *ASHRAE Transactions*, **98**, Part 1, 3–18.

Englander, S. and Norford, L.K. (1992) Saving fan energy in VAV systems – Part 2: Supply fan control for static pressure minimization using DDC zone feedback. *ASHRAE Transactions*, **98**, Part 1, 19–32.

Fairhall, D. (1988) Rubber damper dumps decibels. *Heating and Air Conditioning Journal*, **57**, July, 20–21.

Fiddian-Green, A. (1987) Intelligent controls for fan coil units and variable air volume boxes. In *Proceedings of the Far East Conference on Air Conditioning in Hot Climates*, held in Singapore, 3–5 September 1987, American Society of Heating, Refrigerating and Air Conditioning Engineers, Atlanta, 224–241.

Fiegen, B. and Starke, D. (1989) VAV for light commercial buildings. *Heating, Piping and Air Conditioning*, **61**, July, 89–91.

Filardo, M.J., Jr. (1993) Outdoor air – how much is enough? *ASHRAE Journal*, **35**, January, 34–38.

Finkelstein, W. and Riegel, H. (1979) Varyset-bypass-system voor Trox plafondroosters voor toepassing in VAV-systemen met een grote luchthoevelheid (Varyset bypass system for Trox ceiling grilles in high capacity variable volume air conditioning systems). *Verwarming en Ventilatie*, **36**, October, 663–670.

Finkelstein, W. and Riegel, H. (1980) Varyset-bypass system til anvendelse i VAV-klimaanlaeg med stort volumenstromrade (Varyset bypass system for use in VAV air conditioning plant with wide flow range). *VVS (Denmark) (Tidsskrift for Varme, Ventilation, Sanitet)*, **16** (11), 51–52.

Finkelstein, W. and Haaz, J. (1980) Flow pressure controller. *Heating, Air Conditioning and Refrigeration*, **12**, May, 14–19.

Fisher, A. (1987) Local control for VAV systems. *Building Services CIBSE Journal*, **9**, January, 29–30.

Fowler, D.T. (1979) Performance of motors in variable volume design. *Building Systems Design/Energy Engineering*, **76**, June/July, 4–9.

Gardner, T.F. (1988) Ventilation deficiencies in VAV systems. *Heating, Piping and Air Conditioning*, **60**, February, 89–100.

Gazzi, L. (1977) I controlli di tipo fluidico nei terminali a portata d'aria variabile (Fluidic type controls in variable volume air terminals). *Condizionamento dell'Aria*, **21**, December, 951–957.

Geake, L.W. (1980) Controls for single-duct variable air volume terminal units. *ASHRAE Transactions*, **86**, Part 2, 825–838.

Ghidoni, D.A. and Jones, R.L., Jr. (1994) Methods of exhausting a biological safety cabinet (BSC) to an exhaust system containing a VAV component. *ASHRAE Transactions*, **100**, Part 1, 1275–1281.

Goli, D. (1978) Problemi ugradnje termostata u VAV sisteme (Problems of installing thermostats in VAV systems). *Klimatisacija Grejanje Hladenje*, **7**, November, 30–32.

Goode, E. (1986) Bringing zone control to smaller buildings. *Building Services and Environmental Engineer*, **8**, May, 12–13.

Gorchev, D. (1966) Variable volume reheat system. *ASHRAE Journal*, **8**, January, 98–101.

Gorchev, D. (1972) Variable air volume systems and hardware: systems evolution. *Building Systems Design*, **69**, August, 35–38.

Gosling, C.T. (1978) The design and application of VAV systems, Part 1: System concept. *Heating and Air Conditioning Journal*, **48**, March, 14–20.

Gosling, C.T. and Williams, T.M. (1978) The design and application of VAV systems, Part 2: System design. *Heating and Air Conditioning Journal*, **48**, April, 12–16.

Gosling, C.T. and Williams, T.M. (1978) The design and application of VAV systems, Part 3: Terminals. *Heating and Air Conditioning Journal*, **48**, May, 44–50.

Gosling, C.T. and Williams, T.M. (1978) The design and application of VAV systems, Part 4: Automatic control and heat reclaim. *Heating and Air Conditioning Journal*, **48**, July, 36–42.

Gosling, C.T. and Williams, T.M. (1978) The design and application of VAV systems, Part 5: Energy comparison of air conditioning systems. *Heating and Air Conditioning Journal*, **48**, August, 20–25.

Gosling, C.T. and Williams T.M. (1978) The design and application of VAV systems, Part 6: System survey. *Heating and Air Conditioning Journal*, **48**, November, 38–48.

Goswami, D. (1986) VAV fan static pressure control with DDC. *Heating, Piping and Air Conditioning*, **58**, December, 113–117.

Goswami, D. (1989) Common VAV system problems for design stage handling. *Energy Engineering*, **86** (6), 6–25.

Graves, L.R. (1986) Evolution of intermittent fan terminal. *ASHRAE Transactions*, **92**, Part 1B, 511–518.

Graves, L.R. (1995) VAV mixed air plenum pressure control. *Heating, Piping and Air Conditioning*, **67**, August, 53–55.

Gray, B. (1986) Variable air volume concepts, Part 1. *Australian Refrigeration, Air Conditioning and Heating*, **40**, January, 34–45.

Gray, B. (1986) Variable air volume concepts, Part 2. *Australian Refrigeration, Air Conditioning and Heating*, **40**, February, 26–27.

Guntermann, A.E. (1986) VAV system enhancements. *Heating, Piping and Air Conditioning*, **58**, August, 67–78.

Gupta, V.K., Int-Hout, D., Roberts, M.M., Wessel, D.J., Brickman, H. and Waeldner, W. (1987) A forum on variable air volume. *ASHRAE Journal*, **29**, August, 22–31.

Haessig, D. (1994) Variable air volume controls for VAV fan terminals. *AIRAH Journal*, **48**, November, 29–32.

Haessig, D. (1995) A solution for DX VAV air handlers. *Heating, Piping and Air Conditioning*, **67**, May, 83–86.

Haines, R.W. (1981) Supply and return air fan control in a VAV system. *Heating, Piping and Air Conditioning*, **53**, February, 75–76.

Haines, R.W. (1983) Supply fan volume control in a VAV system. *Heating, Piping and Air Conditioning*, **55**, August, 107–111.

Haines, R.W. (1984) Fan energy – P versus PI control. *Heating, Piping and Air Conditioning*, **56**, August, 107–111.

Haines, R.W. (1984) Control strategies for VAV systems. *Heating, Piping and Air Conditioning*, **56**, September, 147–148.

Haines, R.W. (1986) Outside air volume control in a VAV system. *Heating, Piping and Air Conditioning*, **58**, October, 130–132.

Haines, R.W. (1994) Ventilation air, the economy cycle and VAV. *Heating, Piping and Air Conditioning*, **66**, October, 71–73.

Handel, C., Lederer, S. and Roth, H.W. (1992) Energy consumption and comfort of modern air-conditioning systems for office buildings. In *Ventilation for Energy Efficiency and Optimum Indoor Air Quality: Proceedings of the 13th AIVC Conference* held in Nice, 15–18 September 1992, Air Infiltration and Ventilation Centre, Coventry, 423–430.

Hartmann, K. (1986) Variables volumenstromsystem mit digitaler elektronischer regelung der luftanlasse (variable volume flow system with digital electronic air outlet control). *Klima Kalte Heizung*, **11**, November/December, 1125–1138.

Hartman, T. (1988) Dynamic control of typical air systems. *Heating, Piping and Air Conditioning*, **60**, July, 87–92.

Hartman, T. (1989) TRAV: A new HVAC concept. *Heating, Piping and Air Conditioning*, **61**, July, 69–73.

Hartman, T. (1990) Stand-alone panels and terminal controllers. *Heating, Piping and Air Conditioning*, **62**, November, 41–47.

Hartman, T. (1991) Terminal input and output devices. *Heating, Piping and Air Conditioning*, **63**, January, 91–95, 100–101.

Hartman, T. (1993) *Direct Digital Controls for HVAC Systems*, McGraw-Hill, New York, 110–120.

Hartman, T. (1993) Terminal regulated air volume systems. *ASHRAE Transactions*, **99**, Part 1, 791–800.

Hartman, T. (1995) Global optimization strategies for high-performance controls. *ASHRAE Transactions*, **101**, Part 2, 679–687.

Hartman, T. (1996) Library and museum HVAC: new technologies/new opportunities – Part 2. *Heating, Piping and Air Conditioning*, **68**, May, 63–68, 72, 103.

Haupt, J. (1980) Regeltechnische problemlosungen in klima-anlagen mit variablen luftmengen (Solutions to control problems in variable volume air conditioning installations). *Schweiz. Bl. Heiz. Luft.*, **47** (2), 21–24.

Heating and Air Conditioning Journal (1974) Air conditioning with energy conservation, **44**, November, 14–17.

Heating and Air Conditioning Journal (1977) New fluidics operated control for variable flow air systems, **47**, October, 20–23.

Heating and Air Conditioning Journal (1987) Space savings from low temperature air, **56**, July, 27–28.

Heating and Air Conditioning Journal (1989) Do away with dumping, **58**, February, 27–28.

Heating, Air Conditioning and Refrigeration (1979) Air conditioning in the Sanlamsentrum: an energy-conscious design, **12**, September, 12–27.

Heating, Piping and Air Conditioning (1967) Variable volume air supply selected to air condition triangular office building, **39**, May, 112–114.

Heating, Piping and Air Conditioning (1967) Variable volume air systems cool computers, people in bank's new operations centre, **39**, December, 77–81.

Heck, W.C. (1970) Environmental control of US Steel corporate center. *Air Conditioning, Heating and Ventilating*, **67**, October, 49–55.

Hemsher, F.J. (1980) Static point control method used in high-rise VAV. *Specifying Engineer*, **44**, August, 62–65.

Hill, A. (1981) Air flow synchronization controls in VAV systems. *Heating, Piping and Air Conditioning*, **53**, August, 47–53.

Hittle, D.C. (1987) Control of air conditioning systems, problems and solutions. In *Advances in Air Conditioning: Proceedings of the Chartered Institution of Building Services Engineers (CIBSE) Conference*, held at Heathrow, October, 1987, CIBSE, London, 63–72.

Holmes, M.J. (1974) *Designing Variable Volume Systems for Room Air Movement*, Building Services Research and Information Association, Bracknell, Applications Guide 1/74.

Holmes, M.J. (1974) *Room air distribution with variable air volume supply systems*. HVRA Project Report 15/107.

Holmes, M.J. (1976) *Backward Curved Centrifugal Fans in VAV Systems – Selection and Energy Consumption*, Building Services Research and Information Association, Technical Note 2/76.

Holmes, M.J. (1977) Air systems balancing – regulating variable flow rate systems. In *Proceedings of the Symposium on Testing and Commissioning of Building Services Installations*, held at the Institute of Marine Engineers, London, 3 November 1977, Chartered Institution of Building Services (CIBS), London, G1-9, GD1-2.

Hopkin, P. (1992) Duct design and air distribution. *Australian Refrigeration, Air Conditioning and Heating*, **46**, September, 29–35.

Hospital boasts VAV, heat recovery (1971) *Air Conditioning, Heating and Ventilating*, **68**, April, 29–30.

Humphries, T. (1988) Honing zoning for comfort and cost. *Heating and Air Conditioning Journal*, **58**, October, 61–62.

Inhelder, P. (1982) Die elektronische Regelung von klimaanlagen mit variablem luftvolumenstrom (electronic control of variable air volume air conditioning installations). *Klima Kalte Heizung*, **10**, May, 183–186.

Inhelder, P. and Juen, H. (1983) De elektronische regeling van variabel-luchtvolume systemen: Deel 2 (Electronic control of variable air volume systems: Part 2). *Verwarming en Ventilatie*, **40**, December, 849–854.

Inoue, U. and Matsumoto, T. (1977) Measurement of single duct variable air volume systems. *Transactions of the Society of Heating, Air Conditioning and Sanitary Engineers of Japan*, June.

Inoue, U. and Matsumoto, T. (1979) A study on energy savings with variable air volume systems by simulation and field measurement. *Energy and Buildings*, **2**, January, 27–36.

Intelligent Fan Coil and VAV Terminal Control, Trend Application Guide SD102812, Issue 1/0.

Int-Hout, D. (1986) Microprocessor control of zone comfort. *ASHRAE Transactions*, **92**, Part 1B, 528–538.

Int-Hout, D. (1992) Low temperature air, thermal comfort and indoor air quality. *ASHRAE Journal*, **34**, May, 34–35, 38–39.

Int-Hout, D. and Berger, P. (1984) What's really wrong with VAV systems. *ASHRAE Journal*, **26**, December, 36–38.

Ip, K.C.W. (1991) Dynamic modular simulation of variable water and air volume flow systems in Buildings, Ph.D thesis, University of Bristol.

Jakobczyk, J.S. (1982) Direct digital control for VAV terminals. *Heating, Piping and Air Conditioning*, **54**, February, 77–81.

Janisse, N.J. (1977) Simple controls cut variable volume energy costs. *ASHRAE Transactions*, **83**, Part 1, 598–605.

Janu, G.J., Wenger, J.D. and Nesler, C.G. (1995) Strategies for outdoor airflow control from a systems perspective. *ASHRAE Transactions*, **101**, Part 2, 631–643.

Janu, G.J., Wenger, J.D. and Nesler, C.G. (1995) Outdoor air flow control for VAV systems. *ASHRAE Journal*, **37**, April, 62–68.

Johnson, C.M. (1988) Comparison of variable pitch fans and variable speed fans in a variable air volume system. In *Proceedings of the CIBSE Technical Conference*, held at Brunel University, 1–2 June 1987, Chartered Institution of Building Services

Engineers, London, 120–149. Also (1988) *Building Services Engineering Research and Technology*, **9** (3), 89–98.

Johnson Controls (1988) *Building Pressure Control for VAV Systems*, Engineering Report 447, Johnson Controls.

Johnson, G.A. (1984) Retrofit of a constant volume air system for variable speed fan control. *ASHRAE Transactions*, **90**, Part 2B, 102–211.

Johnson, G.A. (1985) From constant air to variable. *ASHRAE Journal*, **27**, January, 106–114.

Johnson, M. (1988) Variable volume, variable temperature. *Australian Refrigeration, Air Conditioning and Heating*, **42**, January, 23–33.

Jones, E.C. (1968) Variable volume air systems. In *Proceedings of the Heating, Piping and Air Conditioning (HPAC) Conference on Air Systems*, held in Chicago, 15–17 October 1968, 20–24.

Jones, E.C. (1969) Variable volume air systems. *Heating, Piping and Air Conditioning*, **41**, January, 172–176.

Jones, R.S. and Sturz, D.H. (1981) Make sure VAV box noise meets your design specs. *Specifying Engineer*, March.

Jones, W.P. (1994) *Air Conditioning Engineering*, 4th edn, Edward Arnold, London.

Jones, W.P. (1976) Comparative systems for air conditioning office blocks: Part 6. *Heating and Air Conditioning Journal*, **45**, December 1975/January 1976, 42–45.

Jones, W.P. (1976) Comparative systems for air conditioning office blocks: Part 7. *Heating and Air Conditioning Journal*, **45**, February, 14–23.

Jones, W.P. (1980) Energy consumption by variable air volume systems. *Building Services and Environmental Engineer*, **2**, July, 16–19.

Jones, W.P. (1981) *Air Conditioning Applications and Design*, Edward Arnold, London.

Kadziela, P. (1994) Understanding cost-effective air diffusion. *AIRAH Journal*, **48**, October, 27–29.

Kalasinsky, C.C. (1988) The economics of relief fans vs return fans in variable volume systems with economizer cycles. *ASHRAE Transactions*, **94**, Part 1, 1467–1476.

Kasprzak, K. (1992) Selecting electric heaters for VAV systems. *Heating, Piping and Air Conditioning*, **64**, December, 65–66.

Kelley, R. (1995) Room air circulation: the missing link to good indoor air quality. *Heating, Piping and Air Conditioning*, **67**, September, 67–69.

Kettleman, J.E. (1972) Variable air volume systems and hardware: the experience record. *Building Systems Design*, **69**, August, 29–32.

Kettler, J.P. (1972) Variable air volume systems and hardware: system control. *Building Systems Design*, **69**, August, 19–23.

Kettler, J.P. (1980) System-powered variable air volume terminals. *ASHRAE Transactions*, **86**, Part 2: 839–847. Also (1980) *ASHRAE Journal*, **22**, August, 40–42.

Kettler, J.P. (1987) Efficient design and control of dual-duct variable-volume systems. *ASHRAE Transactions*, **93**, Part 2, 1734–1741.

Kettler, J.P. (1988) Field problems associated with return fans on VAV systems. *ASHRAE Transactions*, **94**, Part 1, 1477–1483.

Kettler, J.P. (1995) Minimum ventilation control for VAV systems: fan tracking vs. workable solutions. *ASHRAE Transactions*, **101**, Part 2, 625–630.

Kettler, J. and Noll, R.W. (1974) Central fan system control for VAV application. *ASHRAE Transactions*, **80**, Part 1, 499–504.

Khoo, I., Levermore, G.J. and Letherman, K.M. (1996) Findings of a UMIST VAV research project. In *Variable Flow Control*, General Information Report 41, BRECSU, Garston/Watford.

Kloostra, L.M. (1980) Comparison and advantages of dual-duct variable-volume control assemblies in controlling the perimeter of large buildings. *ASHRAE Transactions*, **86**, Part 2, 848–858.

Kloostra, L. (1990) VAV retrofit slashes utility energy costs. *Heating, Piping and Air Conditioning*, **62**, November, 51–54.

Kostrz, B. (1980) Temvar, eine energiesparende sulzer-klimaanlage mit variablem volumenstrom (temvar, an energy-conserving sulzer variable volume air conditioning installation). *Schweiz. Bl. Heiz. Luft.*, **47** (1), 6–10.

Knight, J. (1979) Variable air volume. *Building Services CIBSE Journal*, **1**, June, 38–40.

Knight, J. (1980) Variable air volume air conditioning design and application – Part 1. *Australian Refrigeration, Air Conditioning and Heating*, **34**, May, 29–34.

Knight, J. (1980) Variable air volume air conditioning design and application – Part 2. *Australian Refrigeration, Air Conditioning and Heating*, **34**, June, 37–42.

Knight, J. (1980) Variable air volume air conditioning design and application – Part 3. *Australian Refrigeration, Air Conditioning and Heating*, **34**, July, 21–24.

Kohonen, R. and Heimonen, I. (1991) Supervisory control of variable air volume (VAV) air-conditioning system. In *Proceedings of the 3rd International Conference on System Simulation in Buildings*, held in Liege, 2–5 December 1990, by the Laboratory of Thermodynamics at the University of Liege and the International Building Performance Simulation Association, Inc., Laboratory of Thermodynamics at the University of Liege, 725–745.

Krajnovich, L. and Hittle, D.C. (1986) Measured performance of variable air volume boxes. *ASHRAE Transactions*, **92**, Part 2A, 203–214.

Lack, C. (1978) VAV: a realistic look at fan energy costs. *Heating and Ventilating Engineer*, **52**, June, 23–24.

Lane, W.Z. (1996) The commissioning procedure of a variable air volume (VAV) system. *AIRAH Journal*, **50**, June, 30–39.

Langbein, C.E., Jr. (1978) Variable air volume design. *Building Systems Design*, **75**, June/July, 30–36.

Laux, H. (1978) Variable-volumenstrom-klimasysteme (Variable volume air conditioning). *Heizung, Luftung, Haustechnik*, **29**, November, 411–418.

Leary, J. (1979) *Energy Consumption of Variable Air Volume Systems*, Electricity Council, Environmental Engineering Section, London.

Lentz, M.S. (1991) Adiabatic saturation and VAV: a prescription for economy and close environmental control. *ASHRAE Transactions*, **97**, Part 1, 477–485.

Levine, A.Z. (1974) Unique all-air system conserves energy in United California Bank. *Heating, Piping and Air Conditioning*, **46**, January, 52–55.

Lewis, J.R. (1988) Application of VAV, DDC and smoke management to hospital nursing wards. *ASHRAE Transactions*, **94**, Part 1, 1193–1208.

Lim, D. (1992) Model example. *Building Services and Environmental Engineer*, **14**, October, 6.

Lindberg, P.R. (1983) VAV and heat recovery in a medical centre. *Heating, Piping and Air Conditioning*, **55**, August, 82–88.

Linford, R.G. (1987) Dual-duct variable air volume: design/building viewpoint. *ASHRAE Transactions*, **93**, Part 2, 1742–1748.

Linford, R.G. and Taylor, S.T. (1989) HVAC systems: central vs. floor-by-floor. *Heating, Piping and Air Conditioning*, **61**, July, 43–49, 56–57, 84.

Liston, T.L. (1970) Low pressure VAVs. *Air Conditioning, Heating and Ventilating*, **67**, August, 39–45.

Lo, L. (1990) VAV system with inverter-driven AHU for high-rise office building in tropical climates: a case study. *ASHRAE Transactions*, **96**, Part 1, 1209–1217.

Lorenzetti, D. (1993) Modelling adjustable speed drive fans to predict energy savings in VAV systems. In *Energy Impact of Ventilation and Air Infiltration: Proceedings of the 14th AIVC Conference*, held in Copenhagen, 21–23 September 1993, Air Infiltration and Ventilation Centre, Coventry, 269–277.

Lorenzetti, D. and Norford, L.K. (1992) Measured energy consumption of variable-air-volume fans under inlet-vane and variable-speed drive control. *ASHRAE Transactions*, **98**, Part 2, 371–379.

Lorenzetti, D. and Norford, L.K. (1993) Pressure reset control of variable air volume ventilation systems. *Proceedings of the American Society of Mechanical Engineers (ASME) International Solar Energy Conference*, April 1993, American Society of Mechanical Engineers, New York, 445–453.

Love, M. (1995) VAV enters a new era. *Building Services and Environmental Engineer*, **18**, June, 20–21.

Love, M. and Smith, P. (1983) Variable air volume air conditioning. *Journal of the Chartered Institute of Building Services*, **5**, February, 35–40.

Lujan, P. (1977) Variable air volume: where are we? *ASHRAE Transactions*, **83**, Part 1, 581–584.

Lynn, M. (1989) Balancing DDC-Controlled Boxes. *Heating, Piping and Air Conditioning*, **61**, July, 79–84.

Mather, C. (1985) Innovative services design for 26 Flinders Street, Adelaide. *Australian Refrigeration, Air Conditioning and Heating*, **39**, June, 17–22.

Mayhew, F.W. (1993) Application of direct digital temperature control systems for maximum system efficiency. *ASHRAE Transactions*, **99**, Part 1, 801–807.

McComb, J. (1992) Reasons for the development of VAV. *Australian Refrigeration, Air Conditioning and Heating*, **46**, September, 36–37.

McGregor, A. (1991) Variable air volume systems for laboratories with fume hoods. In *Proceedings of the CIBSE National Conference*, held at the University of Kent, Canterbury, 7–9 April 1991, Chartered Institution of Building Services Engineers, London, 352–363.

McPherson, R. (1992) Maintenance of VAV systems. *Australian Refrigeration, Air Conditioning and Heating*, **46**, September, 24–28.

Meckler, M. (1994) Carbon dioxide prediction model for VAV system part-load evaluation. *Heating, Piping and Air Conditioning*, **65**, January, 115–118, 123–124.

Meckler, M. (1993) VAV/bypass filtration system controls VOCs, particulates. *Heating, Piping and Air Conditioning*, **66**, March, 57–61.

Milewski, L. (1977) VAV systems controls, problems and solutions. *Heating, Piping and Air Conditioning*, **49**, July, 75–81. Also (1978) Australian Refrigeration, Air Conditioning and Heating, **32**, May, 37–44.

Milewski, L. (1980) VAV for fire management in high-rise buildings. *Specifying Engineer*, **43**, May, 84–89.

Milewski, L. (1981) Fan powered air terminal units. *Heating, Piping and Air Conditioning*, **53**, November, 95–98.

Miller, P. (1991) Diffuser selection for cold air distribution. *ASHRAE Journal*, **33**, September, 32–36.

Monger, S. (1989) Trouble shooting the VAV system. *Energy Engineering*, **86**, 47–54.

Morley, T. (1992) Variable volume air conditioning systems – design considerations. *Australian Refrigeration, Air Conditioning and Heating*, **46**, September, 17–22.

Moschandreas, F.J., Choi, S.W. and Meckler, M. (1995) Testing of VAV/BPFS for reduced energy consumption and improved IAQ. In *Proceedings of IAQ95, Practical Engineering for IAQ*, 31–35.

Mumma, S.A. and Bolin, R.J. (1994) Real-time, on-line optimization of VAV system control to minimize the energy consumption rate and to satisfy ASHRAE Standard 62-1989 for all occupied zones. *ASHRAE Transactions*, **100**, Part 1, 168–179.

Mumma, S.A. and Wong, Y.M. (1990) Analytical evaluation of outdoor airflow rate variation vs. supply airflow rate variation in variable-air-volume systems when the outdoor air damper position is fixed. *ASHRAE Transactions*, **96**, Part 1, 1197–1208.

Mutammara, A.W. and Hittle, D.C. (1990) Energy effects of various control strategies for variable-air-volume systems. *ASHRAE Transactions*, **96**, Part 1, 98–102.

Nelson, B.W. (1995) Turning the page on HVAC retrofit. *Consulting-Specifying Engineer*, **18**, September, 28–34.

Nevins, R.G. and Rohles, F.H. (1974) Thermal comfort and variable air volume systems. *Australian Refrigeration, Air Conditioning and Heating*, **28**, August, 40–45.

Nolli, A. and Ciceri, C. (1980) La regolazione automatica negli impianti a portata variabile (Automatic control of variable air volume installations). *Condizionamento dell'Aria*, **24**, December, 982–988.

Noon, C. (1972) Aerofoil fans in variable volume systems. *Heating, Air Conditioning and Refrigeration*, **5**, January, 9–23.

Noon, C. (1987) Energy savings from variable duty fans. *Heating and Air Conditioning Journal*, **56**, March, 17–20.

Norell, L., Strindehag, O. *et al.* (1989) VAV systems with air quality control. *Klimatisacija Grejanje Hladenje*, **18**, November, 61–63 (in Serbo-Croat).

Noren, A. (1991) Getting the most from VAV. *Heating, Piping and Air Conditioning*, **63**, September, 34–35.

Norford, L.K. and Little, R.D. (1993) Fault detection and load monitoring in ventilation systems. *ASHRAE Transactions*, **99**, Part 1, 590–602.

Norford, L.K., Rabl, A. and Socolow, R.H. (1986) Control of supply air temperature and outdoor airflow and its effect on energy use in a variable air volume system. *ASHRAE Transactions*, **92**, Part 2B, 30–44.

Norman, L.D. (1991) A new VAV valve. In *Interklima 91 (Energy, Ecology and Economy): Proceedings of the 11th International Symposium of Heating, Refrigerating and Air Conditioning*, held in Zagreb, 12–13 June 1991, Faculty for Mechanical Engineering and Naval Architecture, Zagreb University, 201–210.

Nungester, B. and Rhodes, T. (1989) Analysis of flow in small, multi-zone HVAC systems. *ASHRAE Journal*, **31**, October, 17–23.

Obler, H. (1978) Converting double duct to variable air volume. *Heating, Piping and Air Conditioning*, **50**, December, 58–63.

Obler, H. (1979) VAV system eliminates overcooling. *Heating, Piping and Air Conditioning*, **51**, August, 75–80.

Okada, T., Yoshikawa, T., Seshimo, Y. *et al.* (1992) Research and development of a home-use VAV air conditioning system. *ASHRAE Transactions*, **98**, Part 2, 133–139.

Ostrander, W.S. (1975) VAV energy analyses. *Building Systems Design*, **72**, June/July, 16–20, 37.

Oughton, D.R. (1987) Recent developments in VAV system design and control. In *Advances in Air Conditioning: Proceedings of the Chartered Institution of Building Services Engineers (CIBSE) Conference*, held at Heathrow, October 1987, CIBSE, London, 1–18.

Oughton, R.J. (1990) Speed control for VAV fans. *Heating and Air Conditioning Journal*, **60**, July, 34.

Pannkoke, T. (1980) Air terminal units. *Heating, Piping and Air Conditioning*, **52**, December, 49–52.

Patterson, N.R. (1977) Fan selection and control in high-velocity VAV systems. *ASHRAE Journal*, **19**, April, 17–24. Also (1977) *ASHRAE Transactions*, **83**, Part 1, 585–597. Also (1978) *Australian Refrigeration, Air Conditioning and Heating*, **32**, January, 18–28.

Patterson, N.R. (1980) Variable air volume systems: types, modulation and selection of system fans. *Plant Engineering*, **34**, February, 67–70.

Patterson, N.R. (1980) Variable air volume systems: a typical installation. *Plant Engineering*, **34**, February, 75–77.

Patterson, N.R. (1981) Variable air volume systems. *Plant Engineering*, **35**, October.

Paul, L. (1991) Fan reduces the quantity of air. *Heizung, Luftung, Haustechnik*, **42**, August, 486–488.

Peach, J.W. (1974) Twin fans convert dual-duct decks to parallel VAV systems. *Heating, Piping and Air Conditioning*, **46**, April, 64–66.

Percival, G. (1985) *PI control of a single-duct VAV HVAC system*. U.S. Army Corps of Engineers Construction Engineering Research Laboratory, Champaign, IL, CERL-TM-E-85/05, AD-A159 049, Technical Manuscript E85/05.

Peterson, K.W. and Sosoka, J.R. (1990) Control strategies utilizing direct digital control. *Energy Engineering*, **87** (4), 30–35.

Pinnella, M.J. (1985) Modeling, tuning and experimental verification of a fan static pressure control system, Master's thesis, University of Illinois at Urbana-Champaign.

Procel, C.J. (1974) Variable air volume systems: loads and psychrometrics. *ASHRAE Transactions*, **80**, Part 1, 473–479.

Rakoczy, T. (1982) Erfahrungen bei RLT-Anlagen mit variablem volumenstrom (experience with variable volume air conditioning installations). *Gesundheits Ingenieur*, **103**, April, 57–69.

Rakoczy, T. (1987) Instationares rechenverfahren fur variable luftvolumenstromsysteme (non-steady calculation procedure for variable air volume systems). *Klima Kalte Heizung*, **15**, March, 155–160.

Reid, M.A. (1983) Control changes enable VAV conversion. *Heating, Piping and Air Conditioning*, **55**, March, 109–110.

Rees, K., Wenger, J. and Janu, G. (1992) Ventilation airflow measurement for ASHRAE Standard 62-1989. *ASHRAE Journal*, **34**, October, 40–45.

Richardson, G. (1994) Commissioning of VAV laboratories and the problems encountered. *ASHRAE Transactions*, **100**, Part 1, 1641–1645.

Rickelton, D. (1972) Energy conservation aspects of variable air volume systems. *Building Systems Design*, **69**, June, 21–24. Also (1972) *Australian Refrigeration, Air Conditioning and Heating*, **26**, December, 50–55.

Rickelton, D. and Becker, H.P. (1972) Variable air volume. *ASHRAE Journal*, **14**, September, 31–55.

Roberts, D.V. (1987) Controls and VAV technology. *Refrigeration, Air Conditioning and Heat Recovery*, **90**, November, 43–44.

Roberts, J.W. (1991) Outdoor air and VAV systems. *ASHRAE Journal*, **33**, September, 26–30.

Roberts, M.M. (1981) The variable air volume approach to plant air conditioning. *Plant Engineering*, **35**, October, 139–142.

Rowe, D. (1984) Wollongong Government Office Building: energy conservation measures in design. *Australian Refrigeration, Air Conditioning and Heating*, **38**, December, 42–45.

Rowe, D. (1986) Design for air handling and distribution systems. *Australian Refrigeration, Air Conditioning and Heating*, **40**, November, 34–40.

Ross, D.E. (1974) VAV system evaluation for commercial office buildings. *ASHRAE Transactions*, **80**, Part 1, 480–485.

Rowntree, H.E. (1991) VAV and constant-volume systems – a time and place for each. *Consulting Specifying Engineer*, **9**, January, 40–44.

Sauer, H.J., Jr. and Howell, R.H. (1991) *Final Report on ASHRAE RP-590, Control of Outside Air and Building Pressurization in VAV Systems*, American Society of Heating, Refrigerating and Air Conditioning Engineers, Atlanta.

Sauer, H.J., Jr. and Howell, R.H. (1992) Estimating the indoor air quality and energy performance of VAV systems. *ASHRAE Journal*, **34**, July, 43–49.

Sauer, H.J., Jr., Delp, W.W. and Anantapantula, S. (1994) Effect of component size on VAV system performance. In *Proceedings of IAQ94, Engineering Indoor Environments* (E.L. Besch, ed.), held in St Louis, Missouri, October 31–November 2 1994, ASHRAE, Atlanta, 175–181.

Sauer, H.J., Jr., Delp, W.W. and Howell, R.H. (1993) Maintaining indoor air quality and building pressurization with variable air volume systems. In *Proceedings of CLIMA 2000*, held in London, 1–3 November 1993, Chartered Institution of Building Services Engineers, London.

Schlein, R.L. (1982) Saving energy with VAV: a case history. *Specifying Engineer*, **48**, August, 104–105.

Schmidt, R.D. (1977) Variable air volume terminals and diffuser performance. *ASHRAE Transactions*, **83**, 606–612.

Schuurmans, H. (1977) Variable volume all-air systems. *Heating, Air Conditioning and Refrigeration*, **9**, March, 9–16.

Schuurmans, H. and Weill, J.R. (1978) Volume d'air variable: etude d'application et de rentabilite en espace a zones non differenciees (Variable air volume: study of its application and economic competitiveness in non-zoned spaces). *Chauffage, Ventilation, Condionnement*, **54**, September/October, 15–19.

Scofield, C.M. (1991) Low-temperature air with high IAQ. In *Proceedings of the Far East Conference on Environmental Quality*, held in Hong Kong, 5–8 November 1991, American Society of Heating, Refrigerating and Air Conditioning Engineers, Atlanta, 9–15.

Scofield, C.M. (1993) Low temperature air with high IAQ for tropical climates. *ASHRAE Journal*, **35**, March, 52–59.

Scofield, C.M. (1994) California classroom VAV with IAQ and energy savings, too. *Heating, Piping and Air Conditioning*, **66**, January, 89–93.

Scofield, C.M. and Des Champs, N.H. (1995) Low-temperature air with high IAQ for dry climates. *ASHRAE Journal*, **37**, January, 34–40.

Scofield, C.M. and Fields, G. (1989) Joining VAV and direct refrigeration. *Heating, Piping and Air Conditioning*, **61**, September, 137–140, 147–152.

Scott, R. and Hunt, J. A comparative commercial appraisal of chilled ceilings and displacement ventilation with conventional AC systems. *H and V Engineer*, **68**, 727, 5–6.

Sehgal, S. (1972) Variable air volume systems and hardware: the Euram building. *Building Systems Design*, **69**, August, 23–29.

Sessler, S.M. and Shiver, W.D. (1981) Energy can be reduced in air handling systems. *Specifying Engineer*, March.

Shataloff, N.S. (1972) Variable air volume systems and hardware: comfort installations. *Building Systems Design*, **69**, August, 16–19.

Shavit, G. (1981) Study evaluates fan system design options. *Heating, Piping and Air Conditioning*, **53**, August, 69–74.

Shepherd, K.J. (1995) Variable-air-volume systems for optimised design and control, M.Sc. thesis, UMIST.

Shepherd, K.J., Levermore, G.J., Letherman, K.M. and Karayiannis, T.G. (1993) VAV modelling and control. In *Energy Efficient Buildings (Design, Performance and Operation): Proceedings of the CIB International Symposium* (Erhorn *et al.* eds), held in Stuttgart, 9–11 March 1993, Conseil International du Patiment pour la Recherche, l'Etude et la Documentation (CIB), 551–556.

Shepherd, K.J., Levermore, G.J., Letherman, K.M. and Karayiannis, T.G. (1993) Analysis of VAV networks for design and control. In *Proceedings of CLIMA 2000*, held in London, 1–3 November 1993, Chartered Institution of Building Services Engineers, London.

Shepherd, K.J., Levermore, G.J., Letherman, K.M. and Karayiannis, T.G. (1994) VAV design and control. In *Proceedings of the BEPAC Conference*, held at the University of York, 6–8 April 1994, Building Energy Performance Analysis Club (BEPAC).

Shuper, A. (1964) A study in variable volume air conditioning. *Air Conditioning, Heating and Ventilating*, **61**, November, 57–60.

Smith, R.B. (1990) Importance of flow transmitter selection for return fan control in VAV systems. *ASRAE Transactions*, **96**, Part 1, 1218–1223.

Spitler, J.D., Hittle, D.C., Pedersen, C.O. and Johnson, D.L. (1986) Fan electricity consumption for variable air volume. *ASHRAE Transactions*, **92**, Part 2B, 5–18.

Stein, H.L. (1982) Variable air volume air conditioning. *Specifying Engineer*, **48**, 129–133.

Steketee, N.F. (1972) Variable volume air conditioning systems. *Australian Refrigeration, Air Conditioning and Heating*, **26**, January, 39–42.

Straub, H.E. (1970) Room air distribution with a variable volume system. *ASHRAE Journal*, **12**, April, 52–58.

Straub, H.E. (1972) Variable air volume systems and hardware: air distribution. *Building Systems Design*, **69**, August, 38–40.

Straub, H.E. (1986) Terminal requirements based on designers' terminal choices. *ASHRAE Transactions*, **92**, Part 1B, 519–527.

Stubblefield, R.R. (1977) Variable volume-constant temperature/constant volume-variable temperature. *Heating, Piping and Air Conditioning*, **51**, December, 43–45.

Subbarao, B. Analysis of flow rates and building pressurization in a variable air volume system. Master's thesis T6053, University of Missouri-Rolla.

Tallant, D. (1981) Inverters chosen for VAV retrofit. *Heating, Piping and Air Conditioning*, **53**, August, 79–81.

Tamblyn, R.T. (1983) Beating the blahs for VAV. *ASHRAE Journal*, **25**, September, 42–45.

Tamblyn, R.T. (1988) Getting high on low temperature air. *Heating, Piping and Air Conditioning*, **60**, January, 101–108.

Tamblyn, R.T. (1994) Supplying 100% outside air with less energy and providing individual occupant temperature control with less initial cost. In *Proceedings of IAQ94, Engineering Indoor Environments* (E.L. Besch, ed.), held in St Louis, Missouri, October 31–November 2 1994, ASHRAE, Atlanta, 175–181.

Teji, D.S. (1987) Controlling air supply for energy conservation. *Energy Engineering*, **84** (3), 4–13.

Terry, K.B. (1965) Dual duct and dual conduit systems. *Australian Refrigeration, Air Conditioning and Heating*, **19**, September, 22–34.

Terry, K.B. (1969) Variable volume air conditioning systems and devices. *Australian Refrigeration, Air Conditioning and Heating*, **23**, October, 34–48.

Thomas, L.H. (1988) Effective VAV begins with effective fan-volume control. *Consulting Specifying Engineer*, **4**, August, 62–64.

Tinsley, W.E. (1991) Hospital saves energy with VAV system and DDC control. *ASHRAE Journal*, **33**, March, 22–24.

Tisdale, R.F. (1981) Air flow controls: how much accuracy can we afford? *Heating, Piping and Air Conditioning*, **53**, August, 57–64.

Transactions of the Society of Heating, Air Conditioning & Sanitary Engineers of Japan. Actual Measurements and Energy Saving of VAV Systems, June 1980.

Tseytlin, A.A. (1993) Making VAV more efficient and flexible. *Heating, Piping and Air Conditioning*, **65**, February, 55–57.

Urban, R.A. (1970) Design considerations and operating characteristics of variable volume systems. *ASHRAE Journal*, **12**, February, 77–84.

Van Aken, G.J. (1993) Control of outdoor air and exhaust air flow rates for variable air volume air conditioning systems. In *Proceedings of the AIRAH International Conference*, held in Sydney, April 1993, AIRAH, Sydney, Vol. 3.

Ventresca, J.A. (1992) Economizer operation and maintenance for indoor air quality. *ASHRAE Journal*, **34**, January, 26–32.

Vivien, L.E. (1972) Variable volume air meets high rise needs. *Heating, Piping and Air Conditioning*, **44**, January, 84–86.

Walker, C.A. (1984) Application of direct digital control to a variable air volume system. *ASHRAE Transactions*, **90**, Part 2, 846–855.

Walker, C.A. (1984) DDC: variable air volume application demonstrates key advantage. *ASHRAE Journal*, **26**, September, 23–27.

Warren, M. and Norford, L.K. (1993) Integrating VAV zone requirements with supply fan operation. *ASHRAE Journal*, **35**, April, 43–46.

Watt, J.R. (1972) Heat reclaim, variable volume air serve new office. *Heating, Piping and Air Conditioning*, **44**, April, 86–89.

Webster, J. *et al.* (1988) Keeping control of VAV systems. *Building Services and Environmental Engineer*, **10**, June, 18–29.

Wendes, H. (1989) Supply outlets for VAV systems. *Heating, Piping and Air Conditioning*, **61**, February, 67–71.

Wendes, H. (1991) *Variable Air Volume Manual*, Fairmont Press.

Williams, K. (1985) Direct digital control gains interest with VAV. *Specifying Engineer*, **53**, January, 58–63.

Williams, V.A. (1988) VAV system interactive controls. *ASHRAE Transactions*, **94**, Part 1, 1493–1499.

Woods Ballard, W.R. (1980) Why variable pitch fans save energy. *Plant Engineering*, **34**, October.

Woods Ballard, W.R. (1984) Comparative energy savings in fan systems. *Australian Refrigeration, Air Conditioning and Heating*, **38**, May, 18–26.

Yamamori, H. (1991) Integration of a VAV system incorporated with the IAQ concept and a VRV system. In *Proceedings of the Far East Conference on Environmental Quality*, held in Hong Kong, 5–8 November 1991, American Society of Heating, Refrigerating and Air Conditioning Engineers, Atlanta, 47–53.

Yoke, J.H. (1972) Variable air volume systems and hardware: systems for schools. *Building Systems Design*, **69**, August, 33–35.

Yu, H.H.S. and Raber, R.R. (1992) Air-cleaning strategies for equivalent indoor air quality. *ASHRAE Transactions*, **98**, Part 1, 173–181.

Zaheer-uddin, M. and Goh, P.A. (1991) Transient response of a closed-loop VAV system. *ASHRAE Transactions*, **97**, Part 2, 378–387.

Zaheer-uddin, M. and Zheng, G.R. *Static Pressure Control for Balancing VAV Systems and Minimizing Energy Consumption: A Computer Simulation Study*. (No further details available.)

Zaheer-uddin, M. and Zheng, G.R. (1994) A dynamic model of a multizone VAV system for control analysis. *ASHRAE Transactions*, **100**, Part 1, 219–229.

Zheng, G.R. and Zaheer-uddin, M. (1996) Optimization of thermal processes in a variable air volume HVAC system. *Energy*, **21** (5), 407–420.

Zhivov, A.M. (1990) Variable-air-volume ventilation systems for industrial buildings. *ASHRAE Transactions*, **96**, Part 2, 367–372.

Zieger, H.R. (1993) Variable air volume systems – new trends in measurement and controls. In *LSH Jahrbuch*, 25–29 (in German).

Index